RESTORATION ECOLOGY

Related Titles from Jones & Bartlett Learning

Botany: An Introduction to Plant Biology, Fourth Edition
James D. Mauseth

Climatology, Second Edition
Robert V. Rohli & Anthony J. Vega

The Ecology of Agroecosystems
John H. Vandermeer

Environmental Oceanography: Topics and Analysis
Daniel C. Abel & Robert L. McConnell

Environmental Science, Eighth Edition
Daniel D. Chiras

Environmental Science: Systems and Solutions, Fourth Edition
Michael L. McKinney, Robert M. Schoch, & Logan Yonavjak

Essentials of Geochemistry, Second Edition
John V. Walther

Introduction to the Biology of Marine Life, Tenth Edition
John F. Morrissey & James L. Sumich

Invitation to Oceanography, Fifth Edition
Paul R. Pinet

Outlooks: Readings for Environmental Literacy, Second Edition
Michael L. McKinney

Plants, Genes, and Crop Biotechnology, Second Edition
Maarten J. Chrispeels & David E. Sadava

Principles of Environmental Chemistry, Second Edition
James E. Girard

Tropical Forests
Bernard A. Marcus

RESTORATION ECOLOGY

Sigurdur Greipsson, PhD
Kennesaw State University

JONES & BARTLETT
LEARNING

World Headquarters
Jones & Bartlett Learning
40 Tall Pine Drive
Sudbury, MA 01776
978-443-5000
info@jblearning.com
www.jblearning.com

Jones & Bartlett Learning
Canada
6339 Ormindale Way
Mississauga, Ontario L5V 1J2
Canada

Jones & Bartlett Learning
International
Barb House, Barb Mews
London W6 7PA
United Kingdom

Jones & Bartlett Learning books and products are available through most bookstores and online booksellers. To contact Jones & Bartlett Learning directly, call 800-832-0034, fax 978-443-8000, or visit our website, www.jblearning.com.

> Substantial discounts on bulk quantities of Jones & Bartlett Learning publications are available to corporations, professional associations, and other qualified organizations. For details and specific discount information, contact the special sales department at Jones & Bartlett Learning via the above contact information or send an email to specialsales@jblearning.com.

Copyright © 2011 by Jones & Bartlett Learning, LLC

All rights reserved. No part of the material protected by this copyright may be reproduced or utilized in any form, electronic or mechanical, including photocopying, recording, or by any information storage and retrieval system, without written permission from the copyright owner.

Production Credits
Chief Executive Officer: Ty Field
President: James Homer
SVP, Chief Operating Officer: Don Jones, Jr.
SVP, Chief Technology Officer: Dean Fossella
SVP, Chief Marketing Officer: Alison M. Pendergast
SVP, Chief Financial Officer: Ruth Siporin
Publisher, Higher Education: Cathleen Sether
Acquisitions Editor: Molly Steinbach
Senior Editorial Assistant: Caroline Perry
Senior Editorial Assistant: Jessica S. Acox
Production Manager: Louis C. Bruno, Jr.
Senior Marketing Manager: Andrea DeFronzo
V.P., Manufacturing and Inventory Control: Therese Connell
Cover Design: Kristin E. Parker
Assistant Photo Researcher: Rebecca Ritter
Illustrations: Elizabeth Morales
Composition: CAE Solutions Corporation
Cover Image: © Lensara/Dreamstime.com
Printing and Binding: Malloy, Inc.
Cover Printing: Malloy, Inc.

Library of Congress Cataloging-in-Publication Data
Greipsson, Sigurdur.
  Restoration ecology / Sigurdur Greipsson. — 1st ed.
     p. cm.
  ISBN 978-0-7637-4219-5 (alk. paper)
 1. Restoration ecology—Textbooks. I. Title.
  QH541.15.R45G745 2011
  639.9—dc22   2010011824

6048

Printed in the United States of America
14 13 12 11 10   10 9 8 7 6 5 4 3 2 1

Brief Contents

Chapter 1   Introduction to Restoration Ecology   1

Chapter 2   Ecosystem Functioning   33

Chapter 3   Biodiversity   55

Chapter 4   Succession   79

Chapter 5   Assembly   101

Chapter 6   Landscape   131

Chapter 7   Invasive Species   150

Chapter 8   Soil   176

Chapter 9   Sand Dunes   206

Chapter 10   Mines and Polluted Sites   236

Chapter 11   Forest   265

Chapter 12   Endangered Animals   290

Chapter 13   Aquatic Ecosystems: Wetlands, Lakes, and Rivers   314

Chapter 14   Management of Restoration Projects   352

              Glossary   382

              Index   387

Contents

Preface   xiii

Chapter 1 **Introduction to Restoration Ecology**   1
  1.1 Degradation of Ecosystems   3
    Global Warming   4
    Energy Consumption   6
    Ecosystem Management Models   10
  1.2 Value of Ecosystem Services   12
  1.3 Outlook for Ecological Restoration   13
    Different Restoration Approaches   15
    Varying Scales of Restoration   16
    Future of Restoration   18
  Case Study 1.1: Carbon Sequestration of Soil   21

Chapter 2 **Ecosystem Functioning**   33
  2.1 Ecosystem Disturbances   34
    Stress   35
    Abrupt Change   36
    Frequency   36
    Magnitude   37
    Duration, Abruptness, and Return Interval   37
    Mega-Disturbances   37
    Severe Disturbances   38
  2.2 Fire Disturbances   39
    Fire-Adapted Ecosystem   40
    Prescribed Burning   42
    Fire Suppression   42
  2.3 Fragmentation   43
  2.4 Nutrient Cycling   44

2.5 Hydrological Cycling   47
2.6 Functional Groups and Ecosystem Engineers   48
    Drivers and Passengers   48
    Ecosystem Engineers   48
2.7 Keystone Species   49
    Top-Down Control Mechanism   49
    Predators   50
    Mutualistic Species   50
Case Study 2.1: Importance of Fire and Grazing in Grassland Biomes   51

Chapter 3 Biodiversity   55
3.1 Threats to Biodiversity   57
    Habitat Fragmentation   57
    Invasive Species   58
    Chemical Pollution   58
    Hybridization   59
    Overhunting   59
3.2 Extinction   60
    Rate of Extinction   60
    Species Vulnerability   61
    Extinct Species: The Great Auk   61
3.3 Genetic Diversity   64
    Restoration of Genetic Diversity   64
3.4 Restoration of Species Diversity   66
    Seed Harvesting   67
    Seed Production   67
    Seed Processing and Storage   67
    Seed Evaluation   68
    Germination   68
    Seed Dormancy   69
    Seed Priming   69
    Seed Sowing   70
    Nursery-Raised Plants   70
3.5 Ecosystem Diversity   71
    Habitat Fragmentation   71
    Ecosystem Stability   72
Case Study 3.1: Ecological Genetics and Restoring the Tallgrass Prairie   74

Chapter 4 Succession   79
4.1 Theories of Succession   79
4.2 Successional Processes and Restoration   82
    Primary Succession   83

        Secondary Succession   88
        Animals in Succession   89
    4.3 Management of Succession   91
        Reducing or Inducing Disturbances   91
        Introduction or Removal of Species   92
        Nutrient Management   93
        Hydrological Manipulations   94
    4.4 Monitoring Succession   95
    4.5 Inferring Succession   96
    Case Study 4.1: Restoration of Primary Sites   97

Chapter 5   Assembly   101
    5.1 Equilibrium Theory of Island Biogeography   102
    5.2 Ecosystem Resilience and Stability   104
        Resilience   104
        Resistance   105
        Stability   105
        Ecological Constraints   105
    5.3 Alternative Stable States   106
        Regime Shift   107
        Restoring Alternative Stable States   110
    5.4 Assembly Rules   111
        Sequence Introduction   111
        Species Compatibility   113
        Ecosystem Thresholds   113
        Ecosystem Filters   113
    5.5 Unified Neutral Theory of Biodiversity and Biogeography   114
    Case Study 5.1: Resilience and Ecosystem Restoration   116
    Case Study 5.2: Phylogenetic Structure of Plant Communities Provides Guidelines for Restoration   119

Chapter 6   Landscape   131
    6.1 Connectivity   133
        Matrix Restoration   133
        Corridors   134
        Stepping Stones   135
    6.2 Metapopulation   136
        Metapopulation Networks   136
        Metapopulation Dynamics   137
        Metapopulation Restoration Projects   138
    6.3 Landscape Restoration   143
        Passive Restoration   143
    Case Study 6.1: A Metapopulation Approach to Restoration of Pitcher's Thistle in Southern Lake Michigan Dunes   146

Chapter 7	**Invasive Species** 150
	7.1 Process of Invasion 152
	7.2 Effects of Invasion on Ecosystems 154
	7.3 Methods of Control 157
	Prevention 157
	Eradication 158
	Containment 161
	Chemical and Biological Control 161
	7.4 Restoration to Constrain Invasion 162
	Niche Preemption 162
	Fire Management 163
	Increasing Biotic and Abiotic Resistance 164
	Case Study 7.1: Kudzu—The Notorious Invader of the Southern United States 165
	Case Study 7.2: Pale Swallow-Wort: An Emerging Threat to Natural and Seminatural Habitats in the Lower Great Lakes Basin of North America 169
Chapter 8	**Soil** 176
	8.1 Soil Erosion 178
	Global and Local History of Soil Erosion 178
	Wind Erosion 179
	Water Erosion 180
	Factors Affecting Soil Erosion 181
	8.2 Desertification 181
	Regional Effects of Desertification 183
	How Desertification Affects Us 183
	How We Affect Desertification 184
	Efforts to Slow Desertification 185
	8.3 Soil Conservation 185
	Conservation Methods 186
	Irrigation and Saline Soils 189
	8.4 Restoration of Soil Nutrients 190
	Nitrogen 190
	Phosphorus 191
	8.5 Soil Microorganisms 191
	Case Study 8.1: Importance of Soil Microbial Communities 197
	Case Study 8.2: Role of Arbuscular Mycorrhizal Fungi in Restoration of Mine Tailings 201
Chapter 9	**Sand Dunes** 206
	9.1 Dune Formation 207
	9.2 Ecological Processes 208
	9.3 Disturbances 210

x   Contents

         Loss of Vigor   210
         Human Disturbances   211
         Invasion of Non-Native Plant Species   211
         Rising Sea Levels   212
   9.4 Restoration Strategies   213
         Beach Nourishment   214
         Dune-Building Fences   214
         Native Plants   216
   9.5 Long-Term Management   219
   Case Study 9.1: Coastal Erosion at Dauphin Island, Alabama   222
   Case Study 9.2: Ecological Effects of Sandy Beach Restoration in Northeast Norfolk, United Kingdom   225

Chapter 10   **Mines and Polluted Sites**   **236**
   10.1 Mine Waste Restoration   239
         Surface Stabilizers   240
         Metal-Tolerant Genotypes   240
         Stockpiled Soil   241
         Commercial Grasses   242
         Native Plants   242
         Monitoring   243
   10.2 Phytoremediation   245
         Hyperaccumulators   246
         Metal Chelates   246
   10.3 Bioremediation   248
         Impeding Factors   249
         Requirements for Bioremediation   249
         Landfarming   250
   Case Study 10.1: Restoration of Gold Mines in Ghana, West Africa   252
   Case Study 10.2: Phytoremediation of Lead-Contaminated Soil   257
   Case Study 10.3: Bioremediation of a Pesticide: Hydroxylation of Bensulide   261

Chapter 11   **Forest**   **265**
   11.1 Forest Degradation   266
         Clear-Cutting and Selective Cutting   266
         Human Settlement and Ranching   267
         Land Mismanagement   267
         Pollution   270

11.2 Forest Restoration   271
   Passive Restoration   271
   Active Restoration   272
   Regeneration Niche   275
11.3 Tropical Rain Forest   280
   Threats to the Rain Forest   280
   Restoration   282
   Obstacles to Restoration   283
Case Study 11.1: Recovery of Forested Ecosystems After Management in Nova Scotia, Canada   284

Chapter 12   **Endangered Animals**   **290**
12.1 Restoration of Critical Habitats   292
   Area Threshold   294
   Carrying Capacity   294
   GIS Technology   294
12.2 Captive Breeding   294
   Genetic Structure and Genetic Variation   295
   Giant Panda   295
   Black-footed Ferret   296
   California Condor   298
   Role of Zoos   298
12.3 Translocation and Reintroduction   299
   Translocation   299
   Reintroduction   299
   Release Strategies   300
   Reintroduction of the Gray Wolf into Yellowstone Park   301
   Complex Reintroduction   304
Case Study 12.1: Hybridization Between Introduced Walleye and Native Sauger in Montana: Implications for Restoration of Montana Sauger   306

Chapter 13   **Aquatic Ecosystems: Wetlands, Lakes, and Rivers**   **314**
13.1 Degradation of Aquatic Ecosystems   316
   Degradation of Wetlands   316
   Degradation of Lakes   318
   Degradation of Rivers   319
13.2 Restoration of Wetlands   319
   Passive Restoration   320
   Active Restoration   320
   Restoration of Florida's Everglades   321

13.3 Restoration of Lakes   324
    Restoration of the Great Lakes   325
    Restoration of the Broads in Norfolk, United Kingdom   327
13.4 Restoration of Rivers   330
    Active Restoration   330
    Restoration of the North Atlantic Salmon   332
Case Study 13.1: Restoration of Coastal Salt Marshes in Brazil Using Native Salt Marsh Plants   333
Case Study 13.2: Biomanipulation as a Tool for Shallow Lake Restoration in the Norfolk Broads, United Kingdom   338
Case Study 13.3: Restoration of the Kissimmee River, Florida   347

Chapter 14 Management of Restoration Projects   352
14.1 Setting Goals   353
14.2 Planning   355
14.3 Action Plan   357
14.4 Adaptive Management   359
14.5 Monitoring   360
14.6 Aftercare and Final Assessment   362
14.7 Legal Framework and International Agreements   363
    Environmental Legislation   363
    Wetlands   363
    Mined Lands   364
    Invasive Non-Native Species   365
    Soil Erosion   365
    International Conventions   366
    Human Population Overgrowth   366
    Desertification   366
    Biodiversity   367
Case Study 14.1: Integrated Restoration Efforts of Severely Degraded Subarctic Heathland Ecosystem   368
Case Study 14.2: Decision Making in Ecological Restoration   375

Glossary   382
Index   387

PREFACE

The new field of restoration ecology is quickly gaining momentum. At the time of my first professional restoration ecology meeting in 1985, a few mine reclamation journals were in print, but not a single scientific journal existed to unify the multidisciplinary facets of restoration ecology. The Society of Ecological Restoration International was founded in 1988 to promote a sustainable and healthy relationship between nature and culture. Now, the organization has grown to over 2,300 members operating in 37 countries. It hosts several conferences every year to exchange information and ideas among members, and publishes two journals of peer-reviewed scientific articles. Over the past 25 years, the field of restoration ecology has expanded to address problems in a greater variety of degraded ecosystems worldwide as well as to incorporate new strategies that reflect our growing understanding of complex ecology —and the need could not be greater.

Today, the Earth's biodiversity is at risk, as delicate ecosystems struggle to overcome global warming, rain forest destruction, acid rain, overfishing, eutrophication, erosion, and a whole host of other interconnected—and largely anthropogenic—environmental problems. Fortunately, as the severity of these problems has escalated globally, so has the prominence of restoration ecology, which offers practical and economical solutions. Ambitious projects are underway worldwide; large-scale examples within the United States include the Longleaf Pine Initiative, which is working to restore native forests across nine southern states, and the Comprehensive Everglades Restoration Plan, which provides a long-term framework for protecting the depleted freshwater resources and damaged ecosystem of southern Florida. In 2010, of course, the *Deepwater Horizon* oil leak in the Gulf of Mexico turned public attention to the major ecological risks of offshore drilling; the cleanup effort for the disaster requires a thorough understanding of marine, wetland, and coastal restoration ecology.

Many challenges lie ahead for scientists in our field. Effective resource management is key to the long-term success of restoration projects, but this relies on cooperation from legislators, property owners, and businesses, whose interests do not always coincide with the ecological good. New tools that can assign an accurate economic value to healthy ecosystems will be essential. Further complicating the situation, among scientists, opinions differ as to the most effective strategies. Although the theory of community succession has played an important role in restoration ecology, other approaches that emphasize niche assembly or put more emphasis on neutral and/or random processes need to be tested vigorously under various ecological conditions in different biomes. Strategies that facilitate restoration of high biodiversity need to be designed and tested as well. All of this requires careful planning, access to economic, political, and natural resources, and—above all—time.

Teaching a course on such a young and fluid subject is not an easy task. Restoration ecology is multidisciplinary, using techniques derived from molecular biology, geography, oceanography, soil sciences, environmental chemistry, botany, resource management, and—of course—ecology. The first time I taught a course on restoration ecology, it was impossible to find a textbook that covered appropriate topics for the discipline. As a result of that challenge, I started gathering information and writing the textbook that is in your hands.

Restoration Ecology is written to engage the upper-level undergraduate or graduate student in this increasingly relevant topic. A vast majority of the material in the book was tested in the classroom with great success, and the textbook covers both the theoretical background and restoration approaches of various ecosystems. Divided into five logical parts, the text opens with a look at ecological perspectives of restoration (Chapter 1), including nutrient cycling and factors that regulate ecosystem function (Chapters 2 and 3). Chapters 4 and 5 discuss the ecological theories that have shaped restoration ecology. Chapters 6 through 13 are devoted to restoration in practice, providing accounts of restoration efforts in actual ecosystems. The text includes vivid examples of recent restoration projects, many of which are still ongoing and hopefully will stimulate students to consider restoration ecology as a professional career. Chapter 14 delves into the planning, implementation, monitoring, and appraisal of restoration work.

Special Features

End-of-Chapter Tools

This text is intended to be more than just a reader or a reference book, and accordingly it incorporates several pedagogical elements that will help students understand and retain the material. **Key Terms** highlight the most important ideas and information introduced in each chapter, and a comprehensive list of terms is found at the end of the reading. **Key Questions** are designed to help students review and assess their comprehension of the material. For students with the motivation and capacity for independent research, a **Further Reading** section provides relevant texts and articles for deeper exploration of the topics covered in the text.

Case Study Boxes

The text provides a global perspective on restoration ecology with examples of restoration problems worldwide. These examples are highlighted in **Case Study** boxes, written by experts in the field, which can be found throughout the text. The Case Study boxes, in chapter order, are:

- Carbon Sequestration of Soil
- Importance of Fire and Grazing in Grassland Biomes
- Ecological Genetics and Restoring the Tallgrass Prairie
- Restoration of Primary Sites
- Resilience and Ecosystem Restoration
- Phylogenetic Structure of Floridian Plant Communities Provides Guidelines for Restoration
- A Metapopulation Approach to Restoration of Pitcher's Thistle in Southern Lake Michigan Dunes
- Kudzu—The Notorious Invader of the Southern United States
- Pale Swallow-Wort: An Emerging Threat to Natural and Seminatural Habitats in the Lower Great Lakes Basin of North America
- Importance of Soil Microbial Communities
- Role of Arbuscular Mycorrhizal Fungi in Restoration of Mine Tailings
- Coastal Erosion at Dauphin Island, Alabama
- Ecological Effects of Sandy Beach Restoration in Northeast Norfolk, United Kingdom
- Restoration of Gold Mines in Ghana, West Africa
- Phytoremediation as a Reclamation Strategy of Lead-Contaminated Soil
- Bioremediation of a Pesticide: Hydroxylation of Bensulide
- Recovery of Forested Ecosystems After Management in Nova Scotia, Canada
- Hybridization Between Introduced Walleye and Native Sauger in Montana: Implications for Restoration of Montana Sauger
- Restoration of Coastal Salt Marshes in Brazil Using Native Salt Marsh Plants
- Biomanipulation as a Tool for Shallow Lake Restoration in the Norfolk Broads, United Kingdom
- Restoration of the Kissimmee River, Florida
- Integrated Restoration Efforts of Severely Degraded Subarctic Heathland Ecosystem
- Decision Making in Ecological Restoration

Ancillaries

The online *PowerPoint® ImageBank* provides a library of all the art, tables, and photographs in the text to which Jones & Bartlett Learning holds the copyright or has permission to reproduce digitally. With the Microsoft® PowerPoint software you can easily project images from the text in the classroom or insert images into your existing PowerPoint lecture presentations. The ImageBank is available for qualified instructors to download from http://www.jblearning.com/catalog/9780763742195/.

Acknowledgments

The writing of the textbook was time-consuming but enjoyable. Many colleagues and friends have helped in developing the material in the book, and their steady interest has been motivating. I would also like to thank the following reviewers, who were generous with their time and comments:

John H. Brock, Arizona State University Polytechnic
Timothy L. Dickson, Washington University in St. Louis
Jerald Dosch, Macalester College
Molly Hunter, Northern Arizona University
Gary A. Lamberti, University of Notre Dame
Jeffrey W. Matthews, University of Illinois
Michael L. McInnis, Oregon State University
Stephen D. Murphy, University of Waterloo
Justin Podur, York University
Melanie Riedinger-Whitmore, University of South Florida—St. Petersburg
Keith S. Summerville, Drake University
Thomas H. Whitlow, Cornell University

I welcome any comments and suggestions on the book, and I hope it will be used in classrooms worldwide to promote the field of restoration ecology.

Sigurdur Greipsson
Kennesaw State University

1

INTRODUCTION TO RESTORATION ECOLOGY

Chapter Outline

1.1 Degradation of Ecosystems
    Global Warming
    Energy Consumption
    Ecosystem Management Models
1.2 Value of Ecosystem Services
1.3 Outlook for Ecological Restoration
    Different Restoration Approaches
    Varying Scales of Restoration
    Future of Restoration
Case Study 1.1: Carbon Sequestration of Soil

Humans have throughout the centuries drastically impacted their environments. Since their mastery of fire, humans have modified their environment to a large extent. The greatest change resulting from intensive use of fire has been the transformation of forests to grasslands. Such transformation was important for the advance of early hunting and agrarian societies. Also, grazing pressure from domesticated animals has played a critical role in worldwide ecosystem degradation. In addition, direct hunting has resulted in the extinction of many species and at the same time has impacted the ecological role these species played. When extinction or significant reduction occurs in a keystone species, it can result in drastic changes to ecosystem functioning. Reduction in biodiversity can also induce major shifts in ecosystem structure and function. More recently, widespread pollution has impacted the environment on a global scale.

Humans are one of the few species that have built settlements in all biomes on Earth. Human settlements stretch from the high arctic to equatorial rain forests. Human dominance has brought about large-scale modification to Earth's landscape. Today, humans have transformed about 40% of terrestrial land, typically changing complex native ecosystems into monocultures (including rangelands and agroforests), industrial, commercial, and residential landscapes. Mainly through agriculture, humans dominate (directly or indirectly) about one-third of the net primary terrestrial productivity.

A common misconception is that nature can provide humans with inexhaustible resources that are free of charge. Although humans have not yet reached their carrying capacity in food production, they do use about 55% of the potable water that some predictions show becoming a scarce resource in the near future. This scarcity of potable water is already alarming in many countries. In rural China more than 500 million people use contaminated drinking water. Shortages of food and water are some of the most pressing problems that humans will face in the near future.

The chief underlying reason for the intensified use of natural resources is the recent exponential growth of the human population. In 1798 AD Thomas Malthus warned in his book, *An Essay on the Principle of Populations*, about the danger of exponential growth of the human population while resources (food) increase only arithmetically. A few milestones in the growth of the human population indicate its explosive nature: 70,000 years ago the human population was only about 2,000, in 1200 AD it was 400 million, by 1900 AD it had increased to 1.6 billion, and currently it is roughly 6.5 billion. The exponential growth of the human population has doubled its size since 1960 AD. The growth rate of the population is about 1.2% per year, which adds about 80 million people to Earth each year. The vast majority of this growth (90%) takes place in developing countries. Currently, most of the human population live in Asia. About 1.3 billion people live in China, and about 1.1 billion live in India. In the United States there are about 300 million people, representing less than 5% of the world's population.

The carrying capacity of humans on Earth is rapidly reaching its threshold. The growth of the human population rose sharply around 1850 AD due to advances in agriculture (chemical fertilizers), industry (mechanized production of goods), and medical sciences (vaccination and antibiotics). It is predicted that human population will reach about 9.4 billion by the year 2050, peak at 11 billion by 2150, and thereafter begin to level off. The fastest growth in the human population is currently taking place in Asia and Africa. And, though growth of the human population has stabilized in some developed countries like Sweden, Switzerland, Russia, and Japan, population growth in many developing countries like Ethiopia, Nigeria, Congo, Yemen, and Pakistan is expected to double or triple before it begins to level off. Future perspectives for many countries are grim. For example, Nigeria is Africa's most populous country, with more than 135 million people and projected to increase to 258 million by 2050. Nigeria's livestock population has grown from about 6 million in 1950 to about 66 million in 2005 and is projected to be about 126 million in 2050. The forage needs of this rapidly expanding livestock population will help to turn most of the country into a desert by that time.

Social problems are expected to grow worldwide, because many affluent countries are experiencing zero population growth, whereas the population in some of the poorest countries is growing out of control. Disparity in wealth and job opportunities will result in mass emigration from developing countries. For example, the population in Nepal, a small developing country, has grown from about 2 million to more than 20 million in just a few decades. This growth has not only put enormous stress on the environment (including deforestation and soil erosion) but has also resulted in civil unrest, war, and mass emigration.

Although it is difficult to come to terms with the challenges associated with providing enough food for a population reaching 11 billion people, it is clear that such population pressure will have a huge impact on the environment. A sizeable increase in the human population will require a greater food supply that will, in turn, require more land for agriculture. The increased need for agricultural land and lumber will increase deforestation. Increases in waste production may result in large-scale pollution problems. Examples of the pressures the growing human population put on the environment can be found on isolated islands that became overpopulated. These islands serve as models of population growth pressures for much larger continents. For example, colonization of humans on isolated Pacific islands was usually followed by extinction of large animals and detrimental alteration of vegetation on the islands. Some environmental changes resulting from human colonization were even more intensive. For example, the Pacific Easter Islands lost their entire forest after colonization, mainly due to over-logging. The destruction on this island was so intensive it caused the extinction of several once-abundant tree species. A much larger island, Haiti, is now going through a similar transformation where massive logging is causing soil erosion, land degradation, and a shrinking food supply, among other social problems.

Processes of ecosystem degradation and the efforts to mitigate these are outlined in this chapter. Incentives for ecological restoration and the varied strategies for such work are also discussed.

1.1 Degradation of Ecosystems

Environmental problems abound, many of which have led to large-scale degradation of biomes. Today, environmental degradation is experienced globally and is the result of, among other causes, acid rain, massive soil erosion associated with agriculture, increased ultraviolet radiation due to stratospheric ozone depletion, excess addition of nutrients (eutrophication), and global warming. This list can also include more localized environmental degradation causes such as tropical rain forest destruction and desertification. Global environmental problems, including acid rain, ozone depletion, and global climate change, need to have a legislative framework and international effort to be solved. These issues are discussed further in Chapter 14.

Large-scale environmental degradation can induce negative feedback loops in the ecosystem. For example, large-scale deforestation of rain forests can interrupt hydrological cycles in the tropics. Deforestation in West Africa is probably contributing to the observed drought in the sub-Sahara (Sahel) region. Rainfall in tropical rain forests depends on moisture that comes from vegetation (via a process called evapotranspiration). Deforestation may, with time, therefore lead to a conversion of rain forest to a much drier savanna. Such changes are predicted to take place in the near future in the Amazon basin and to be intensified in tropical West Africa.

Global Warming

Increased levels of CO_2 in the atmosphere, mainly due to the burning of fossil fuels (coal, oil, and natural gas) and to large-scale forest clearings in the tropics, are mainly causing the current global warming. Today, CO_2 emission is highest in China. Carbon dioxide levels in the atmosphere have risen from about 290 ppm in 1860 AD to more than 370 ppm in 2000. In May 2008 the value had risen to 388 ppm and in March 2010 it was 391 ppm (**Figure 1.1**). During the same time the CO_2 level has risen, an increase in average global temperature has occurred (**Figure 1.2**). Future predictions on the impacts of **global warming** are grim. These include a multitude of problems including extreme weather events such as more frequent and stronger hurricanes and tornadoes, excessive heat waves, droughts, and torrential rain.

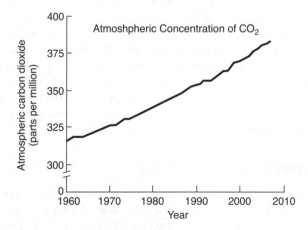

Figure 1.1 Increase of CO_2 with time in the atmosphere. (Data from Scripps Institute of Oceanography.)

Degradation of Ecosystems

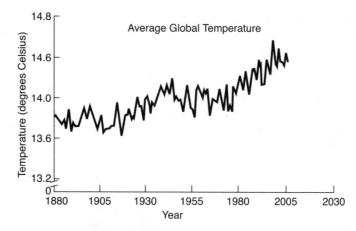

Figure 1.2 Increase of average temperature with time on Earth. (Data from GISS, BP, IEA, CDIAC, DOE, and Scripps Institute of Technology.)

One of the first signs of global warming came with the melting of ice caps on mountains and the shrinking of many large glaciers. Some of Asia's largest rivers (India's Ganges, Myanmar's Irrawaddy, Vietnam's Mekong, and China's Yellow, to name a few) originate in ice caps atop high mountain ranges such as the Himalayas. As the ice caps of the Himalayas and other mountains disappear, it is predicted that the water flow in these rivers will decline. The long-term consequences of declining water in these large rivers will be disastrous. Global warming will have an even greater impact on regions that are already facing chronic drought and desertification, such as in the countries bordering sub-Sahara (Sahel) and northern regions of China.

The sea level is rising due to the melting of glaciers and thermal expansion of the sea. It is expected that sea levels will rise about 20 cm over this century. This predicted sea level rise may destroy rice-producing deltas and floodplains in populous countries of Asia, including China, India, Myanmar, Vietnam, Thailand, and Bangladesh. This alone could create hundreds of millions of climate refugees bringing unprecedented consequences. As the sea waters rise, coral reefs will also be adversely affected. It is likely that more than 60% of existing reefs will be destroyed worldwide within the next 30 years. And, because coral reefs are important as spawning grounds for a myriad of fish species, their destruction could have unforeseen consequence for marine life. The impact of sea level rise on coastal environments will be discussed further in Chapter 9.

Global warming will most likely cause displacement of most ecosystems on Earth. Many species will in the near future simply find themselves in a warmer climate than the one in which they have adapted to live. Although migration of vegetation away from the equator and toward colder temperatures is possible, it is unlikely that whole ecosystems will be able to move hundreds of kilometers within a short time period. Also, global warming will affect delicate biological processes, such as vernalization, reproduction of plants, and seed germination, with unforeseen consequences. Animal populations are already suffering from global warming, as is shown by the declining moose (*Alces alces*) population in the Agassiz National Wildlife Refuge, Minnesota (from about 4,000 animals in 1985 to just 237 animals today). Similarly, the Arctic's polar bears (*Ursus maritimus*) have been in rapid decline. As the polar ice has been affected by global warming, the polar bear's very survival has been threatened. In 2007 the Arctic sea ice was 39% below the long-term average from 1979 to 2000. This progressive decline in sea ice in the Arctic has resulted in the polar bear's inclusion in May 2008 on the threatened species list under the Endangered Species Act.

Global warming has also been linked to unusual outbreaks of diseases. For example, global warming has been linked to a drastic increase (90%) in the larvae of pine beetles of lodgepole pines in Colorado, increasing the number of dead forest trees. This in turn, and especially when combined with the excess heat and drought caused by global warming, increases the risk of wildfires.

Energy Consumption

To solve the imminent problems of global warming, society needs to drastically reduce its dependency on fossil energy (coal, oil, and natural gas) and harness the power of renewable energy resources. Currently, the global energy need is about 16 TW (terawatts, 1 TW = one trillion watts) per year. This need is expected to increase to 26 TW in 2050 and rise even further to 39 TW by 2100. Today, most of this energy comes from burning fossil fuels, and the vast majority is used in developed countries (the United States uses about 25% of the world's energy).

Energy consumption in the United States changed drastically over the twentieth century. During the first four decades, most of the energy came from coal and biomass (logs). After the Second World War (1945) energy consumption increased and the proportion of oil and natural gas usage became greater than that of coal. Today, most energy consumption in the United States is by industry (33%) and transportation (28%). These percentages of energy consumption are followed by residential (21%) and commercial (18%).

Most of the electrical energy in the United States is derived from coal (45%) and natural gas (20%) firing plants; this is followed by nuclear power (19%) and hydropower, a renewable resource that accounts for only 7% of the total electrical energy. Wind and solar power, other sources of renewable energy, currently account for less than 1% of the total electrical energy.

Nuclear Power Although the United States does not rely heavily on nuclear power for generating electricity, other countries do. For instance, France produces most of its electricity (80%) by nuclear power. Germany has about 27% of its electricity produced by nuclear power and the United Kingdom about 21%. Compared with burning fossil fuels, which release CO_2 into the atmosphere, nuclear power facilities are more environmentally friendly with zero carbon emission. The downside, however, is that uranium mining is devastating to the environment and requires on-site restoration measures. Also, permanent storage of radioactive waste (the byproduct of nuclear power) has not yet been solved. In addition, there is a general opposition from the public against building new nuclear power facilities in the United States.

Oil Oil consumption has increased steadily in the world and, concurrently, so too has oil's price. In just the last few years the price of crude oil has risen from $30 per barrel (2004) and peaked to about $140 per barrel (2008). Increased demand for oil in China and India without any significant increase in worldwide oil production is the main reason behind this sharp rise in price. Oil consumption in China increased from 165 million barrels in 1996 to 1,064 million barrels in 2006. Although oil production in the United States peaked in the early 1970s, it has since declined sharply. During this same time oil prices have increased steadily. The world has not yet reached the peak of oil production, but it is getting close. Most of the oil reserves of the world still in the ground are located in the Middle East (Saudi Arabia, Iraq, and Iran). Political instability in this region only exacerbates the many oil dependency issues that exist today.

Because of rising oil prices and increased awareness of the relationship between CO_2 emission and global warming, it is predicted that renewable energy resources will increase in proportion and perhaps even phase out fossil fuel usage in the near future. **Renewable energy** resources include hydropower, wind energy, and solar power.

Hydropower Global electrical production by hydropower is currently only about 17%. However, some countries get most of their electrical energy through hydropower; these include Norway (90%), Brazil (81%), and Canada (60%). Large hydroelectric installations are among the world's largest electric power stations; these include the Three Gorges, China (18.2 GW, 1 GW = one billion watts), Iaipu, Brazil (14 GW), and Guri, Venezuela (18 GW). In the United States the largest hydropower installations are found on the West Coast (Washington, Oregon, and California). Although large hydropower installations might not increase worldwide in the near future, medium size (around 100 MW, 1 MW = one million watts) hydropower installations could add to the total contribution of hydropower. The negative impact of hydropower dams on rivers is discussed in Chapter 13.

Wind Energy Wind energy is another renewable energy source that has great potential. Although it currently makes up less than 1% of the total electricity source in the United States, wind energy supplies more than 20% to the total electrical needs of Denmark and more than 5% in Spain and Germany. Wind energy is the fastest-growing renewable energy source in the world. One reason for the rapid increase in wind-generated electricity is its relatively low cost. The cost of producing wind turbines dropped about 90% between 1980 and 2000. The electricity produced by wind turbines is also competitive to other energy sources. The cost of electricity produced by wind turbines is about

4 cents per kilowatt hour, which is about the same as for coal-firing electrical plants. And, just as the price to produce wind-generated electricity has decreased, the potential power of turbines has increased. Today, gigantic wind turbines such as the E-126 in Germany can produce more than 7 MW. Wind turbines are typically installed in locations that meet critical levels of wind velocity (7–7.5 msec at 50-m height). Such wind farms can typically generate about 100 MW. One of the largest wind farms in the United States is the Horse Hollow Wind Energy Center in Texas. Here, 421 wind turbines generate enough electricity to power 230,000 homes. Another benefit of wind farms is they can improve economic opportunity in rural areas. For example, a 250-MW wind farm in Iowa provides more than $2 million in property taxes to the local rural community.

Wind energy has a great potential to supply the world with electricity. The total electrical need of the United States could be matched by installing wind turbines in only three states: Texas, North Dakota, and Kansas. To supply 20% of the electrical need of the United States with wind power turbines would only cover about 0.6% of the total land area. Agricultural land can, however, still be used around such wind turbines. The location of wind farms, however, must be selected carefully to avoid negatively impacting the environment. Offshore wind farms such as the Cape Wind Project (Massachusetts) have met with huge opposition mainly due to the adverse aesthetic impacts. Other detrimental impacts of wind turbines include noise and risk to wildlife.

Solar Power Solar power is harnessed by the use of photovoltaic (PV) cells that convert light into electricity (**Figure 1.3**). Recently, the capacity of PV cells to generate electricity has improved greatly. Currently, the cost of PV electricity is competitive with nuclear-generated electricity (8–12 cents per kilowatt hour).

It is critical to locate PV cells where the annual solar energy is high. The annual solar energy in the United States is highest in the southwest (southern California, New Mexico, and Arizona). To generate enough electricity for the United States, PV cells would have to cover about 20,600 km^2. This required area will probably be reduced in the near future as PV cells become more efficient in generating electricity. Solar power farms (such as the 9-MW Tucson, Arizona Electric Power) are already in operation. Solar PV cells can also be installed on the roofs of residential and business buildings. As the PV cells become more efficient, these options are probably going to be used widely.

Future Impacts of Renewable Energy As generating renewable electrical energy becomes more economical, such energy can be used in the transportation system to run electrical trains and electrical cars, thereby reducing oil consumption even further. Other energy sources for vehicles, including ethanol, biodiesel, and hydrogen, can also help in reducing oil consumption.

It is not unrealistic to expect renewable energy resources to rapidly increase their proportion of generated electricity in the United States. In the near future wind energy could account for 30% and solar energy for another 30% of the total

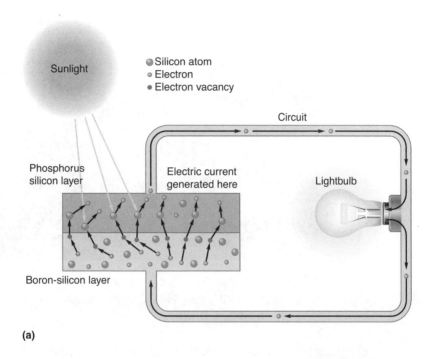

Figure 1.3 (a) Solar photovoltaic (PV) cells convert sunrays into electricity. (b) Solar PV cells installed on roofs provide homes with renewable energy. The two darker panels on the left of the garage roof are solar hot water panels. (Courtesy of the Department of Energy.)

(c)

Figure 1.3 *Continued* **(c)** Solar power farms using PV cells generate renewable energy. (© Stangot/Dreamstime.com.)

electrical production. Hydroelectricity along with geothermal power could add another 10%. Nuclear power (not renewable but zero CO_2 emitting) could contribute another 20%. The remaining 10% could come from fossil fuel (natural gas and coal) ideally if emission of CO_2 into the atmosphere is prevented. Carbon dioxide that is emitted from such power plants could, for instance, be injected into safe underground sediments for storage. Or, alternatively, it could be linked to algal-biomass production to be turned into ethanol or biodiesel.

Ecosystem Management Models

Appropriate land management, including restoration, is essential to halt land degradation. Models dealing with land management, however, have shown that public ownership of land has discouraged people from taking direct action to prevent serious land degradation. A well-known model describing this phenomenon, called the "tragedy of the commons," was introduced by Garrett Hardin in 1968. The model is based on an actual situation where farmers had access to common land where they could graze their livestock without any restriction. It showed that each person would only act in their own best interest with regard to access to natural resources. Such behavior, in turn, leads to an escalation in land degradation. This model has been extended to apply to any situation where commonly held natural resources are used by many people without any individual responsibility. Such situations lead to increasing pressure on finite natural resources as the human population increases exponentially. If property rights and responsibilities are not clearly outlined in the use of natural resources, an inevitable overexploitation and dysfunction of these systems follow. What this model and its expanded application provided, ultimately, was a link among social, political, and biological sciences into one framework to illustrate how they relate to the management of natural resources.

Ecosystems have limited capabilities to regenerate from disturbances. In this context it is important to consider the scale of the disturbances, which is discussed further in Chapter 2. It is, however, questionable whether or not biomes can regenerate from large-scale degradation. Ecosystems may, however, possess "self-regulating" mechanisms that counteract destructive disturbances. One of these self-regulating mechanisms suggests an inherent control on global warming, and a model of this has been proposed. The Gaia Model, as it is known, has gained popularity because of its great explanatory power. It predicts that global warming will result in increased evaporation from the sea, leading to an increase in cloud formation. This, in turn, will act like a shutter on a window facing the sun; sunrays will be reflected by the clouds and will be partially blocked, causing a cooling of Earth's surface. Although the Gaia Model has been given some importance, even its proponents have recently warned that the current trend of global warming could already be out of control and that our reliance on fossil fuel as an energy source should be rapidly curtailed.

Just as land degradation has worsened over time, so too have marine fisheries been degraded substantially worldwide. Most commercial fish species are currently intensively exploited. Management tactics such as quotas on fisheries have shown mixed success. Among alternative management strategies is one that would establish large marine reserves. Marine reserves are protected oceanic areas where fisheries are prohibited, as is the exploration for minerals, oil, and gas. Among the first marine reserves was the Goat Island Bay Reserve in New Zealand, established in 1971. As the marine ecosystem recovered within the reserve, a "spillover" was noticed where species within the reserve began to migrate to adjacent areas and, consequently, improved local fisheries. Today, New Zealand has 31 marine reserves. Marine reserves are now viewed as natural fish hatcheries that replenish surrounding oceans. Areas for marine reserves should be strategically selected where high biodiversity and ecosystem integrity is of importance. The world's largest marine reserve (the Northwestern Hawaiian Islands Marine Natural Monument) was established in 2006 when more than 300,000 km^2 of the Pacific Ocean was protected from commercial and sport fishing under the Antiquities Act. An even larger marine reserve (900,000 km^2) has been proposed for the Central Pacific that would include the uninhabited islands of Palmyra, Howland, and Baker. These islands are surrounded by coral reefs that are rich in biodiversity.

The near future will most likely see three possible scenarios in response to the unprecedented global environmental degradation that is taking place worldwide:

1. Ecosystem degradation will continue at an escalating rate with unforeseen consequences.

2. An effective strategy of no further degradation will be developed. In this case a loss of an ecosystem is counterbalanced by restoration.

3. A global strategy will be developed in which the rate of ecosystem restoration exceeds the rate of degradation. (Such a strategy is discussed later in this chapter.)

The renowned naturalist Edward O. Wilson at Harvard University has predicted that this century will be devoted to ecological restoration. Let's hope his vision comes true.

1.2 Value of Ecosystem Services

Ecosystem services are defined as life-supporting ecosystems. These processes include a continuous supply of materials (food, potable water, fiber, and lumber), energy (biomass), and new information (particularly for the agricultural and pharmaceutical industries) that benefits humans. Furthermore, ecosystem services provide humans indirectly with maintenance of soil fertility, hydrological control, carbon fixation (leading to climate stabilization), oxygen production, pollination of crops, and natural pest control. In addition, ecosystem services also provide subjective benefits having aesthetic, recreational, or cultural value. Such ecosystem services contribute to our standard of living but do not necessarily have a market price. It is, however, paramount that ecosystem services are recognized as life supporting; humans cannot live without a constant supply of food, oxygen, and potable water.

The total value of ecosystem services provided by functional ecosystems worldwide has been estimated to be worth at least $33 trillion. This amount is nearly double the combined gross national product of the entire world. Although it is hard to come to terms with such an enormous amount, it is obvious that it will be very costly, and perhaps impossible, to replace such services. And, given that some vital ecosystem services, such as oxygen production, are not included in this estimation, it is easy to see that the value of integral ecosystems will likely continue to grow rapidly. Estimations on ecosystem services are important because services that are free of charge are often treated inappropriately, as outlined by the "tragedy of the commons" model.

It is critical to realize that ecosystem services are generated by functional and integral ecosystems. Degraded ecosystems do not provide the same quality of ecosystem services as fully functional and integral ecosystems. Therefore, the value of ecosystem services that functional and integral ecosystems provide should be considered in large-scale restoration. The cost of restoring a damaged ecosystem, unfortunately, does not typically reflect the value of the ecosystem services. High economic values for ecosystem services are a tremendous incentive for restoring degraded ecosystems to become integrated and fully functional.

Although estimated economic value, as shown above, can be assigned to ecosystem services, it is often difficult to do so because the value is often influenced by our current needs. For example, it can be challenging to put economic value on information that benefits agriculture in the future. Humans still need the services that are provided by integral ecosystems characterized by genetic diversity. Because most of the caloric consumption of humans is provided by only a few grass species (i.e., rice, wheat, corn, barley, and oats) there is an ongoing pressure on improving (through breeding and genetic engineering) the production and adverse environmental tolerance of these species. New varieties of crops that are tolerant to soil salinity or drought are being developed. The global agricultural system is vulnerable to pathogenic attacks. One way of dealing with this problem is to breed pathogen-resistant varieties of crops, and such work is continuously taking place in agricultural research stations worldwide. For example, viral-resistant strains of corn (*Zea mays*) are provided by introducing resistant genes from a wild grass, teosinte (*Euchlaena mexicana*). The only native habitat of this wild grass (in the area of Jalisco, Mexico) was recently purchased to conserve this genetic material. It is estimated that the value of this genetic material was approximately $330 million. Similarly, other crop plants may benefit from genetic varieties found in related wild plant species.

Another approach in the evaluation of ecosystem services is to compare the function of an intact ecosystem with the function provided by a comparable artificial (engineered) system. For example, the cost of operating water purification facilities can be compared with the services provided by a functional watershed, including intact wetlands and forests. The city of New York compared two options in providing potable water to its residence: restoration of nearby watershed or building a water treatment facility. In 1996 the city of New York invested about $1 billion to restore a watershed in the Catskill Mountains, lying north of the city. After the restoration project was completed, the clean water from the Catskills was piped directly to New York City. The alternative option to build a water treatment facility would have cost $6 to 8 billion, followed by about $300 million annually for maintenance. Large-scale restoration and conservation of watersheds is certainly an economical option and should be compared as an alternative to the cost of constructing water treatment facilities. Conservation and restoration of the watershed in the Catskills was therefore the most cost-effective option. In addition, management of this watershed provided other benefits, such as highly valued recreation and flood control areas.

1.3 Outlook for Ecological Restoration

Restoration ecology has recently been established as a field of science. Although it is an interdisciplinary field of science, it is founded on ecological principles. Restoration ecology is an applied discipline that provides pragmatic solutions to environmental problems. Restoration is therefore "an acid test of our ecological

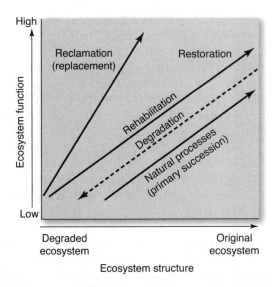

Figure 1.4 Model of ecosystem degradation and restoration options. Ecosystem function includes primary productivity and nutrient cycling. Ecosystem structure includes ecosystem complexity and biodiversity. (Reproduced from A. P. Dobson, A. D. Bradshaw, and A. J. M. Baker, *Science* 277 (1997): 515–522. Reprinted with permission from AAAS [http://www.sciencemag.org/].)

understanding," as elegantly phrased by the late Anthony D. Bradshaw. It can be implemented at the species, the community, the ecosystem, and even the large-scale biome level. **Restoration** is the process of rebuilding a degraded ecosystem until it reaches its original state (the state before any major disturbance). This includes its predisturbance structure (including taxonomic composition and complexity) as well as its functions (including primary productivity and nutrient cycling). To reach the original ecosystem state, different restoration approaches might be used (**Figure 1.4**). These include revegetation, reclamation, and rehabilitation (each discussed below).

It may, however, be difficult to reach the predisturbance ecosystem state for several reasons:

- Often, very little is known about the predisturbance state of an ecosystem.

- Many ecosystems have been disturbed by humans for such a long time it is impractical and often unrealistic to outline a prehuman disturbance state.

- Restoration efforts aiming at a "pristine" ecosystem state may be prohibitively complicated and consequently expensive.

- The predisturbance state of the ecosystem might not truly represent a "pristine" state because the ecosystem might already have been affected by other types of anthropogenic disturbances. (This situation is exemplified by the intensive ecosystem disturbances that took place in North America after the arrival of European settlers and the previous less-intensive disturbances caused by Native Americans.)

The question, then, is should restoration aim at an ecosystem state as it was before any human disturbances or a state affected by moderate human disturbances? Also, degradation may have led to the loss (or, in the worst case, extinction) of a keystone species on which the whole ecosystem depended.

Restoration of complex ecosystems can also be difficult to accomplish due to historical contingencies. For example, the assembly of natural ecosystems has been influenced by geological events such as the Ice Ages, where colonization of a large swath of land followed primary succession (see Chapter 4). Such ecosystem assembly was essentially a "tabula rasa" (new beginning). The ecosystem assembly was therefore influenced by an historical contingency affecting immigration trajectories of the flora and fauna. It would be nearly impossible to re-create similar immigration events and reach the complexity of such ecosystems.

Different Restoration Approaches

Selecting the appropriate restoration approach for each case mainly depends on the severity of the ecological disturbances. Another factor to consider is the final state of the ecosystem for which the work is aiming.

A strict restoration aims at attaining the "pristine" state of the ecosystem. The endpoint of the restored ecological state therefore should be an exact replica of the pristine state. Strict restoration has been criticized in the past for aiming at an almost impossible endpoint. Similarly, active restoration involves continuous human interactions to keep the succession on the preferred trajectory toward an endpoint that is an exact replica of the pristine state.

Another approach, the liberal or passive restoration, does not necessarily consider the predisturbance state of the ecosystem as the ultimate goal. Instead, this approach allows succession to follow its own trajectory, without active interference. This approach therefore considers several different endpoints of the final ecosystem.

Restoration of a severely degraded ecosystem often starts with revegetation, a process where the initial aim is to establish vegetation cover on a site. **Revegetation** is an initial attempt to restore any part of the structure and/or function of an ecosystem. This work is often accomplished by the use of commercial plants, some of which may be non-native. Revegetation involves, for instance, establishment of grass cover on denuded sites. It also can include seeding or transplanting hardy shrubs and trees to promote wildlife. Revegetation is typically applied to large areas after considering ecological, economic, and sociological constraints and demands. After initial revegetation efforts, different succession trajectories are sustained.

Reclamation of severely degraded sites aims at making such sites more habitable to wildlife. Often, the aim is to reach the predisturbance state. Examples of severely degraded sites typically used in reclamation include landfills, mines, gravel pits, and road works.

Rehabilitation is another approach that aims at rebuilding essential structures and functions of degraded ecosystems. This work does not necessarily aim at "pristine" or predisturbance states of the ecosystem. The final state of the rehabilitated ecosystem should, however, be stable and functional. Rehabilitation of forests is common where a different species combination is used instead of the predisturbance species assemblage.

Re-creation and **ecological engineering** are two other approaches. In each of these, rebuilding certain physical components of an ecosystem is involved. This work could, for instance, include installing hydrological devices (dams, levees, and drainage ditches) to improve ecosystem functioning. Strict engineering solutions include building structures (from rock or cement) that can, for instance, reduce coastal erosion. These include building seawalls, groins, breakwaters, and revetments. Ecosystem engineering aims at designing, building, and operating ecosystems to solve environmental problems. For example, wetlands have been designed to deal with wastewater. Species commonly used in such wetlands include water hyacinth (*Eichhornia crassipes*), common reed (*Phragmites australis*), and cattail (*Typha latifolia*). These species are all non-native and can become invasive. They grow rapidly and respond quickly to eutrophic waters by massive biomass production. Recently, there has been a trend in using native species for this purpose.

Acid mine drainage is commonly treated with constructed wetlands. Today, reducing and alkalinity producing wetland systems are commonly constructed to deal with this problem. Such constructed wetlands consist of layers of limestone rock, compost, and standing water. A network of perforated pipes is placed below a limestone layer to facilitate drainage. Water moves uniformly down through the organic layer and the limestone layer. As water moves through the compost substrate, dissolved oxygen is removed by the decomposition of organic material, iron is reduced, and the compost acts as a sink for aluminum hydroxide. A series of ponds is typically connected to the reducing and alkalinity producing wetland system to improve quality of the flowing water. Such constructed wetland system is referred to as a successive alkalinity producing system.

Varying Scales of Restoration

Whatever approach is used, ecological restoration is performed on different scales. Restoration can be implemented successfully on a small scale, for instance, in restoring mine tailings. Also, restoring fragments of habitats that represent a vast historical ecosystem is a common practice. Such an approach has been compared with creating "ecomuseums" or "living museums." The immediate benefits of such efforts are numerous and have a positive impact on the society. Ecological restoration can also involve more complex, much larger communities such as an integral watershed, forest, or wildlife.

Large-scale restoration programs involving whole biomes are rare. Examples include the Great Lakes (discussed in Chapter 13), the Florida Everglades (discussed in Chapter 13), forests in the northeastern United States and Canada (discussed in Chapter 11), and reintroduction of the gray wolf in the Yellowstone Park and bighorn sheep in the Rocky Mountains (discussed in Chapter 12). Other large-scale restoration projects include Chesapeake Bay, Columbia River, Coastal Louisiana, and San Francisco Bay and Delta.

Ecological restoration must play a pivotal role in abating global warming. This should be done chiefly by restoring biomes with high primary productivity (such as the tropical rain forest) to abate excess amounts of atmospheric CO_2. Other vast ecosystems could also sequester large amounts of CO_2. In fact, large-scale projects dealing with carbon **sequestration** into soil are already being implemented (see **Case Study 1.1** on page 21).

Acid Rain **Acid rain** is one example of air pollution that leads to widespread degradation of ecosystems that requires large-scale restoration effort and international mitigation (discussed further in Chapter 14). Acid rain occurs when gaseous emissions of sulfur oxides (SOx) and nitrogen oxides (NOx) interact with air moisture and sunlight. These chemicals are converted to strong acids such as sulfuric acid (H_2SO_4) and nitric acid (HNO_3), which get diluted with rain. Sources of acid rain are predominantly coal-burning electric facilities. Today, China is the leading emitter of SOx. In the United States acid rain is greatest in the eastern part of the country, mainly because of prevailing wind direction and the high number of coal-firing plants along the Appalachian mountains. In the northeastern United States the average pH of acid rain is about 4.6; however, the pH of rainfall can occasionally drop to 4.0.

Acid rain can directly affect vegetation and soil, which become gradually acidified. Such acidification can cause loss of native vegetation and soil microorganisms. Acid rain can also cause toxic substances to leach from soil and pollute groundwater. In Sweden, mercury (Hg) pollution of lakes is a widespread problem due to acid rain. Shallow lakes with low buffering capacity are also impacted rapidly by acid rain. With such potentially devastating consequences, environmental impacts are only worsened by the environment's almost complete inability to neutralize acid rain. As a result, acid rain has already damaged many shallow lakes, forests, and agricultural fields. It has even had direct effects on human health.

The simplest way to reduce acid rain is to use less energy from fossil fuels. Conventional methods of reducing NOx and SOx from smokestacks include installing scrubbers that spray a mist of water and powdered limestone into the fumes and recapturing most of the sulfur. Vehicle emissions of NOx and SOx can also be limited by using catalytic converters. However, areas that are already polluted by acid rain are typically restored by spreading powdered limestone on the ground to neutralize acids. Wide scale use of this method though would be prohibitively expensive.

Ozone Layer Restoration of the **stratospheric ozone** (O_3) layer is another example of large-scale restoration. The stratospheric ozone layer blocks almost all ultraviolet radiation from reaching the surface of Earth (**Figure 1.5**). These restoration efforts have immediate benefits for human health (i.e., reduced skin cancer) and also provide ecosystem benefits. The culprit behind stratospheric ozone destruction is the emission of chlorofluorocarbon (CFC) gas. By thinning the ozone layer considerably, all terrestrial life could be endangered due to excess ultraviolet radiation. Ozone is naturally produced, mainly when thunderstorms strike in the tropics, leading to a recharge of the stratospheric ozone layer.

In 1987 the international community moved to curtail detrimental impacts on the ozone layer by prohibiting the release of CFCs through the Montreal Protocol. This protocol called on 24 industrialized nations to stop production and use of CFCs. Unfortunately, several developing countries were allowed to phase out their CFC production over a longer time period than the more industrialized nations. The purpose was to alleviate any undue economical hardship, but the result is that, at present, CFC production is still ongoing. Today, more than 180 countries have signed a treaty on banning production and use of CFCs. As the amount of CFC production has been drastically reduced, it does appear that the ozone layer has been able to regenerate itself to some extent (**Figure 1.6**). Although it will probably take up to one century for the ozone layer to regenerate to its former concentration, in this case it was enough to curtail the degenerative factors to promote restoration.

Future of Restoration

Is restoration of major ecosystems worldwide possible? Lester R. Brown at the Earth Policy Institute has estimated how much it will cost to restore major ecosystems worldwide. His estimation included reforestation, soil conservation, rangeland restoration, restoration of hydrology, biodiversity, and also fisheries. He estimates the total annual budget for such a mega-enterprise at $93 billion. This price pales, however, in comparison with the $1 trillion that is spent annually around the world on military activities. Restoration efforts at this scale are needed on Earth for at least a century to ensure proper functioning of the major ecosystems that support our lives.

Today and in the near future, ecological restoration will play a pivotal role in regeneration of damaged ecosystems worldwide. Conservation efforts have recently been changing direction with emphasis on various aspects ranging from preservation and protection of pristine systems to the restoration of fragmented and degraded ecosystems. Ecological restoration is particularly practical because degraded land can be purchased at a much lower price than a comparable area of land in a pristine state from the same ecosystem. Therefore, much larger areas of degraded land can be purchased by conservation organizations. Restoring large swaths of land is essential for the conservation of many animal populations (see Chapter 12) and for the functioning of whole biomes.

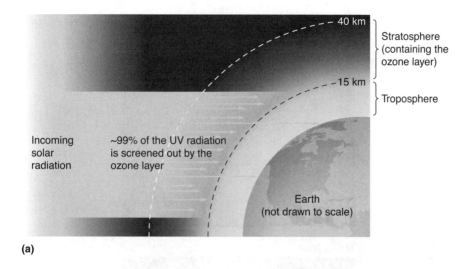

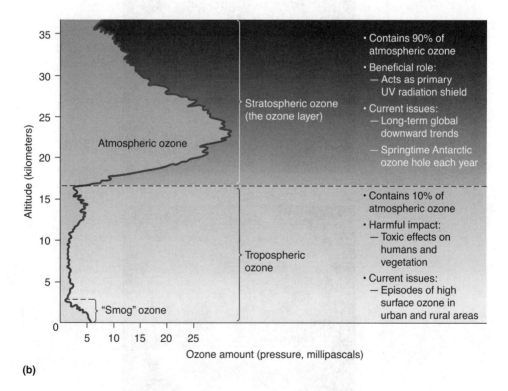

Figure 1.5 The atmospheric ozone (O_3) layer. **(a)** The stratospheric ozone layer screens almost all UV radiation from the sun. **(b)** Ozone concentration in the stratosphere and the troposphere.

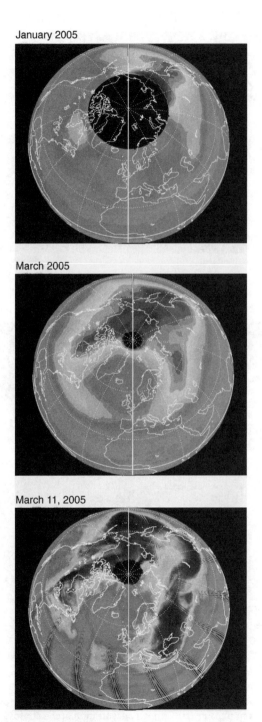

Figure 1.6 Changing size of the ozone hole above the Arctic. These images were computer generated. (Photos courtesy of NASA/JPL/Agency for Aerospace Programs (Netherlands)/Finnish Meteorological Institute.)

Summary

Humans have greatly modified their surrounding environment, often leading to drastic changes in ecosystem structure and function. The rapid growth of the human population is the underlying reason for today's increased pressure on the environment, which has led to unprecedented global degradation. Global pollution such as carbon dioxide emissions from burning fossil fuel is resulting in global warming. Global warming, in turn, is likely to impact almost all ecosystems worldwide in the near future. Most ecosystems will be displaced due to the warmer climate. Acid rain, also resulting from burning fossil fuel, has already acidified soil in large areas on Earth and drastically affected shallow lakes. Land management models emphasize property rights and responsibilities to manage natural resources. Ecosystem services that support our lives have enormous monetary value. This fact puts an even greater importance on proper management, conservation, and restoration of these life-supporting systems. Restoration aims at achieving an ecosystem as it was before disturbance and degradation. It can be challenging to define the predisturbance state of an ecosystem. Restoration efforts can involve passive restoration, where almost no human interference is needed, or active restoration, where continuous efforts are needed. Restoration can have different components such as revegetation, reclamation, or rehabilitation. Each approach aims to rebuild particular structural and/or functional components of the damaged ecosystem.

1.1 Case Study

Carbon Sequestration of Soil

Rattan Lal, Carbon Sequestration and Management Center, The Ohio State University, Columbus, OH

The atmospheric concentration of CO_2, which contributes about 50% of the total radiative forcing due to greenhouse gases (GHGs), has increased from 280 ppmv since the late 1700s to 377 ppm in 2004. It is presently increasing at the rate of about 0.47% or 1.9 ppm yr^{-1}. About 35% increase in CO_2 concentration in the post–industrial era is primarily due to two activities: fossil fuel combustion at about 7 Pg C yr^{-1} and deforestation and soil cultivation at 0.6–2.5 Pg C yr^{-1}. Of the total source of 8.5 Pg C yr^{-1} (range, 7.6–9.5 Pg C yr^{-1}), atmosphere absorbs 3.3 Pg, ocean 2.2 Pg, and an unknown terrestrial sink about 3.0 Pg C yr^{-1} (2.1–3.9 Pg C yr^{-1}). Similar to CO_2, concentration of CH_4 has also increased from about 700 ppb in the pre-industrial era to 1,783 ppb in 2004, an increase of 155%. The annual growth rate has been

(continued)

Case Study (continued)

13 ppb y^{-1} during the late 1980s and 5 ppb between 2000 and 2005. Principal sources of CH_4 are fossil fuel exploitation, wetland, rice cultivation, ruminant animals, and biomass burning. Well-drained soils oxidize CH_4. Atmospheric abundance of N_2O has increased from 270 ppb in the pre-industrial era to 319 ppb in 2004 and is presently increasing at the rate of 0.8 ppb yr^{-1}. Principal sources of N_2O are fertilizer use, biomass burning, combustion of fuels, and industrial processes. Land use, land use change, soil degradation, and soil management strongly impact fluxes of GHGs between terrestrial/aquatic ecosystems and the atmosphere.

Increase in atmospheric concentration of all three GHGs is considered responsible for the observed and projected increase in global temperature. Although fossil fuel combustion has been a major source since the industrial era, land use conversion, biomass burning, and soil cultivation have been the dominant sources of GHGs ever since the dawn of settled agriculture about 10,000 years ago. World soils are a sink for atmospheric CO_2 under most natural ecosystems and mostly a source under agricultural ecosystems.

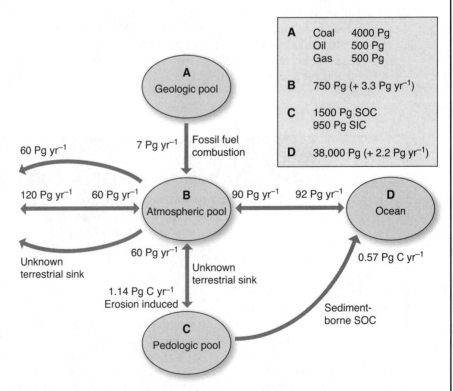

Figure 1.7 Global C pools and fluxes among them. (Courtesy of Rattan Lal, Ohio State University. Illustration adapted from original source.)

World soils constitute the third largest among four global C pools (**Figure 1.7**). The pedological pool, estimated at 2,500 Pg (Pg = petagram = 1×10^{15} g = 1 billion tonne = 1 gigatonne or Gt), has two distinct components: soil organic carbon (SOC) and soil inorganic carbon (SIC) pools. The SOC pool consists of highly active humus and some inert charcoal and is derived from remains of plants and animals at different stages of decomposition. The SIC pool includes elemental C and carbonate material such as calcite, dolomite, and gypsum. The pedologic pool interacts with (1) the atmospheric pool through soil respiration; (2) the biotic pool through transfer of biomass C in roots, detritus material, and leaf litter into the subsoil; and (3) oceanic pool through transfer of sediment-borne and dissolved organic carbon via rivers into oceans. The pedological pool is strongly influenced by land use, land use change, soil management, and farming/cropping systems.

Most agricultural soils contain lower SOC pools than their counterparts under natural ecosystems. Thus, conversion of agricultural soils and degraded ecosystems to a restorative land use can restore the SOC pool while improving soil quality. This chapter, therefore, describes the basic principles of carbon (C) sequestration in soil as a strategy of mitigating the climate change due to atmospheric enrichment of CO_2 and outlines some land use and management options that can restore the SOC pool.

Soil as a Source for Atmospheric CO_2

The SOC pool in undisturbed natural ecosystems is at a steady-state equilibrium, because the biomass C input into the system is balanced by losses through mineralization. Conversion of natural to agricultural ecosystems disturbs this equilibrium and leads to reduction in the SOC pool. Depletion of the antecedent pool is caused by numerous factors, including a decrease in addition of the biomass C to the soil, an increase in rate of mineralization caused by changes in moisture and temperature regimes, and an increase in losses of SOC pool by erosion and leaching. The rate of SOC depletion is more drastic in the tropics than in temperate region soils. Over time, soils under agricultural ecosystems attain a new equilibrium at a much lower level than that under the natural ecosystems (**Figure 1.8**). The rate of SOC depletion

$$d = \frac{\Delta y}{\Delta x}$$

is generally high in soils prone to accelerated erosion and of coarse texture containing low concentrations of low-activity clays (e.g., 1:1 clay minerals). In contrast, the rate of SOC depletion is low in clay soils with high concentration of high activity clays, high water retention capacity, and low internal drainage; relatively less prone to accelerated erosion; and in cool/temperate climates. The rate of depletion also depends on the antecedent pool, being more in soils with high than low SOC pool. In general, soils of the temperate climates lose 30% to 50% of the antecedent pool within 50 to 100 years of land use change (see Figure 1.8). In contrast, soils of the tropics may lose 50% to 75% within 10 to 20 years after conversion from natural to agricultural ecosystems.

(continued)

Case Study (continued)

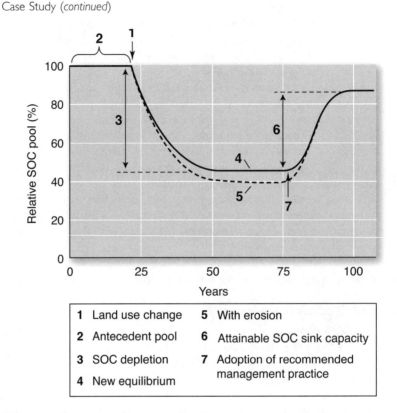

Figure 1.8 A schematic model showing the magnitude and rate of SOC depletion by land use conversion from natural to agricultural ecosystem and SOC sink capacity and rate of sequestration upon adoption of recommended management practices. (Courtesy of Rattan Lal, Ohio State University. Illustration adapted from original source.)

Soil Degradation and SOC Pool Depletion

The magnitude of depletion of the SOC pool is accentuated by soil degradation processes. The rate of SOC depletion is especially high in soils prone to erosion, which involves preferential removal of the light (low-density) humus or soil organic matter fractions concentrated in vicinity of the soil surface. In contrast to erosional processes that remove SOC by wind or water movement, other degradative processes deplete the SOC pool by decreasing the amount of biomass C returned to the soil. Several degradative processes are nutrient depletion, acidification, elemental imbalance leading to toxicity (e.g., Al, Mn, Fe) or deficiency (e.g., N, P, K, Zn), structural decline, compaction, and crusting. In contrast to other degradative processes, higher rate of SOC depletion with erosion is attributed to two factors: preferential removal of SOC fraction by flowing water and blowing wind and decrease in biomass production and addition to the soil because of severe decline in soil quality by accelerated erosion.

Table 1.1 Extent and Severity of Global Soil Degradation by Different Processes (in million hectares)

Degradation Process	Light	Moderate	Strong	Total
Water erosion	343	527	224	1,094
Wind erosion	269	254	26	549
Chemical degradation	93	103	43	239
Physical degradation	44	27	12	83
Total	749	911	305	1,965

Reproduced from D.J. Greenland and I. Szabolcs, (Eds.); 1994; *Soil Resilience and Sustainable Land Use*; CAB International, Wallingford, UK.

Globally, there are large areas of degraded soils, especially in regions of predominantly extractive farming practices (e.g., sub-Saharan Africa, South Asia) and harsh climate (tropics) and steep terrain (Andean region, Himalayan regions, Loess plateau of China). Of the estimated 1,965 million hectares (Mha) affected by different degradation processes, 1,094 Mha are affected by water erosion, 594 Mha by wind erosion, 239 Mha by chemical degradation, and 83 Mha by physical degradation (**Table 1.1**). Of the total degraded area, strongly degraded soils comprise 224 Mha (20.6%) by water erosion, 26 Mha (4.4%) by wind erosion, 43 Mha (18%) by chemical degradation, and 123 Mha (14.5%) by physical degradation. The magnitude of SOC depletion is generally higher in strongly degraded than in slightly or undegraded soils. Severe depletion of the SOC pool, with attendant decline in activity and species diversity of soil flora and fauna, is called "biological degradation." Soils managed with extractive farming practices, based on low input and removal of crop residues and biomass, are prone to biological degradation and are a source of atmospheric CO_2. A strategy of restoration of degraded soils and ecosystems can therefore reverse the degradation trends and lead to large sequestration of atmospheric CO_2. Soils with strongly depleted SOC pools respond positively with significant increase in SOC pool proportional to increase in C input through roots, crop residues, and other biosolids.

Historic Loss of Soil Carbon Pool

Estimates of the historic loss of SOC pool since the industrial era around the 1700s range from 40 to 537 Pg C. **Table 1.2** shows that world cropland soils have lost 66 to 90 Pg (78 ± 12 Pg) of C from the root zone, most of it emitted into the atmosphere through mineralization. Of this amount, 19 to 32 Pg has been lost by erosional processes, and some of this (20% or more) was also emitted into the atmosphere. Over and above the loss of SOC, soils of the arid and semiarid regions may have lost additional 10 to 20 Pg of C from the SIC pool. The SOC loss since the onset of agriculture 10 millennia ago is estimated at 320 Pg. In contrast, estimates of fossil fuel emissions are about 270 ± 55 Pg. Thus, world soils have historically been a principal source of enrichment of atmospheric CO_2. Such estimates of historic soil C loss, made

(continued)

Case Study (continued)

Table 1.2 Estimates of the Historic Loss of SOC Pool from World Cropland Soils

Soil Order	Total Area (Mha)	Cropland Area (Mha)	Estimated Historic Loss (Pg)
Alfisols	1,330	290	15–18
Andisols	106	50	5–7
Aridisols	1,556	40	0.2–0.3
Entisols	2,168	80	0.8–1.3
Inceptisols	946	150	8–13
Mollisols	925	290	7–11
Oxisols	1,012	300	22–27
Spodosols	348	50	1–3
Ultisols	1,175	130	6–7
Vertisols	320	60	1–2
Gellisols	1,120	0	0
Others	1,870	110	0.2–0.3
Total	13,035	1,550	69–90

Reproduced from *Progress in Environmental Science* by R. Lal. Copyright 1999 by Elsevier Science & Technology Journals. Reproduced with permission of Elsevier Science and Technology Journals in the format Textbook via Copyright Clearance Center.

at a soilscape and farm or watershed level, provide a reference point with regards to the soil C sink capacity (see Figure 1.8). The latter comprises potential and attainable capacity. The potential sink capacity equals the historic C loss, 66 to 90 Pg on global scale since the industrial era. The attainable sink capacity equals the actual amount that can be sequestered. As a rule of thumb, the attainable sink capacity is about two-thirds of the historic C loss.

Soil Carbon Sequestration

Carbon sequestration implies transfer of atmospheric CO_2 into another long-lived C pool. There are a wide range of technological options of sequestration of atmospheric pool into long-lived pools (**Figure 1.9**). Geological sequestration implies capture, purification, transport, and injection of atmospheric CO_2 into a stable geological strata (e.g., coal mines, old oil wells, or saline aquifer). Oceanic sequestration is of two types: biotic sequestration involves increase in photosynthesis by plankton and other flora through Fe fertilization, and abiotic sequestration is the injection of industrial CO_2 deep into the ocean where it stays in a liquid form. Chemical sequestration involves conversion of purified CO_2 into carbonates and other stable minerals.

Soil C sequestration is one of the three forms of terrestrial sequestration, each of which depend on the natural process of photosynthesis (**Equation 1**):

$$106CO_2 + 16NO_3^- + HPO_4^{2-} + 122H_2O + 181H^- = C_{106}H_{263}O_{110}N_{16}P + 138O_2 \quad \text{(Eq. 1)}$$

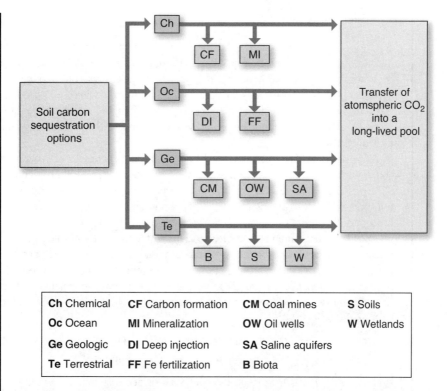

Figure 1.9 Strategies of sequestration of atmospheric CO_2. (Courtesy of Rattan Lal, Ohio State University. Illustration adapted from original source.)

This reaction governs the primary production by chlorophyll-bearing plants. Annually, 120 Pg of $CO_2 - C$ is converted into biomass by the photosynthesis reaction outlined in **Equation 1**. About half of this is returned back to the atmosphere through respiration and half transferred to soil and eventually returned to the atmosphere through soil organic matter decomposition. A small fraction of the photosynthetic material returned to the soil, however, is converted into SOC pool through the process of "humification." Humus is a dark brown or black amorphous material. It has a large surface area and high charge density and is bound to clay minerals as organomineral complexes. A hypothesis of formation of aggregates or organomineral complexes is shown in **Equation 2**:

$$xy[(Cl\text{-}P\text{-}H)] \underset{\text{Dispersion}}{\overset{\text{Aggregate}}{\rightleftharpoons}} [(Cl\text{-}P\text{-}H)_y] \underset{\text{Dispersion}}{\overset{\text{Aggregate}}{\rightleftharpoons}} [(Cl\text{-}P\text{-}H)_y]_x \quad \text{(Eq. 2)}$$

$$\text{(Domain)} \qquad \text{(Micro-aggregate)} \qquad \text{(Aggregate)}$$

where Cl is a clay particle, P represents a polyvalent cation (e.g., $Ca^{+2}, Al^{+3}, Mn^{+4}, Fe^{+3}$), and H is a long-chain humic polymer. Aggregation begins with precipitation of xy number

(continued)

Case Study (continued)

of organomineral complexes as domains into x number of microaggregates and eventually into a few complex aggregates. Carbon sequestered in stable microaggregates is encapsulated for a long time and is physically protected against decomposition by microbial processes. Formation of organomineral complexes with polyvalent cations leads to chemical protection and to long-chain complex polymers to biological protection. These are some of the protective mechanisms responsible for long residence time of the SOC pool.

Transfer of SOC into subsoil, away from the zone influenced by climatic elements and anthropogenic perturbations, is another mechanism of physical protection of the SOC pool. Transfer of SOC into the subsoil, either through a deep tap root system or transport as dissolved organic carbon and its re-precipitation, is an important pedological process of SOC sequestration called "illuviation of clay-borne SOC."

Soil C sequestration also involves changes in the SIC pool, which is an important constituent in subsurface horizons of soils of the arid and semiarid regions. The two types of carbonates in soil are the primary or lithogenic carbonates and secondary or pedogenic carbonates. Primary carbonates are derived from the weathering of parent rock. In contrast, secondary carbonates are formed through the reaction of atmospheric CO_2 with Ca^{+2} or Mg^{+2}:

$$CO_2 + H_2O \leftrightarrows CaCO_3 + H_2CO_3 \leftrightarrows CCa^{2+} + 2HCO_3 \qquad \text{(Eq. 3)}$$
gas   C   Aquous   Aquous   Aquous

An increase in the CO_2 concentration in soil air or decrease in pH drives the reaction in **Equation 3** to the right, thereby dissolving the carbonate and translocating Ca^{+2} and HCO_3^- into the subsoil along with the percolating water. The reaction moves to the left, leading to precipitation of carbonates (chemical or C) if the partial pressure of CO_2 is lowered, pH is increased, or soil moisture content is decreased. The presence of CO_2 in soil air is increased by addition of compost, mulch, other biosolids and root respiration. Water percolating down the profile, either by irrigation or rainfall, can translocate the material in the subsoil where it may precipitate as $CaCO_3$. The rate of SIC sequestration as secondary carbonates is only 5 to 10 Kg C ha^{-1} yr^{-1}. However, the process occurs in large areas of arid and semiarid regions, which cover about 4 to 4.9 billion hectares of Earth's land area.

Pedogenic carbonates formed from base-rich bedrocks or noncarbonate sediments are a net sink of atmospheric CO_2. In contrast, those formed from calcareous parent materials are neither sink nor source because for each mole of CO_2 sequestered, a mole of CO_2 is released upon reprecipitation of $CaCO_3$:

$$CaCO_3 + CO_2 + H_2O \longrightarrow Ca^{+2} + 2HCO_3 \longrightarrow CaCO_3 + CO_2 + H_2O \qquad \text{(Eq. 4)}$$

On geological time scale this process simply transfers carbonates from land via rivers to the ocean.

Leaching of bicarbonates is another mechanism of SIC sequestration. In a system of partial or complete soil leaching, such as in irrigation with good-quality water, the process can transfer as much as 0.25 to 1.0 Mg C ha^{-1} yr^{-1}.

Whereas most of the above processes can be managed/altered through soil, water, or vegetation manipulation, SIC sequestration can also occur through natural process of chemical weathering. For example, weathering of Ca-Mg silicates (e.g., Plagioclese, Olivine, pyroxene, volcanic gloss) into 1.1 clays (e.g., kaolinite, halloysite) can lead to fixation of atmospheric CO_2:

$$CaAl_2Si_2O_8 + 2CO_2 + 3H_2O \longrightarrow Al_2Si_2O_5(OH)_4 + Ca^{+2} + 2HCO_3^- \longrightarrow CaCO_3 + CO_2$$
(Eq. 5)

The reaction depicted in **Equation 5** causes net removal of atmospheric CO_2 because of the transport of Ca^{+2}, Mg^{+2}, and HCO_3^- from the weathering site via rivers into the ocean where it precipitates as carbonates. Although Ca^{+2} and Mg^{+2} are transported from the weathering site, Al and Si remain as clay minerals. On a geological time scale, this reaction is a net C sink of atmospheric CO_2.

Rates of Soil Carbon Sequestration

Adoption of a restorative land use and recommended management practices can restore the depleted SOC pool. Restorative processes are set in motion more readily in soils with high than with low resilience characteristics. Rates of SOC sequestration vary widely among soil types (texture, clay minerals), landscape position (foot slope vs. shoulder slope), climate (temperate vs. tropics), and precipitation (humid vs. dry) (**Table 1.3**). High rates of SOC sequestration (about 1 Mg C ha^{-1} yr^{-1}) are observed in cool and humid climates. Irrigation and water conservation can enhance SOC rates in soils of arid and semiarid climates. Adoption of those recommended management practices that enhance the quantity and quality (C/N ratio) of biomass returned to the soil generally increases the SOC sequestration rate. Incorporation of a cover crop in the rotation cycle also increases the rate of SOC sequestration. Relative duration of the cover cropping versus cultivation period can be estimated through

Table 1.3 Potential Rates of Soil Organic Carbon Sequestration (millions of grams of carbon per hectare per year)

Land Use and Management	Temperate Climate		Tropical Climate	
	Humid	Dry	Humid	Dry
Conservation tillage	0.5–1.0	0.25–0.5	0.25–0.5	0.1–0.25
Cover cropping/cropping systems	0.2–0.5	0.1–0.2	0.1–0.2	0.05–0.1
Integrated nutrient management/manuring	0.3–0.6	0.2–0.4	0.2–0.4	0.1–0.2
Water management/irrigation	−0.1 – −0.2	0.25–0.5	−0.1 – −0.2	0.2–0.4
Agroforestry	0.3–0.6	0.2–0.4	0.4–0.8	0.2–0.4
Improved grazing	0.2–0.4	0.1–0.2	0.4–0.8	0.1–0.2
Soil restoration	0.5–1.0	0.4–0.6	0.8–1.2	0.2–0.4

A negative sequestration in humid climates is caused by drainage of poorly drained/wetland soils.

(continued)

Case Study (*continued*)

use of simple models used to express the SOC dynamics:

$$\frac{dC}{dt} = -KC + A \qquad \text{(Eq. 6)}$$

where K is the decomposition constant, C is SOC concentration, and A is the accretion constant and reflects the amount of biomass C added to the soil. Soil C sequestration occurs when the quantity (−KC + A) is positive and depletion when it is negative. The numerical value of K is high for practices based on plow tillage, crop residue removal or burning, drainage of wetlands, summer fallowing, extractive farming practices, high stocking rate, and cropping systems that exacerbate soil erosion. In contrast, K is low for practices based on afforestation, crop residues/biomass returned to soil, conservation tillage, use of cover crops, integrated nutrient management including manuring, controlled grazing with improved pastures, and agroforestry systems. To assess the time (duration) of cover cropping (t_{cc}) versus that of cropping (t_c) in the rotation cycle, **Equation 6** can be written as follows assuming a steady-state condition when $dC/dt = 0$:

$$(-K_c C_m + A_c)t_c + (-K_{cc} C_m + A_{cc})t_{cc} = 0 \qquad \text{(Eq. 7)}$$

where C_m is the mean SOC concentration, K_c and K_{cc} are decomposition constants, and A_c and A_{cc} the accretion constants during the cropping and cover cropping phase, respectively. Terms in **Equation 7** can be rearranged to calculate the ratio of cultivation period (t_c) to that of the cover cropping (t_{cc}):

$$\frac{t_c}{t_{cc}} = (A_{cc} - K_{cc} C_m)/(K_c C_m - A_c) \qquad \text{(Eq. 7)}$$

Constants K and A are determined experimentally, and C_m is chosen as the desired level of SOC concentration with regards to the improvement in soil quality. Experimental assessment of the SOC sequestration rates, for diverse soils and management systems in major global ecoregions, is a high priority.

Implications to Global Food Security

Improvement in soil quality through carbon sequestration has numerous ancillary benefits. Important among these are decrease in non–point source pollution through reduction in runoff and erosion, improvement in water quality through increase in the effectiveness of soil as a biomembrane for filtering and denaturing pollutants, increase in biodiversity especially that of soil biodiversity, reduction in net CO_2 emission through off-setting emissions due to combustion of fossil fuel, increase in use efficiency of petroleum-based inputs (e.g., fertilizers, irrigations), and increase in agronomic productivity. Indeed, success in achieving the U.N. Millennium Goals of eradicating poverty and eliminating world hunger is closely linked to success in enhancing SOC sequestration in soils of developing countries.

A strong and a direct relationship exists between the SOC pool and agronomic yields, especially in mineral soils managed by the resource-poor farmers. All other edaphological factors remaining the same, yield of crops grown on mineral soils increases with increase in SOC pool up to certain limit. Increase in SOC pool by 1 Mg C ha^{-1} can increase grain yield by 150 to 300 Kg ha^{-1} of corn, 30 to 60 Kg ha^{-1} of wheat, 20 to 30 Kg ha^{-1} of soybeans. Global increase in food production by increasing SOC pool by 1 Mg C ha^{-1} yr^{-1}, estimated at 24 to 40 million Mg yr^{-1}, is enough to meet the current and projected food deficit globally and in regions characterized as hunger hot spots of the world.

It is in this context that restoration of degraded soils and ecosystems is a high priority. Reduction in the net CO_2 emission and the attendant mitigation of climate change is an ancillary benefit of SOC sequestration. The principal benefits of SOC sequestration include eliminating hunger, advancing food security, and eradicating poverty in rural population of Asia, Africa, and Latin America. Carbon sequestration of soils, through adoption of recommended management practices and restoration of degraded soils, is an important strategy to achieve these goals.

Key Terms

Acid rain   17
Carbon sequestration   17
Ecological engineering   16
Ecosystem services   12
Global warming   4
Reclamation   15

Re-creation   16
Rehabilitation   16
Renewable energy   7
Restoration   14
Revegetation   15
Stratospheric ozone   18

Key Questions

1. How does the increase in human population result in environmental degradation?
2. How can appropriate land management counteract land degradation?
3. Describe ecosystem services.
4. Define the term ecological restoration.
5. Compare strict restoration to liberal restoration.

Further Reading

1. Brown, L. R. 2006. *Plan B 2.0. Rescuing a planet under stress and a civilization in trouble.* New York: W.W. Norton.
2. Diamond, J. 2005. *Collapse. How societies choose to fail or succeed.* New York: Viking Penguin.
3. Gates, D. M. 1993. *Climate change and its biological consequences.* Sunderland, MA: Sinauer Associates.

4. Gilbert, O. and Anderson, P. 1998. *Habitat creation and repair.* Oxford, UK: Oxford University Press.
5. Harris, J. A., Birch, P., and Palmer J. 1996. *Land restoration and reclamation. Principles and practice.* Harlow, Essex, UK: Longman.
6. Kangas, P. C. 2004. *Ecological engineering.* Boca Raton, FL: Lewis Publishers.
7. Jordan III, W. R., Gilpin, M. E., and Aber, J. D. 1987. *Restoration ecology. A synthetic approach to ecological research.* Cambridge, UK: Cambridge University Press.
8. Mitch, W. J. and Jorgensen, S. E. 2004. *Ecological engineering and ecosystem restoration.* New York: John Wiley & Sons.
9. Perrow, M. R. and Davy, A. J. 2002. *Handbook of ecological restoration*—Volumes 1 and 2. Cambridge, UK: Cambridge University Press.
10. Urbanska, K. M., Webb, N. R., and Edwards, P. J. 1997. *Restoration ecology and sustainable development.* Cambridge, UK: Cambridge University Press.

2

ECOSYSTEM FUNCTIONING

Chapter Outline

2.1 Ecosystem Disturbances
  Stress
  Abrupt Change
  Frequency
  Magnitude
  Duration, Abruptness, and Return Interval
  Mega-Disturbances
  Severe Disturbances
2.2 Fire Disturbances
  Fire-Adapted Ecosystem
  Prescribed Burning
  Fire Suppression
2.3 Fragmentation
2.4 Nutrient Cycling
2.5 Hydrological Cycling
2.6 Functional Groups and Ecosystem Engineers
  Drivers and Passengers
  Ecosystem Engineers
2.7 Keystone Species
  Top-Down Control Mechanism
  Predators
  Mutualistic Species
Case Study 2.1: Importance of Fire and Grazing in Grassland Biomes

Ecosystem functioning is regulated by various processes such as nutrient cycling, primary productivity, and energy flow between trophic levels. This functioning can be controlled by various mechanisms, including a top-down feedback mechanism. For instance, a top trophic level predator controls a population of herbivores. On the other hand, primary productivity of an ecosystem can induce a bottom-up feedback mechanism that controls the population size of large herbivores. The ultimate goal of restoration efforts is to enhance functioning of degraded ecosystems. This involves installing control mechanisms to enhance ecosystem function.

Generally, natural disturbances affect ecosystem functioning and may increase available niches in the ecosystem and therefore indirectly increase biodiversity. Infrequent and drastic disturbances, however, are more likely to lead to massive extinction of species on a local scale.

Disturbances can have cascading effects on the ecosystem. For example, excessive wind that damages trees can facilitate insect and fungal attack on broken limbs. This in turn gives negative feedback and increases a forest's susceptibility to further disturbances.

Various types of disturbances cause ecosystem degradation that can lead to serious disintegration or ecosystem collapse. Such disturbances are likely to impact community composition and ecosystem functioning because they can have widespread effects. For example, large-scale deforestation can lead to reduced precipitation and even changes in climate patterns that can intensify drought. Disturbances such as overgrazing by livestock on marginal lands can also have widespread effects by altering the albedo (reflection of sunlight) of the land surface and, consequently, changing regional patterns of temperature and precipitation. Different scales of disturbances and their influences on ecosystems are discussed further below, as is the importance of managing disturbances in restoration. This involves both installing a natural disturbance regime and curtailing anthropogenic disturbances.

Restoration efforts aim at building up critical budget of macronutrients (nitrogen [N], phosphorous [P], and potassium [K]) to ensure nutrient cycling. Effective nutrient cycling is essential for ecosystem function. Factors that influence nutrient cycling are discussed further in this chapter. Restoration of the hydrological cycle is also critical for functioning of ecosystem. Factors that affect the hydrological cycle are discussed further in this chapter. Few species or group of species may be critical for the functioning of the ecosystem. The role of keystone species, functional groups, drivers, ecosystem engineers, predators, and mutualistic species in ecosystem functioning are also discussed.

2.1 Ecosystem Disturbances

A **disturbance** is defined as any event that disrupts ecosystem structure or function. Natural disturbances are ubiquitous but vary tremendously in their ecological impacts. They are the main factors behind spatial heterogeneity in ecosystems. Natural disturbances therefore play a crucial role in maintaining different niches within the ecosystem, which in turn increases biodiversity.

An ecosystem can experience different regimes of disturbances. The **disturbance regime** is the sum of all disturbances that impact an ecosystem. Total disturbance regime combines the effects of natural and introduced disturbances by humans. Examples of human-caused disturbances include

- Changes in fire regime
- Overgrazing by livestock
- Logging
- Clear-cutting
- Various agricultural practices (i.e., slash/burning and fallow)
- Surface mining
- Altered hydrology (i.e., draining or flooding)
- Large-scale pollution (i.e., acid rain or eutrophication)

Among disturbances that are considered to be natural are

- Fire induced by thunderstorm lightning
- Any abnormal weather patterns (i.e., hurricanes, tornadoes, and flash floods)
- Gap dynamics in forests
- Coastal processes
- Landslides
- Natural erosion
- Volcanic activities

In addition to these natural disturbances, there are those that are caused by ecosystem factors, including insect outbreaks, diseases, predations, impacts of ecosystem engineers, and invasion of non-native species.

Stress

Ecosystems can be disturbed through continuous stress. Continuous stress may affect ecosystem functioning with a direct or abrupt influence on ecosystem structure. One example of the impact of stress on organisms can be found in the sharp population decline of California's Chinook salmon. In 2002 about 804,000 Chinooks were counted in California's Sacramento River as they returned from sea to spawn. In 2006 this number had declined to 277,000. In 2008 just 90,000 Chinooks were counted. One cause of this sharp decline in the Chinook population is declining water quality of the river due to huge water diversion for agriculture and human consumption.

Abrupt Change

Disturbances may also affect ecosystems abruptly. Abrupt changes in ecosystem function or structure is also called regime shift and is discussed further in Chapter 5. Abrupt changes might take an ecosystem to a certain critical state that can result in an alternative ecosystem state. Alternative ecosystem states are usually characterized by a shift in dominance among species (discussed further in Chapter 5). Shifts in an ecosystem state are usually not desired because they can lead to a loss in ecosystem services that can, in turn, have an adverse economical impact. Restoring a degraded ecosystem to another desired state is not a straightforward process and is one that may require drastic and expensive approaches (see Chapter 5). There is a considerable interest therefore in identifying factors that act as switches between ecosystem states. It would be helpful to identify such switches before they take place, but so far that has been a challenge.

Frequency

In addition to what causes them and to how they affect ecosystems, disturbances are also characterized by their frequency. The frequency of disturbances can vary tremendously. Disturbances are typically infrequent and vary in magnitude from subtle local ecological changes to widespread catastrophic ones. Disturbance frequency is usually interrelated to disturbance magnitude. Disturbances of low magnitude usually have high frequency. Conversely, disturbances that have high magnitude usually have low frequency.

Frequent disturbances most likely result in dynamic shifts in ecosystem states. This has been termed the nonequilibrium paradigm, and it has strong implications for restoration. For example, ecosystems that are in nonequilibrium may exist in several alternate states. The restoration work must therefore aim at the final ecosystem as a "moving target." With this in mind, using only one ecological state as a reference site might not be a realistic option in restoring that site. (This phenomenon is discussed further in Chapter 5.)

The frequency of hurricanes in the northeastern United States is low (on average two major hurricanes make landfall every century). Disturbances associated with hurricanes can have an impact on ecosystem structure and function. However, the temperate forests of the northeastern United States demonstrate high resilience to such disturbances.

The impact of hurricanes on temperate forests can be tremendous, leading to more than 70% decline in density of trees. This is caused by the direct blow-down of large trees and also by indirect damage from the impact of falling trees. The survival of damaged trees remains above 40% several years after a hurricane. Regeneration of vegetation in the blow-down understory is usually rapid. After hurricanes, species that typically dominate sites early in secondary succession increase their cover. Because of the high rates of tree survival after a hurricane, biodiversity is not greatly impacted. Hurricanes result, however, in a massive structural alteration of the forest, where the average height of the canopy is reduced from more than 20 m in height to just 1 to 5 m. Even with this, however, ecosystem functions, such as primary productivity and nutrient cycling, remain largely intact. During the initial phases of secondary succession, ecosystem productivity declines. However, as big logs decompose productivity usually recovers rapidly.

The frequency of hurricanes in southern United States is higher; more than three hurricanes are expected to make landfall in Florida each year. Hurricane Katarina, which made landfall in August, 2005, damaged more than 2 million ha of forests in the Gulf states of Alabama, Louisiana, and Mississippi. The native longleaf pine (*Pinus palustris*) was more resistant to storm damage than loblolly (*Pinus taeda*) and slash pines (*Pinus elliottii*). Widespread damage to forests by storms can increase the risk of insect infestation and other pests, and the risk of invasive plants is increased.

Magnitude

Ecological events must contain a minimum magnitude (intensity or physical force) to be considered a disturbance. For example, wind has to reach a minimum velocity to cause damage in a forest and volume in a river water has to reach a certain level volume to cause flooding. Disturbances vary in their magnitude. Disturbances of great magnitude can have large-scale devastating effects on an ecosystem and, in turn, initiate primary successions. Those of lesser magnitude may have minimal disturbance on ecosystems and in turn initiate small-scale dynamics.

Duration, Abruptness, and Return Interval

Disturbances are also characterized by their duration, abruptness, and return interval. The duration of disturbances can vary from few days to years. Hurricanes and fires only last for few days, but acid rain has disturbed various ecosystems for decades. The abruptness of disturbances is measured on the impact they make on the ecosystem. Fire can sweep through the understory of a forest and mostly burn leaves and fallen logs; however, crown fires are usually more damaging. The return interval considers the time period of secondary succession that results in an ecosystem similar to the one before disturbances (**Figure 2.1**). Factors that affect succession are discussed further in Chapter 4.

Mega-Disturbances

Mega-disturbances include events that have abrupt or catastrophic effects on ecosystems, such as volcanic eruptions, tsunamis, meteoric impacts, and ice ages. These events have, over the millennia, greatly impacted life on Earth. Mega-disturbances like meteoric impacts have even played huge roles in the

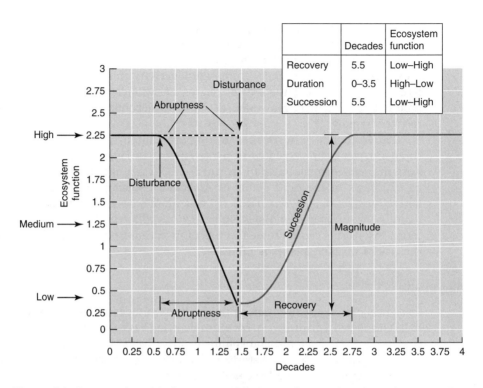

Figure 2.1 Conceptual model of ecosystem disturbance showing abruptness, duration, and magnitude. The recovery period through succession is also shown. (Adapted from P. S. White and A. Jentsch. *Progress in Botany: Volume 62*. New York: Springer, 2001.)

evolutionary history on Earth by decimating large number of dominating species. At the same time, however, mega-disturbances have opened new niches for surviving species and allowed evolutionary processes, including large-scale adaptive radiation, to take place.

Severe Disturbances

Severe disturbances can decimate and disrupt ecosystems, result in drastic changes in ecosystem composition, and functioning. Some disturbances may leave only a few survivors behind. However, live mature trees, seedlings, and buried seeds may persist through severe disturbances. Furthermore, remaining biological structures such as dead trees provide refugia that enhance recolonization and improve landscape connectivity involving seed dispersal by animals.

Species that survive severe disturbances are called **biological legacies**. Biological legacies play an important role in ecosystem restoration, especially recolonization and ecosystem succession. Those found in disturbed forests include all surviving species (i.e., plant, animal, and microorganisms). Among these are seed-producing trees, seed banks in soil, and understory species. Biological legacies also incorporate biological structures such as dead trees that can play an important role in the recolonization process by providing habitat for seed dispersing animals. When they exist as large surviving trees, biological legacies can provide large numbers of seeds and therefore influence succession trajectories. They may also play a critical role in sustaining natural disturbance regimes by promoting natural fire regime. Biological legacies also modify harsh microclimatic conditions and increase niches in the ecosystem.

An important aspect of restoration effort is to halt severe disturbances (especially human ones) and to reintroduce a natural disturbance regime to restore a desired ecosystem state. This often involves efforts such as the management of a fire regime, which maintains specific ecosystem assemblages, and the establishment of irregular water flow of a river, which in turn maintains characteristic river assemblages.

2.2 Fire Disturbances

The introduction of fire to ecosystems with no history of this type of disturbance can have devastating effects, especially in cases where plants have not adapted to fire. Intense fire can lead to other disturbances. After a severe fire, soil can begin to erode. Also, nutrient leaching of soil is not uncommon after fire.

Fire disturbances can increase the risk of invasion of non-native species, due to both the actual removal of native species by the fire disturbance and the resulting changes in the physical environment (notably, increased light availability). Invasive plants can also change the fire regime of an ecosystem. For example, invasive trees in the Florida Everglades have increased both fire frequency and fire intensity within the park.

Natural fire regimes typically maintain characteristic structure and function of ecosystems. Forest fires can even enhance complex landscape patterns composed of different plant communities. For example, in vast forests across the United States, frequent low-intensity ground fires were a normal part of the natural disturbance regime of the ecosystem before the European settlement. Ground fires probably blazed in these forests with low frequency (every decade), clearing away flammable debris such as broken limbs and litter (leaves, needles, branches, and trunks). These low-intensity ground fires typically regenerate forests. Although old trees with dead limbs can catch fire, large healthy trees are usually not affected by these ground fires.

Fire regime can also maintain certain landscape characteristics. Fire can, for instance, increase an expanse of grasslands. Conversely, elimination of fire from grasslands can lead to encroachment of forests. Invasion of grasses into forests or shrublands can intensify the frequency of fire. This, in turn, results in increased cover of grasslands. For example, low-intensity fires may allow perennial plant parts to survive in the soil, whereas moderate fires may eliminate these while allowing the seed bank (seed buried in soil) to survive (see **Case Study 2.1** on page 51). Very intense fires may eliminate all living plants so that recovery must occur through dispersal and colonization of plants.

Fire-Adapted Ecosystem

The southeastern United States coastal plain (from eastern Texas to Virginia) was once dominated by a longleaf pine savanna ecosystem. The understory of these forests was dominated by grasses and contained high species diversity. In fact, this community contained one of the highest levels of biodiversity in the United States. Also, this community harbored large numbers of rare and threatened endemic species. Natural disturbances of this ecosystem were of varying scales. Large-scale disturbances included hurricanes, but small-scale disturbances included fire initiated by lightning, windstorms, and outbreaks of parasites, pests, and diseases. Ground fire swept frequently through this ecosystem about every 1 to 3 years. Sites with fire frequency of 3 to 5 years were codominated with other species of pines and/or hardwoods. This high frequency of fire contributed to the high biodiversity in the forest understory. Without fire, hardwood encroachment took place. Forest fire regulates understory vegetation and maintains certain landscape complexity, including large forest gaps. The longleaf pine savanna community has almost been decimated due to fire suppression, silviculture, agricultural practices, and other land conversions. The longleaf pine used to cover more than 36 million ha. Today, about 97% of its cover has been eliminated. Fragments of longleaf pine forest are, however, found throughout its historical range. The once extensive longleaf pine forests have been replaced mainly by loblolly and slash pine plantations. Today, the transition of using loblolly and slash pines in forestry instead of longleaf pines is regarded as a mistake. Longleaf pine forest is more resilient than loblolly and slash pines. Longleaf pine is more resistant to insect infestation, adverse climate, and fire than loblolly and slash pine. Moreover, longleaf pine lumber is more valuable than other pines cultivated in the south. Large-scale restoration of longleaf pine is of national interest and is a key to sustainable forestry in the south. Other benefits include high potential for carbon sequestration and retention of water. An effective strategy for restoration of longleaf pine forest will initially involve compensation to landowners for their long-term commitment. The America's Longleaf Initiative is an umbrella organization that aims at restoring the longleaf pine forest. The initial aim is to triple the cover of longleaf pine forest within 15 years. Restoration of the longleaf pine savanna must involve establishing a natural fire regime. One important aim for this restoration is to enhance and sustain the high biodiversity. Several rare species found in this community must be managed in compliance with the Endangered Species Act. These include species such as gopher tortoise, indigo snake, red cockaded woodpecker, and flat woods salamander. Some of the existing remnants of this once extensive community have been managed in private and public land for recreation hunting and lumber production. Also, some of these sites are located on relatively poor soils for agriculture, which is probably the main reason for the existence of these remnants today.

The largest remaining forest of longleaf pine in the southeast is located on Eglin Air Force Base in the panhandle of Florida. The site contains about 150,000 hectares of pine forest. More than 20% of this forest is dominated by sand pine (*Pinus clausa*) and slash pine (*Pinus elliottii*) plantations. The site is critical for future restoration efforts on longleaf pine forests because it harbors almost 90% of the remaining old-growth longleaf pine forest. Rare plant species are documented on this site (**Figure 2.2**). This forest was managed for turpentine production and grazing by frequent burning before 1947 when it was transferred to the U.S. Air Force. Forest fires were suppressed in the 1940s, causing great impact on the ecosystem. Large-scale logging was practiced in this forest until 1988. Presently, limited timber harvest is still allowed at Eglin Air Force Base, where rare and endangered species have been used as an "umbrella" to manage and protect the whole ecosystem.

Restoration of the longleaf pine forest at Eglin Air Force Base has mainly focused on implementing prescribed burning. In turn, this has encouraged regeneration of the fire-dependent ecosystem. Prescribed burning has therefore been used as a restoration tool for the longleaf pine savanna. On average, less than

Figure 2.2 Fire-adapted wiregrass community in southern United States. (Courtesy of Sigurdur Greipsson, Kennesaw State University.)

21,000 hectares have been burned annually, resulting in more than a 7-year fire return interval. Annual winter burning was particularly effective in increasing species richness. Prescribed burning depends on critical buildup of longleaf pine needles, which serve as a high-quality fuel. The build-up of longleaf pine needles makes it possible to maintain the high frequency of burning needed to maintain this ecosystem.

Prescribed Burning

Prescribed burning has been enhanced by the use of geographical information system, or GIS, technology to identify critical habitats for rare species. Sites are ranked according to the need to burn by using a spatially explicit burn prioritization model. Factors used in such ranking include vegetation characteristics, total number of burns, and presence of endangered species. Such modeling allows managers to incorporate existing monitoring efforts and research data into their priority ranking and decision on when to burn a site.

Fire Suppression

Fire suppression of forests has been practiced for decades in the United States. Although it requires large funds (containing large to medium-sized forest fires can cost tens to hundreds of millions of dollars), the U.S. Forest Service has, since 1910, implemented a strict policy on putting out every forest fire that is detected. Over the years fire-fighting techniques have improved drastically, and helicopters and large aircrafts guided by GPS (global positioning system) technology are now used to haul in water to extinguish fires. Consequently, the areas of forests and grasslands burned by wildfire have declined drastically. The enormous success in fire suppression has, in fact, intensified the situation and increased drastically the fuel load (i.e., accumulated flammable debris) of the forest with unknown ecological consequences.

The high fuel load of fire-suppressed forests makes them more susceptible to high-intensity crown fires, which have a much more destructive effect than do frequent low-intensity ground fires. For example, shortly after Yellowstone Park was established a strict fire control practice was implemented and was followed until the 1970s. Although the practice became lenient in the 1970s whereby small fires that were initiated by lightning were allowed to burn as long as they did not threaten humans or buildings, this was, however, not enough to prevent massive buildup of fuel load. During the summer of 1998 wild fires burned out of control and devastated about 570,000 ha in Yellowstone Park and adjacent areas. The fires were uncontrollable for weeks, finally stopping in the fall when the first snows came to the rescue.

Fire suppression can also promote the transition from open forest to dense thickets. For example, fire suppression has increased the density of ponderosa pines (*Pinus ponderosa*) by a factor of 50 since the early 1900s. Alternative strategy in fire suppression is to thin forests to avoid catastrophic forest fires.

However, logging and the prevention of fire outbreaks by thinning forests on federal land has been met with public opposition.

Fire suppression and climate change can result in widespread intensive fires as was the case in 2007 when more than 240,000 ha of forested land burned in southern Georgia and northern Florida. Such widespread and intensive wild fires had not been recorded there before.

2.3 Fragmentation

Degradation of an intact ecosystem resulting in isolated dysfunctional patches within an otherwise degraded landscape is called **fragmentation**. Although habitat fragments can contain certain characteristics of the original ecosystem, species composition can also change within fragmented habitats. Habitat fragmentation has many detrimental effects on ecosystems. For example, large-scale fragmentation of native habitats is the main cause of species extinction. Fragmentation generally affects most species living in the affected ecosystem, because habitat fragmentation reduces population size of most species simply by drastically reducing the available habitat. In addition, fragmented populations are typically exposed to increased levels of stress and disturbances, including parasites, predators, and competition with weeds and non-native species (**Figure 2.3**). Drastic fragmentation eventually leads to species extinction,

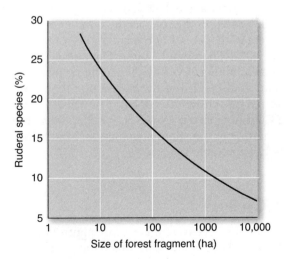

Figure 2.3 Effect of fragmentation on percentage of ruderal species. Fragmentation increases the risk of invasion of weeds and non-native species. (Reprinted from *Biol. Conserv.*, vol. 91, M. Tabarelli, W. Mantovani, and C. A. Peres, Effects of habitat fragmentation..., pp. 119–127, Copyright (1999), with permission from Elsevier [http://www.sciencedirect.com/science/journal/00063207].)

especially of habitat specialists. Fragmentation has adverse impact on ecosystem structure and function. Severe fragmentation may be followed by total ecosystem deterioration.

Fragmentation can both increase and decrease the impact of disturbances. Fragmentation of forests, for example, typically results in smaller fire compartment sizes. This results in reduced fire frequency because the potential area that can burn is reduced significantly in size. Fragmentation can conversely increase disturbance frequency. For example, wind disturbances are more common on fragmented forests where large trees are typically broken down. Also, the magnitude of a landslide can be increased by forest fragmentation, especially in mountainous regions. Also, flooding intensity of rivers and streams generally escalates with an increase in land drainage due to agricultural practices or deforestation.

Fragmentation of forest can impact bird populations. Populations of certain area-sensitive bird species decline as fragmentation increases, especially as core habitat is reduced in size. These may include warblers and flycatchers that require large areas of intact forest or core habitat. On the other hand, fragmented forest provide habitat for birds that are generalist in their habitat selection. These include jays and cardinals.

Rivers have been fragmented worldwide because of the construction of dams and other impoundments. An example of this can be found in the Columbia River, the fourth largest river in North America, where an advanced hydroelectrical system was developed comprising more than 400 dams. These dams have fragmented the river and reduced spawning grounds of salmon. The salmon catch gradually decreased in the river as the number of dams increased. Due to extensive fragmentation, the salmon population in the Columbia River has declined more than 90%. The Grand Coulee Dam alone reduced the original spawning area of the Columbia River by 70%. The fragmented river has been colonized by carp and shad instead of the migrating salmon. In the face of these adverse effects, mitigating efforts have focused on releasing hatchery-raised salmon back into the Columbia River. However, hatchery-raised salmon have decreased the genetic variation of the salmon population. In addition, the released salmon were particularly vulnerable to diseases.

2.4 Nutrient Cycling

The function of an ecosystem is mainly controlled by effective cycling of nutrients. Massive disturbances can affect **nutrient cycling** of ecosystems. Clear-cutting a forest, for example, not only affects the structure of the forest but also leads to increased leaching of soil nutrients, especially N. These soil nutrients can simply leach into the groundwater system or can be permanently lost by increased surface runoff. One of the most detrimental impacts of clear-cutting is how it interrupts the N cycle. This is mainly the result of the interrupted nutrient uptake of plants. It temporarily increases N availability as a result of decomposition and nutrient mineralization. Nitrogen availability can decline further if topsoil is disturbed.

Nutrient availability in the soil usually spikes after disturbances, especially if the litter cover (i.e., leaves, limbs, trunks) is increased. A flush of available soil nutrients is associated with the initial stage of secondary succession. This is usually followed by nutrient limitations as a result of decreased supply of nutrients and increased demand from colonizing vegetation. Then, the usual succession leads to rapid regrowth of fast-growing understory vegetation. After clear-cutting, a rapid regrowth of understory vegetation usually decreases water runoff and, eventually, nutrient leaching stabilizes.

Similarly, ecosystems that are disturbed by topsoil erosion experience reduced soil nutrient levels. On sites that have experienced serious decline in soil nutrient levels, vegetation cover cannot be easily established unless macronutrients, in particular N, P, and K, are added back to the soil. To establish vegetation cover on sites that have lost their topsoil, chemical fertilizers are usually added generously for several years. Fertilizer addition builds up the nutrient pool in the soil and may "jump-start" nutrient cycling. In restoration of mine sites where topsoil is lost, chemical fertilizers are typically added to the soil in quantities of 200 kg per hectare per year (using a 20-10-10 NPK fertilizer). Fertilizers are typically applied only for few years on sites to be revegetated. These must be used judiciously because their application can affect succession trajectory (discussed in Chapter 4).

An alternative method for increasing nitrogen input in poor soils involves using legumes in the revegetation process. Legumes form associations with nitrogen-fixing bacteria on their roots that harness nitrogen (N_2) gas directly from air and converts it to ammonia (NH_4^+) or nitrates (NO_3^-) (see Chapter 8). Legumes can produce more than 100 kg of N per hectare per year. This nitrogen then becomes available for other plants through decomposition of plant tissue.

As levels of N increase in the soil, plants that are demanding for nutrients start to colonize revegetation sites. High doses of fertilizer usually have positive effects on the growth of commercial nurse crops that are typically used in the revegetation. Nurse species are usually commercial grasses that respond well to fertilizer application and are readily available and easy to grow. Nurse crops typically begin to decline in growth as soon as fertilizer application ceases. At the same time, less-nutrient-demanding native plants may begin to colonize revegetation sites.

To build up sustainable ecosystems on restoration sites, special attention must be paid to N and P cycles (**Figures 2.4** and **2.5**). Pools of nutrients in the soil can indicate if an ecosystem has reached a point where it can be regarded as being sustainable. For example, in temperate ecosystems, about 1,000 kg of

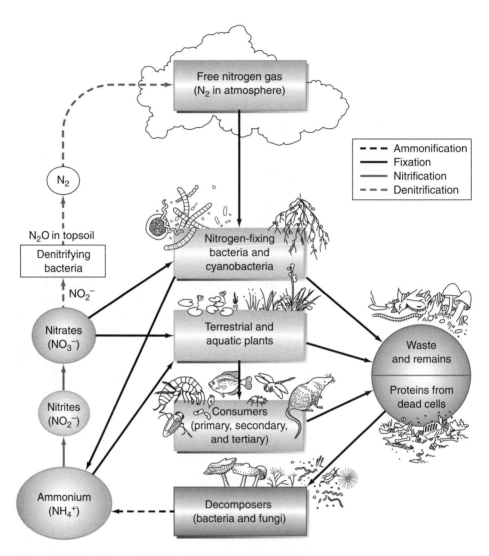

Figure 2.4 Nutrient cycling (nitrogen [N]) within ecosystems.

N per hectare per year should be stored as organic matter in the soil to provide for annual N supply. The annual supply of N in any ecosystem is controlled by the rate of mineralization, which is the breakdown of organic matter into inorganic nutrients (see Chapter 8). Phosphorus cycling on restoration sites is enhanced by fertilizer application and by introducing mycorrhiza, a ubiquitous symbiont of vascular plants that enhance P uptake by plants (discussed further in Chapter 8).

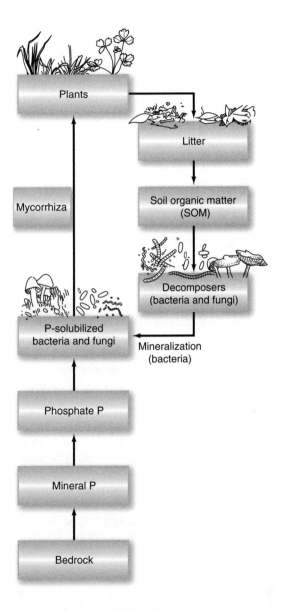

Figure 2.5 Nutrient cycling (phosphorus [P]) within a terrestrial ecosystem.

2.5 Hydrological Cycling

When water falls on the ground of intact ecosystems as precipitation (rain or snow), most of it percolates into the soil. This water enters the soil–water system or the ground–water system. Otherwise, water can flow directly into creeks, rivers, ponds, or lakes as surface runoff. Water that enters the soil–water system is available for plants and other organisms. Water can evaporate directly from the ground or is transpired through vegetation. In this way water reenters the atmosphere. Degraded lands usually have impervious soils compared with intact ecosystems, and therefore surface water runoff is increased under such conditions. Degraded lands usually have compacted soils that increases the risk of surface water runoff. Vegetation cover on the ground helps in intercepting the kinetic energy of the raindrops, therefore reducing their erosive power. Revegetation is critical in protecting the soil from erosion. Adding mulch (chopped straw) on the ground is a temporary measure to prevent soil erosion. This also helps in reducing sedimentation into wetlands, ponds, or lakes. Riparian buffer zones also help in reducing the amount of sediment entering streams. Wetlands can buffer the flow of surface runoff and control floods. Through landscape restoration wetlands provide a great service in storing surface water. Wetlands are also critical in recharging aquifers.

2.6 Functional Groups and Ecosystem Engineers

Functional groups are made up of species that have similar structures, share similar physiological activities, or respond similarly to disturbances. Also, species in the same functional groups typically exploit the same resources. For example, N_2-fixing species form one functional group, understory species can form another, and shrubs and canopy trees yet another. Restoration of different functional groups increases the resilience of an ecosystem.

Drivers and Passengers

For practical purposes, functional groups of species can be classified into drivers and passengers. Drivers can have a determining impact on ecosystem processes, such as succession. On the other hand, passenger species are part of the ecosystem but do not significantly alter ecosystem functioning. The roles of species as ecosystem drivers or passengers can, however, change along succession trajectory. Drivers removed from an ecosystem, however, can have a large impact on its functioning, whereas removing passenger species has little effect.

Ecosystem Engineers

Species that can cause changes in ecosystem structure and, therefore, control the availability of resources for other species or even create new habitats are called **ecosystem engineers**. Moreover, species that are ecosystem engineers usually create or modify mosaics of habitats within the landscape and thereby increase niches and biodiversity within the ecosystem. The beaver (*Castor canadensis*) is an example of an ecosystem engineer. The beaver builds dams in streams and small rivers by selectively cutting down trees in riparian forests and creating complex landscape of ponds, bogs, and flooded meadows. Consequently, the hydrology and geomorphology of rivers and adjacent embankments is modified by the activities of beavers. Beaver ponds reduce the velocity of stream flow and initiate sedimentation. When such sediment is colonized by plants, primary succession is initiated. Beavers also affect their ecosystem when they abandon their dams as soon as food resources decline. Abandoned beaver dams are subject to community succession that can last for decades.

Another example is the activity of the elephant (*Elephantidae* sp.) that creates large gaps within forests by trampling and tearing up shrubs and small trees. Another ecosystem engineer, the gopher tortoise digs long tunnels, altering soil properties. These tunnels also provide habitats for other species such as snakes, frogs, and ants. Other ecosystem engineers are known to affect the soil environment as well. For example, termites, ants, and earthworms mix the soil or modify the environment, including soil hydrology, by their own burrows and create new habitats. Nitrogen-fixing species are also classified as ecosystem engineers. These organisms alter nutrients cycle of soils.

2.7 Keystone Species

Species that are essential for functional and structural integrity of an ecosystem are termed **keystone species**. Keystone species typically influence processes that lead to the formation of a community. It is often impossible to point to a keystone species within functional ecosystems. Only after a keystone species disappears is the role they played revealed. The loss of a keystone species usually has a large effect on ecosystem integrity and can even cause disintegration and collapse. At the same time a population of a keystone species declines, another species can invade the vacant niche and replace the keystone species, thus changing the ecosystem permanently. Restoration of keystone species is, therefore, essential in attaining the original state of the ecosystem. Although it is a cost-effective approach in restoration, it requires great ecological knowledge.

Keystone species can have a great impact on restoration processes by maintaining ecosystem structure and function. During community succession a series of transitional keystone species, also termed "nexus species," can determine succession trajectories of ecosystems. This is discussed further in Chapter 4. Keystone species are often the principle driving force behind succession and are, therefore, of great interest in ecological restoration. Keystone species can determine the composition of the community and regulate the abundance of its component species.

Top-Down Control Mechanism

The top-down control mechanism within ecosystems is provided by predators that act as keystone species by controlling the size of prey populations. Eliminating predators from such ecosystems can lead to an increase in prey population that may deplete critical food resources, leading to a cascade of ecological processes.

An example of this can be found in the complex interaction between the sea otter (*Enhydra lutris*) and the offshore kelp forest (brown algae) of California. For many years the sea otter was almost hunted to extinction for its fur. On top of that, recent over-fishing in the North Pacific has resulted in a migration of killer whales (*Orcinus orca*) closer to the shore, which means an increase in the predation of sea otters.

Submerged kelp forests that grow offshore in California provide diverse subtidal communities with multiple niches. The brown algae can reach 120 m in height and provide shelter, food, and spawning ground for a myriad of fish species. The kelp forests also help in attenuating sea waves that otherwise augment coastal erosion. Sea otters, although low in number, play a critical role in maintaining this complex ecosystem by controlling the population of sea urchins (*Echinodea* sp.), a bottom grazer of the kelp. A sea urchin that grazes on the stem of kelp decimates the whole brown algae, causing tremendous damage to the kelp forest. If sea otters are reduced in number or decimated, the sea urchins population increases drastically and, consequently, has a devastating effect on kelp forests, resulting in "sea urchin barrens."

Restoration of the sea otter population is an important step in the restoration of this offshore ecosystem. Although the sea otter population declined drastically and was absent from large coastal areas, small and remote populations were located. Animals from these remote populations were subsequently translocated to sites of sea urchin barrens and strictly protected. As the sea otter population grew in number, the sea urchin population declined and the kelp forest regenerated.

Predators

Predators can act as keystone species, especially if their presence controls population size of other species. For example, wolves and coyotes represent key predators because they limit the abundance of elk populations. (Restoration of the gray wolf in the Yellowstone Park is discussed in Chapter 12.) Planktivorous fish species are also key predators because they limit the abundance of large-sized zooplankton in lakes. Restoration of lakes may involve management of planktivorous fish species. This method is call biomanipulation and is discussed in Chapter 13.

Mutualistic Species

Mutualistic species are essential for the maintenance of other species. For example, the rate of plant reproduction frequently depends on the presence of mutualistic pollinators (bees, birds, and bats). Also, symbiotic microorganisms such as mycorrhizal fungi and rhizobia bacteria regulate nutrient cycling (P and N) in the ecosystem and, therefore, act as mutualistic keystone species.

Summary

Restoring functional ecosystems involves managing disturbances, which, in turn, includes initiating natural disturbance regime to restore desired ecosystem function and composition. Also, curbing unwanted disturbances is part of restoration efforts. Anthropogenic disturbances can lead to ecosystem disintegration. These include clear-cutting forests, habitat fragmentation, mining, and agricultural practices. Natural disturbances are, however, essential in maintaining certain characteristics of particular communities. Nutrient cycling, especially N and P, is impacted by severe disturbances. Restoring nutrient cycling is critical to enhance succession on drastically degraded sites. Similarly, restoration of the hydrological cycle enhances ecosystem function. In the restoration process it is beneficial to identify keystone species that control ecosystem functioning and composition. Ecosystem engineers are similarly important in reinstating ecological processes and landscape heterogeneity. Restoration efforts aim at installing control mechanisms such as introduction of top-predators to ensure ecosystem function.

2.1 Case Study

Importance of Fire and Grazing in Grassland Biomes
Andy Dyer, Department of Biology and Geology,
University of South Carolina, Aiken, SC

In most ecosystems woody perennial vegetation gains dominance over time because, once established, large plants can competitively reduce resource availability for smaller plants (e.g., by resource depletion through canopy shading or by rapid resource capture below ground). Yet, grasslands are stable and persistent biomes with worldwide distribution in subtropical and temperate regions. The Great Plains of North America, the Pampas of Argentina, the Russian Steppe, the California grasslands, the Serengeti of Africa, and the Inner Mongolian grassland all have something in common. What factors in such different parts of the world and in such different climates act to limit woody plant growth and favor herbaceous dominance?

The key to grassland stability, diversity, and productivity is natural disturbance by both abiotic factors, particularly fire, and biotic interactions with grazing herbivores. Fire and grazing are common throughout many habitat types but are integral to the maintenance and persistence of grasslands. Within and among different grasslands, each disturbance can vary in importance, intensity, frequency, timing, and duration and can interact with each other. For example, grazing intensity is often directly related to fire frequency. Wildfires are caused by lightning and occur as fuels mature and build up, sometimes as often as every year. The dead and unpalatable biomass is removed and

(continued)

Case Study (continued)

the following season sees fresh growth from perennial species and often much higher densities of annual species. Studies in Africa and North America found that grazers are more likely to choose areas that have been burned recently because the freshly growing plants have much higher nitrogen content.

High grazing intensity reduces standing biomass in grasslands. This has the twofold effect of favoring species diversity by reducing the biomass of dominant species and reducing fire frequency by preventing fuel buildup. However, grazing is often patchy in time and space, and senescent biomass increases over time in ungrazed areas. As a result, woody species encroach on the grassland and unpalatable species increase in size and abundance. In a very short time fuel loads will be sufficient to support another fire. Thus, fire creates better grazing conditions for herbivores by favoring herbaceous species and herbivory acts to maintain the diversity of the herbaceous community.

Fire and grazing are complementary, but not equivalent, forces. That is, both reduce biomass and reset successional trajectories, but each has distinct characteristics. Fire removes biomass rapidly and indiscriminately; grazers reduce biomass selectively and based on palatability. Fire is a warm season event; grazing can be year-round. Fire has a return time measured in years; grazers may return within the same season. Additionally, each interacts differently with the seed bank, soil characteristics, nutrient cycling, and activity levels of other organisms.

Grazing can be removed completely from a system, but absence of fire only increases the probability of fire. When grazing is removed, the herbaceous vegetation becomes decadent, litter accumulates, and woody species are no longer suppressed. Grazing-dependent systems undergo a physical transformation toward one or more potential alternate states. In contrast, the removal of fire is never permanent. With an estimated 3 billion lightning strikes on Earth per year, the potential for fire increases in nearly all temperate biomes as vegetation becomes decadent and woodier. Fire in fire-prone systems is inevitable. And because fire is a function of climatic conditions, suppression by humans only intensifies the eventual conflagration.

The plants of the grasslands are well adapted to fire and grazing, which indicates a long evolutionary history with these disturbance types. Grass morphologies, in particular, indicate adaptive responses to the selective pressures of regular biomass loss. For example, meristems and buds are always near to or in the ground. Clonal growth via rhizomes and stolons is characteristic of grasses whether they are turf-formers or bunchgrasses. The low placement of meristems protects theses tissues from both grazers and from the heat of fire. Rapidly moving grassland fires remove dead plant material but often do not kill green tissues, and temperatures at the soil surface may not increase appreciably. Seeds on or slightly beneath the soil surface are not killed, and, in fact, germination can be stimulated by the heat pulse.

The general growth forms of grasses reflect grazing and fire tolerance. The caespitose form (bunchgrasses) produces new tillers at the edges of the bunch as older material collects in the middle and slowly suppresses new growth. Grazers select the newest growth, and the slow buildup of older material invites eventual fire. When finally burned, bunchgrasses show vigorous regrowth, often with increased tiller numbers and seed production. In contrast to bunchgrasses is the prostrate growth form typical of sod-forming or turf grasses. Stoloniferous and rhizomatous grasses are highly resistant to grazing, and even heavily stocked ranges are very resilient. Low biomass accumulation produces low-intensity fires that quickly remove dead material without harming the regenerating portions of the grasses.

In more arid regions, grasslands often contain a significant annual component. Annual species contribute large amounts of very fine and flammable fuels each year. These grasslands are very fire-prone, yet the plants are well adapted to grazing. Annual grasses germinate with winter rains and grow rapidly as spring temperatures rise; they set seed and senesce by mid to late spring. Grazing is common in the winter and spring, whereas fires occur from late spring until the first rains of the fall. Grazers favor the green winter forage, but the maturing grasses become unpalatable as they produce seed heads. Early-season fires can greatly reduce seed survival of annual grasses, but many annual grasses drop their seeds as they mature, and those seeds, once near the soil surface, easily escape the heat of most fires.

Grassland diversity, productivity, and health decline when fire and grazing are removed for even short periods of time. The buildup of decadent biomass and litter shades mature plants, prevents seedling emergence, and lowers survival rates. In bunchgrasses, new tillers are suppressed by shading, and those that do emerge are etiolated. Turf grasses are forced to grow vertically to compete for light. Returning grazing and fire to grasslands after long absence requires careful consideration. With the increased biomass, subsequent fires burn hotter and longer, and increased mortality of basal portions of the plants is common. On the other hand, reducing decadent biomass by introducing livestock is often difficult because of herbivore reluctance to eat low-quality forage. Resource managers must cautiously reintroduce fire and grazing in such as way as to minimize damage. However, the health of the grassland is often quickly restored with careful management of these two important forces.

The management of grasslands, or any biome, for "ecosystem health" is a relatively new endeavor. A few decades ago ecologists assumed that every ecosystem had a mature stage, called a climax condition, at which point the vegetation was more or less static, species dominance was set, and the species composition did not change any further. In other words, the successional endpoint had been reached and gross disturbances such as fires and flooding were undesirable. Today, we understand that reaching a successional endpoint is actually quite rare and often represents unhealthy ecosystem conditions because it implies that natural disturbances may be missing. When fires finally occur under late successional conditions, they are intense and can be catastrophic. Under natural conditions successional trajectories are frequently reset by low- to moderate-intensity disturbance to less "mature" states. In fact, disturbance at a variety of temporal and spatial scales is a common, predictable, and necessary occurrence in every ecosystem. The long-term stability and productivity of grassland biomes is a dependent outcome of a regular disturbance regime even though the particulars of the disturbances may be unpredictable. For grasslands worldwide these disturbances are grazing and fire.

Key Terms

Biological legacy   38
Disturbance   34
Disturbance regime   35
Ecosystem engineers   48
Fragmentation   43
Functional group   48
Keystone species   49
Nutrient cycling   44

Key Questions

1. Describe the importance of natural disturbance regime on ecosystem functioning.
2. List disturbances that are caused by humans.
3. Describe the process of habitat fragmentation.
4. What ecosystem processes are affected by habitat fragmentation?
5. Describe ways of restoring the N cycle on degraded soil.
6. What role do keystone species play in the ecosystem?
7. What is the difference between ecosystem drivers and passengers?

Further Reading

1. Mooney, H. A. and Godron, M. (eds.). 1983. *Disturbance and ecosystems.* New York: Springer Verlag.
2. Pickett, S. T. A. and White, P. S. (eds.). 1985. *The ecology of natural disturbance and patch dynamics.* San Diego: Academic Press.

3

BIODIVERSITY

Chapter Outline

3.1 Threats to Biodiversity
   Habitat Fragmentation
   Invasive Species
   Chemical Pollution
   Hybridization
   Overhunting
3.2 Extinction
   Rate of Extinction
   Species Vulnerability
   Extinct Species: The Great Auk
3.3 Genetic Diversity
   Restoration of Genetic Diversity
3.4 Restoration of Species Diversity
   Seed Harvesting
   Seed Production
   Seed Processing and Storage
   Seed Evaluation
   Germination
   Seed Dormancy
   Seed Priming
   Seed Sowing
   Nursery-Raised Plants
3.5 Ecosystem Diversity
   Habitat Fragmentation
   Ecosystem Stability
Case Study 3.1: Ecological Genetics and Restoring the Tallgrass Prairie

The term **biodiversity** refers to variety of life forms; it includes genetic diversity, species diversity, and ecosystem diversity. In this case biodiversity refers to variety both within and among species.

It is still not known how many species live on Earth. Although about 1.75 million species have been taxonomically identified, it is estimated that between 3 and 100 million species exist. Currently, biodiversity is threatened worldwide, mainly by habitat fragmentation.

Although species extinction is a natural process, its rate has increased rapidly over the last two centuries. Factors that promote species extinction as well as ecological consequences of extinction are discussed further in this chapter.

There is no definitive method for measuring biodiversity. A simple, commonly used measure is that of **species diversity**. Species diversity refers to the number of species within a defined area. Indices used to measure biodiversity include the Simpson Index and the Shannon-Wiener Index. These indices take into account not only the number of species present, but also their relative abundances. Floristic quality indices are commonly used to estimate plant diversity on restoration sites. These indices consider both the number of plant species and their fidelity to certain habitats.

Biodiversity provides great economic values. It is estimated that about 80,000 edible wild plant species could be used by humans. Increasing diversity of crop species could add to food security in the world. Currently, only five grass species (rice, corn, wheat, barley, and oat) account for most human caloric intake.

The pharmaceutical industry derives huge benefits from biodiversity. About 25% of drugs used in the United States contain ingredients derived from plants. Each year values of drugs derived from plants are estimated to be worth at least $84 billion. Harnessing biodiversity in tropical rain forests sensibly (including pharmaceutical uses), without destruction, could be more profitable than the more devastating slash-and-burn type of agriculture or one-time clear-cut logging (see Chapter 11). Also, generic resources such as lumber, fisheries, and rangelands add values to biodiversity. Even wildlife-related recreation is based on biodiversity. It is estimated that in the United States about $104 billion is spent annually on such activities.

Ecological services are provided by functional ecosystems containing high biodiversity. These services include the hydrological cycle, CO_2 fixation, O_2 generation, nutrient storage and cycling, drought and flood mitigation, waste remediation, and soil preservation. Another enormous benefit of functional ecosystems is climate stability. Degradation of ecosystems and eventual failure of their function negatively impact societies (see Chapter 1 for further discussion).

On a global biogeographical scale, biodiversity tends to increase sharply from the polar regions toward the equator. Biogeographical trends in biodiversity are associated with annual rainfall and temperatures. Precipitation and temperatures are also factors that determine ecosystem productivity. The highest ecosystem productivity is found in the tropical rain forest where high temperatures and precipitation prevail. Tropical rain forests also harbor the highest biodiversity of all biomes. It is estimated that between 50% and 90% of all plant and animal species live in the tropical rain forest. Generally, high biodiversity is associated with old communities that occupy heterogeneous habitat over large areas that experience low frequency of natural disturbances. These issues are discussed further in Chapter 11.

Restoration "tools" such as soil amendments, use of chemical fertilizers, soil pH adjustment, and prescribed burning are effective in increasing plant biodiversity. Among these tools, fire can open new niches and bring about the recruitment of new species into an ecosystem (see Chapter 2). Strategies used in restoring biodiversity at the genetic, species, and ecosystem level are discussed further in this chapter. Effective use of native plant materials in restoration efforts is discussed in details in this chapter.

3.1 Threats to Biodiversity

The main factors behind serious decline in biodiversity are

1. Habitat fragmentation and destruction, which affects about 73% of all species
2. Invasive species, which affect about 68% of species
3. Chemical pollutants, which affect about 38% of species
4. Hybridization (e.g., between native and non-native species), which affects about 38% of species
5. Overhunting, which affects about 15% of species

More than one factor could synergistically affect a particular species.

Habitat Fragmentation

Habitat fragmentation is a leading factor in species decline. Fragmentation forces species to live in small patches surrounded by modified landscape such as agricultural fields, industrial parks, or residential areas (see Chapter 6). **Fragmentation** refers to the process where a large swath of native landscape is cut into much smaller and isolated units. The downsized, isolated populations inhabiting fragmented patches face genetic inbreeding and impoverished gene pools, often regarded as the first step toward extinction. Indeed, habitat fragmentation has contributed to species extinction. Fragmentation especially affects species that require large areas of intact habitats. It usually leads to an increased proportion of species that are habitat generalists and a declined proportion of species that are habitat specialists. Fragmentation can result in isolation of populations within the landscape, affecting dispersal patterns of organisms and, therefore, affecting gene flow between populations, which may affect evolutionary processes within species.

Invasive Species

Invasive non-native species, another factor important in the decline of biodiversity, can lead directly to extinction of native species. For example, the Nile perch (*Lates niloticus*), a large, predatory, aggressive, non-native fish, was intentionally introduced in the 1950s into Lake Victoria to improve the lake's fisheries. This introduction has resulted in extinction of about 200 smaller native cichlid fish species, at least half of the original number of such species in the lake. Another example is the brown snake (*Boiga irregularis*), introduced to the island Guam, which has exterminated most native forest birds on the island. Altogether, invasive vertebrates put excessive extinction pressure to about 20% of endangered species worldwide. Invasive non-native species can also out-compete native species from their habitat, altering ecosystem functioning (see Chapter 6).

Not all non-native species, however, are a direct threat to native species. Non-native species that become naturalized have, in fact, lead to increased biodiversity of particular habitats. For example, more than 50% of all plant species in New Zealand are naturalized non-native species.

Chemical Pollution

Chemical pollution related to industry or agriculture can have direct impact on the physiology of sensitive animals. Pollutants can accumulate in animal tissues, causing reproductive failures and other ailments. Chemical pollutants are blamed, in part, for the current worldwide decline in amphibian populations. Amphibians have very permeable skin, so they tend to be highly susceptible to airborne environmental contaminants. These pollutants have been known to cause neurological damage and developmental defects in amphibians. For example, the ubiquitous endocrine-disrupting pesticide atrazine causes slowed gonadal development and hermaphroditism in the leopard frog (*Rana pipiens*). Such irregularities can have drastic effects on the reproductive success of species.

One of the most notorious chemical pollutants, DDT (dichlorodiphenyltrichloroethane), was developed in the 1940s as an insecticide to combat malaria-carrying mosquitos. Its use was implicated in the decline of populations in large, raptorial birds during the mid-twentieth century as well as in the causation of cancer. DDT accumulates in the fat of animals, particularly omnivores that are placed high on the food chain, such as humans and the bald eagle (*Haliaeetus leucocephalus*). It is toxic to embryos and impairs calcium absorption in adult birds, thus causing their eggs to have very thin, brittle shells, further reducing reproductive success. DDT is also toxic to aquatic and amphibious species where it is poisonous to fry and larvae, accumulating in the tissues of adults, modifying behavior, reproduction, and development. These result in decreased survival rates of the affected species.

Some of the most toxic chemicals are now banned (see Chapter 14). Also, public awareness of the damages of many of these chemicals, including DDT, has increased. This increased awareness is in part due to relentless work of activists such as Rachel Carson, who published *Silent Spring* in 1962 and raised general awareness on these issues.

Hybridization

Hybridization between native and non-native species can also contribute to the decline of native species. If, for instance, hybrids are stronger or better adapted to the environment (hybrid vigor), they may out-compete parental species for resources such as food and territory. Also, if parental native species continue to mate outside their own species, fewer viable offspring of the parental species are produced. Hybrids inherit traits from both non-native and native species, sometimes rendering them more fit for their environment and sometimes less fit. New trait combinations in hybrids can even allow individuals to occupy new territories and environments.

Among other causes, hybridization can be induced by environmental disruption. For example, species of swordtails, a freshwater fish native to Central America, use chemical signals (pheromones) to locate mates of the same species. When the water in which they live becomes polluted, their ability to sense these signals is impaired. In this case the hybrids produced are viable and can mate with each other, as well as with the parental species. However, the level of hybridization in such areas can drastically affect species integrity and ultimately lead to decline of the parental native species.

Overhunting

Overhunting of target species can reduce their numbers to dangerously low levels and, in extreme cases, can even cause extinction. One example of this can be found in the story of the American bison (*Bison bison*), a species that many years ago had reached a total population size estimated at 30 million. For centuries these animals were hunted substantially by Native Americans. The bison provided, among other necessities, meat for food, skin for clothing and tents, sinew for bowstrings, and dung for fuel. In the nineteenth century, however, European settlers hunted the bison nearly to extinction. Their hides were highly prized, and massive commercial hunting resulted in near extinction as only a few hundred individual bison remained. Through successful conservation programs, bison populations have recovered somewhat in the past century, although their numbers have not come close to previous levels. Today, about 250,000 bison exist. Many of these carry genes from domestic cattle, indicating they are not pure bison but hybrids. Estimates suggest the number of remaining pure bison is only between 12,000 and 15,000 individuals.

3.2 Extinction

Rate of Extinction

Species **extinction** is simply the unrecoverable loss of all members of a particular species on Earth. The current rate of extinctions worldwide is alarming; over the next half-century it is predicted that one-third to two-thirds of all species worldwide will become extinct. For forest birds over this time span, the estimated extinction figure is 50%. Deforestation leading to habitat fragmentation or loss of habitat is the main factor responsible for these catastrophic figures. In tropical rain forests the extinction rate is estimated to be about three species per hour. This may be an underestimation because unidentified species that become extinct go unnoticed. Although extinction is a natural process, the extinction rate worldwide has been accelerated to 1,000 to 10,000 times the natural extinction rate (without human activities). About 75 species of birds and mammals were driven to extinction, mostly through overhunting, between 1600 and 1900 AD. The extinction rate was even more rapid (one species per year) during the twentieth century. These high rates of species extinction are unprecedented in the natural history of Earth. In fact, scientists predict that we are now facing the sixth mass extinction on Earth, marking the greatest period of mass extinction that life on Earth has ever experienced, and it comes without any catastrophic event, such as a meteorite impact or an ice age.

The rate of species extinction in the United States has been high since European settlement began; it is well documented that about 100 species have become extinct during this period. The status of another 440 species is questionable, and most of them may already be extinct. The extinction rate varies between taxonomic groups. For example, about 22 species of birds have already become extinct. These include the great auk, Carolina parakeet (*Conuropsis carolinensis*), and Labrador duck (*Camptorhynchus labradorius*). Snails have also suffered a great loss, with 26 species already extinct and the status of another 106 species unknown. Plants have been affected as well; 11 species are now extinct with 130 additional species categorized as questionable.

Island species are even more prone to extinction. Hawaii has the highest number of extinct species in the United States (249 species). Of the mainland states, Alabama holds the record with 96 extinct species. Most of these are freshwater species that have suffered from habitat alteration in streams, lakes, and wetlands. Over the entire United States one-third of species are considered at risk of extinction; 1,050 species are endangered and 389 species are listed as threatened. The risk of extinction varies between taxonomic groups. For example, about 70% of freshwater mussels are at risk of extinction, as are 37% of freshwater fish species. Surprisingly, only 14% of birds are at risk of extinction. Only 2.7% of the entire species of flowering plants are are at risk of extinction.

Species Vulnerability

It is important to estimate which species are vulnerable to extinction. Such estimations mainly consider population size (number of individuals) and habitat availability. Population size or density is usually a determining factor for the risk of species extinction. **Minimum viable population** refers to the effect of population size on the likelihood of extinction. Reducing population size can increase the risk of extinction in several ways. When populations reach a critically low limit, they become vulnerable to diseases or environmental disasters that can cause extinction. Small populations also have lower genetic diversity, which can lead to inbreeding depression. The minimum viable population varies between species. A minimum viable population for vertebrates would probably have to be greater than 1,000 individuals for a population to maintain necessary genetic diversity. Degradation of the ecosystem, such as habitat fragmentation, typically leads to a decrease in population size of all species. In this case restoration of a target population is based on landscape restoration that aims at increasing the species' habitat.

Another important strategy in estimating species vulnerability to extinction is to use population viability analysis to assess how the size of a population influences its risk of becoming extinct. This type of analysis is usually performed over a specific time period, typically over one century, using computer simulations. The **population viability analysis** predicts a species' chance for long-term survival.

Extinct Species: The Great Auk

The great auk (*Pinguinus impennis*) provides one of the best documented accounts of species extinction. This bird became extinct in 1844 directly due to overhunting by humans. At one time this bird probably filled a similar niche in the North Atlantic as the penguin (family Spheniscidae) does today in the southern hemisphere. The great auk was a seabird, but human impacts confined the bird to isolated skerries (sea rocks). It was the largest member of the Alicidae family, reaching about 75 cm in height and weighing roughly 5 kg. Related species that still exist are dwarfed in size by comparison. Before any human encounter the bird's main terrestrial predator was the fox. The great auk had adapted to live on the seaside, and, though unable to fly, was an excellent diver. Its wings were small and probably functioned similarly to those of the penguin to aid in rapid diving. The great auk was likely superior to any other bird in the North Atlantic in diving, but its vulnerability was its egg incubation: it laid only one egg per year. The life span of the great auk ranged between 30 and 40 years, with reproduction most likely starting at 4 to 7 years of age.

Between the months of May and July the auks came ashore, usually on skerries, and incubated a single egg for about 40 days. The bird did not lay another egg if for any reason the first egg was lost. During the incubation period the bird was particularly vulnerable to predators. Upon successful incubation a newly hatched chick stayed with its parents for only about 1 week and was then on its own. Not only were the low reproductive rates detrimental to species survival, so too was the low parental investment, which made the chicks vulnerable to predators. Limitation for breeding grounds (the critical habitat) probably also contributed to the reduced population size of the bird.

The maximum population size of the great auk is estimated to have been about 100,000 breeding pairs. The main colonies of this bird were dispersed in Scandinavia, the United Kingdom, the Faroe Islands, Iceland, Greenland, and Newfoundland, Canada. The largest colony was found off the shores in Newfoundland on Funk Island. Skeletal remnants of these birds can be found along the coast in Europe (including Spain and Italy) and on the eastern coast of North America as far south as Florida. It is not known if these were just migrant birds or small satellite populations (**Figure 3.1**).

The extinction of this bird took place gradually, and its disappearance from most of its known colonies is documented. The last pair of great auks to reproduce in the United Kingdom was found on the island Papa Westray in the Orkney

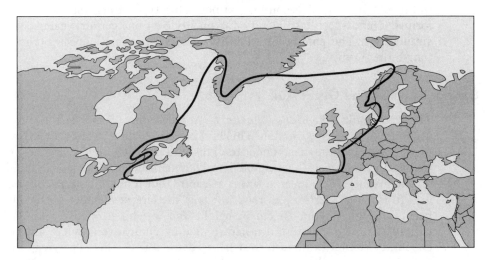

Figure 3.1 Maximum distribution of the great auk. Colonies were mainly confined to coastal areas.

Islands; both birds were killed and are now on display in the British Natural Museum. Other large colonies of the great auk in the United Kingdom, such as the one on St. Kilda, had disappeared in 1821. A decline in the birds' population size was noticed in the Faroe Islands in 1808. The last great auk in Greenland was killed in 1815. The birds were fairly widespread in southern coastal sites of Norway but were last seen there around 1300. The main colonies in Norway had disappeared before that time, leaving only migrant birds.

In 1497 fishermen from Europe began to fish in the Grand Banks of Newfoundland. The meat of the great auk was their sustenance; it was cooked and consumed or salted for storage. The feathers were plucked and used in mattresses, and the oil of the bird had good market value in Europe. It was easy to kill this bird using only a club, especially during its nesting period. The largest colony of the great auk on Funk Island disappeared around 1800.

After this period the last colonies were only found on a few skerries off the coast of southwest Iceland. One of these colonies, close to the Westman Islands, was overhunted around 1800. Another colony was found in Geirfuglasker (skerry of the great auk), which was described as being perfectly made for this bird and hosted about 1,500 breeding pairs. It is estimated that in total only a few thousand breeding pairs could be found off the coast in Iceland before its population size began to decline. These colonies in Iceland played an important role in the metapopulation structure of this species by connecting American colonies with the European colonies. The breeding colonies in Greenland, Iceland, and the Faroe Islands most likely acted as stepping stones, facilitating gene flow between North America and Scandinavian-European populations. Unfortunately, these breeding colonies were frequently visited by hunters, who clubbed down the bird.

It was a natural disaster, however, that played a role in the finale of this bird's existence. In 1830 a submarine volcanic eruption close to Geirfuglasker, followed by tremendous earthquakes, caused the skerry to collapse into the sea. Subsequently, the breeding pairs of the great auk migrated to the small portion of Eldey (Fire Island), which served as the last breeding ground for this species. Seventy-five birds were killed in the years before 1844, and numerous eggs were destroyed by fishermen without any apparent reason. Between 1831 and 1840 about 10 great auks were killed and all were sold to museums or private collectors.

After the submergence of Geirfuglasker, it was obvious that the population of the great auk had declined to a dangerously low level due to the lack of breeding grounds and overhunting. No conservation measures were proposed at that time. Ironically, there was a mounting pressure from museums and private collectors to obtain specimens before this species disappeared altogether. In fact, several expeditions were launched to Eldey for this purpose. The finale took place on Eldey in early July 1844 when three Icelandic sailors killed the last known individuals of the great auk. Interestingly, the viscera of these last great auks are preserved in the Zoological Museum in Copenhagen, Denmark.

With the extinction of the great auk we can only speculate about its role in offshore marine ecosystems. These birds could perhaps have played two important roles in marine ecosystems. First, they may have served as an intermittent host for parasites in seals and, therefore, acted as a control mechanism for the growth of the seal population. Because seals prey on cod, which used to be the main commercial fish species on the Grand Banks, the population size of seals has increased on these coastal sites, likely limiting the regrowth of the cod population. Second, they may have reduced population sizes of larger cod-eating fish, thus allowing for a larger cod population size.

3.3 Genetic Diversity

Biodiversity also refers to the variety of genes that are present in a population. Greater combinations of genes within a population (increased heterogeneity) increase population stability. Increased genetic heterogeneity increases a population's resistance to diseases. For instance, if many individuals in a population are genetically susceptible to a disease, the likelihood of local extinction is increased. In addition, genetic diversity of populations is important for the survival of species. High genetic diversity within a population decreases the risk of developmental and reproductive defects caused by inbreeding.

It is common to find significant genetic variation among plant populations, especially where locally adapted genotypes are in place. This is of concern because locally adapted genotypes are often the focus of conservation and restoration projects. The likelihood of a successful establishment and persistence on restoration sites are increased by using local genotypes. Also, the use of local genotypes ensures maintenance of the genetic integrity of local populations. Introduction of foreign genotypes should be avoided. Information on genetic variation of species used in restoration is usually absent. Ideally, genetic variation within the local population (if still available) should be examined and compared with more distant populations.

It is important to ensure that native species used in restoration projects are adapted to local conditions and contain appropriate genetic diversity. It is, therefore, suitable to use locally adapted populations of native species. This involves collecting seeds or other propagules of native plants. It is critical, then, that surrogate sites contain native species with high genetic diversity. Also, selecting species from adjacent surrogate sites increases the chances of species being properly adapted to the local conditions. The protocol for collecting seeds of native plants is discussed in detail in Section 3.4.

Restoration of Genetic Diversity

Isolated and inbred populations often show rapid decline in fitness, which can lead to extinction. Restoration of genetic diversity of an inbred and isolated population can be promoted by the introduction of a new genotype.

Although it is well documented that inbreeding can result in lower fitness, not many studies show that this trend can be reversed by introduction of novel genes into the population. A study on an isolated population of adder (*Vipera berus*) in Sweden strongly suggests that it is possible to restore the genetic diversity of an inbred population. This particular adder population was isolated by the Baltic Sea and arable land for at least one century and showed very low genetic variability. The adder population showed another sign of severe inbreeding demonstrated by high incidences of malformed or stillborn offspring, a definite warning signal that the population is heading toward extinction. To improve the genetic diversity of the inbred adder population, 20 male adders were captured from another population (that showed high genetic diversity) and released into the inbred population. The surviving foreign adders were removed from the inbred population after four mating seasons. The introduced male adders reproduced successfully with the inbred female adders. Four years later (the time it takes adders to reach sexual maturity) a dramatic increase was noticed in population growth. The genetic variation of the inbred population was examined throughout this study using DNA analysis. It was noticeably improved after the introduction of the new adders (**Figure 3.2**). Moreover, the proportion of stillborn offspring rapidly declined soon after the introduction of the new adders. The introduction of new genetic

Year	Number of individuals	
1982	18	
1983	22	
1984	25	
1985	23	
1986	23	
1987	20	
1988	20	
1989	19	
1990	17	
1991	15	
1992	9	← Introduction of
1993	6	new males
1994	7	
1995	4	
1996	8	
1997	12	
1998	20	
1999	32	

Figure 3.2 Total size of the adder population in Sweden. Arrow indicates introduction of new genetic material. (Adapted from T. Madsen, et al., *Nature* 402 (1999): 34–35.)

material into the inbred adder population by translocating a few unrelated male adders had drastic effects on genetic variability and, consequently, on population survival.

3.4 Restoration of Species Diversity

Restoration of species diversity in severely degraded ecosystems can be accomplished by direct introduction of native plants through the seeding of a mixture of native keystone species (multispecies approach) or by the sequential planting of native keystone species (single-species approach). Rare species should be introduced in a single-species effort focusing on their specific environmental requirements. Another approach is to improve vectors of seed dispersal into the restoration site. This approach is feasible if an intact site is found nearby. Vectors of seed dispersal can be improved by increasing bird dispersal of seed into the restoration site by planting trees or shrubs or by installing artificial perches. Structural heterogeneity and complexity of the ecosystem facilitate biodiversity. Also, animal routes can be established by facilitating connectivity between isolated sites. Animals are effective seed dispersers and can regulate community composition. By restoring habitats containing high biodiversity, animal diversity should increase spontaneously as community complexity is established.

Biodiversity on restoration sites is increased by introduction of native plants, either as seeds or as container-based plants raised in a nursery. Biodiversity is expected to increase with time after initial restoration efforts. Similarly, native species are expected to gradually colonize restoration sites and increase in diversity and complexity with time. High biodiversity is considered to be an important indicator of successful restoration.

To use native plant material in restoration projects, it is necessary to gain information about seed biology and seedling establishment of native plants. Successful restoration projects often depend on the correct choice of native plant genotypes. Additionally, restoration efforts must take into consideration the risk of introducing inappropriate genotypes. Today, environmental laws increasingly require the use of native plants in restoration projects. The use of quality native plant material in restoration projects involves seed collection from native stands, seed processing, storage, in some cases seed coating, and an appropriate seeding method.

Although a variety of native plant material can be purchased for restoration purposes, it is also necessary to use locally adapted genotypes (see **Case Study 3.1** on page 74). Pure seeds of keystone plants can be used alone as well as within seed mixes of a particular community. It is important, however, for the success of restoration projects to gain knowledge on basic biological properties of the species in question.

First, an ideal surrogate site (that provides seed) should be identified near the restoration site. Such a site should harbor healthy populations of desirable native plants. During initial seed collection it is important to distinguish between genotypes of native species. Genetic variation of the surrogate population can be tested using DNA analysis before harvesting.

Seed Harvesting

Harvesting **native seeds** requires careful monitoring in the field because seed maturation may fluctuate between years. Random sampling of seeds in the field at regular time intervals while seeds are maturing can result in improved quality of harvested seeds. Identifying an optimal time for seed collection is also crucial for restoration programs because native seeds often shatter quickly after maturation. This task is challenging as not all plants of one native stand put out mature seeds at the same time. The optimal timing of seed collection, therefore, provides the highest probability of harvesting quality seeds before shattering takes place. Also, seeds can be lost through animal consumption and through random events such as strong gales or heavy rainfall.

Harvesting seeds of native stand requires skilled manual labor. Seeds are usually collected directly from the plant and can be picked by hand or harvested using special harvesting machines. For example, seeds of some species can be harvested using a mechanized seed stripper, which increases drastically the amount of seed collected. Seed collection of native species involves identifying a proper collection method and adjusting or even modifying machinery for this purpose.

Seed Production

Seed production of native plants can be increased in the field by the judicious application of chemical fertilizers. This approach can, however, have a negative environmental impact on the surrogate site in the long run, especially if added nutrients result in increased growth of competitive or invasive plants. Propagation in nurseries and cultivated seedbed establishment are alternative strategies. These are particularly useful alternatives in cases where only a limited amount of native seed can be collected directly from surrogate sites. Native plants can then be cultivated ex situ in seedbeds to increase seed production. This process requires a long-term commitment and can take several years or decades until a reasonable harvest is produced.

Seed Processing and Storage

Seed processing involves drying seeds fairly quickly after harvest to avoid fungal infection. Seeds are usually dried in large bulk by passing over dry and warm airflow. Excessive heat should be avoided because it may kill the seed embryo. Seeds can also be dusted with commercial insecticides and fungicides. The seed can be coated with various materials, such as ground lime or diatomaceous earth, to keep them dry during storage period and for size uniformity for precision seed drilling in the field.

It is also necessary to clean seeds before any long-term storage by removing debris, seed wing, and other seed appendages. These structures are on seeds to optimize their dispersal but are not needed when seeds are dispersed by hand or machines. Cleaning seeds results in decreased bulk and weight, which reduces storage space and storage time; facilitates use of seed in nursery production; and allows precise seed drilling in the field. Proper seed treatment is usually determined by small-scale testing followed by seed germination trial.

Long-term storage of seeds is necessary if seed maturation in the field is sporadic. It is necessary to dry seeds to below 14% seed moisture as soon as possible after harvesting. Seed moisture is controlled by storing seeds hermetically in sealed plastic bags at subfreezing temperatures to avoid fungal or insect damage. Ideal conditions for the long-term storage of seeds (at least 1 year) are identified by small-scale trials. Seeds should also be stored in sealed containers to avoid attacks by animals such as rodents and birds. Seed batches should include information such as species name, collection date, collection site, seed germinability, and life expectancy. Great care should be taken in the separate storage of different genotypes of the same species (e.g., genotypes with varying levels of salt tolerance). It is necessary to apply quality control during seed storage by periodically examining small batches of seed and testing their viability and germination.

Seed Evaluation

Seeds should be evaluated for purity, viability, and total germination. Seed purity indicates the proportion of weeds or other nontarget plants in the seed batch. It is important to know about seed purity because pure seed batches can avoid inappropriate introduction of weeds or non-native plants. Seed purity is increased by separating desired seeds from other species and common weeds that, unfortunately, often contaminate harvests. This is usually accomplished by airflow separators or by a shaking mechanism (seeds are sorted according to size using a series of sieves). Seed viability examines the proportion of live/dead seed embryos of the target species within a seed batch. This involves standardized biochemical tests for seed viability, such as the tetrazolium test. In this test seeds are incubated in a 2,3,4-triphenyl-tetrazolium chloride solution. The seeds are then cut and examined under a microscope. Pink staining of the embryo indicates mitochondria activities and therefore living seeds.

Germination

Basic germination requirements for each native species may vary tremendously and should be identified. For this purpose the pregermination requirements of seeds, along with optimal germination requirements, should be identified. This requires testing seed germination in a controlled environment, usually involving sterile conditions and the use of incubators where temperature and light can be regulated. The effects from various factors, such as light regime and temperature (constant or fluctuating), are carefully examined. Information on seed viability and total germination is important to evaluate seed quality and estimate the rate of seeding in the field (i.e., kilogram of seed per hectare).

One standardized test for seed germination involves placing surface-sterilized seed on moist filter paper in Petri dishes or on sterile water agar (0.7%) plates. The seed may require pretreatments such as scarification or stratification (cold period) followed by water imbibitions. Scarification involves abrading the seed coat (structure that protects the embryo). This can be achieved by mechanical or chemical abrasion (using strong acids). Stratification, on the other hand, requires the placement of seeds in a cold (above freezing) place for weeks or months. After either one of these, water imbibition allows seeds to soak up enough water to initiate germination. Water imbibition is a precursor for germination. Each of these factors must be standardized for each specific species. Petri dishes containing seeds are then placed in closed incubators under controlled dark/light periods and temperature regime, depending on species requirements, for optimal germination. These conditions should simulate the microenvironment in which seeds germinate in the field.

It is often necessary to measure thermofluctuations in the field (using thermo-data loggers) to identify germination triggers. The time required for germination tests vary but usually does not require more than 1 month. Germination is usually accomplished when the seed's radicle (first sign of roots) is visible. Viable seeds that do not germinate readily within 1 month are classified as being dormant.

Seed Dormancy

Seed dormancy is common in seeds of native species. Dormancy is viewed as an obstacle in restoration projects where uniform germination is desired right after sowing. Small-scale lab tests can identify the main cause of dormancy and the most suitable method for treating it. Seeds commonly have a hard coat, which can induce dormancy. However, many methods are available to increase water imbibitions and alleviate seed dormancy, including scarification (chemical or mechanical) of seed coat and application of plant hormones. For example, placing such seeds in sulfuric acid (H_2SO_4) for a specified time period is usually enough to thin or break the seed coat and therefore allow water imbibitions. Machines specifically designed for seed scarification are available commercially but need to be carefully adjusted and monitored for each seed type. Though an effective method, scarification is not the only treatment available. Seeds can also be coated with substances to overcome dormancy. These include the plant growth hormone GA_3 (gibberellic acid).

Seed Priming

Seed priming can improve total germination because it synchronizes seed germination in the field. This technique involves osmotic priming, which controls the amount of water imbibed by the seed. The critical step in priming is to allow seeds to imbibe enough water to start the germination process but not enough to finish it. Primed seeds are then sowed and usually germinate soon if soil moisture conditions are favorable.

Seed Sowing

Seeds can be broadcasted on the ground or sown directly in the ground by mechanized seed drillers. Seed germination and seedling establishment are usually enhanced by preparing the ground before sowing by, for example, plowing or disking. Also, amending the soil with organic (compost) and inorganic (chemical fertilizers) nutrients improves plant establishment. Adjusting soil pH (using lime or elemental sulfur) is also beneficial in this regard. Soil pH requirements do, however, vary among species. Establishment of surface sown seeds is improved by hydroseeding, where seeds are mixed with water-retaining chemicals and fertilizers. The mixture is then sprayed on the ground from a pressurized container that is usually kept on a vehicle. This method is especially beneficial in arid regions or if desiccation is a problem. Establishment can also be improved by simply raking a thin layer of soil over the sown seed to avoid direct exposure to sunlight. Seeds that do not germinate or establish on the surface are drilled into the ground. Such an operation is usually accomplished by specialized seed drillers that can be drawn by tractor.

For seeds that are sown by seed drillers, it is necessary to determine their sowing depth. This is the maximum depth from which seeds can emerge when buried. Sowing depth is determined by simply examining potential emergence capacity from soil burial of seedlings of the species in question. If seeds are sown at too shallow a depth, they face the risk of dessication. If they are sown too deeply, seedlings cannot emerge to the surface. Seeds of different species have different capacities in emerging from soil burial. Seeds of many sand dune plants that are adapted to sand burial usually have great potential for seedling emergence. Seed size is another factor in determining the emergence capabilities of the seedling.

Sowing density (i.e., kilogram of seed per hectare) should be determined to maximize the use of resources. Sowing density is usually inferred from the total germination percentage of a particular species used. This is critical for optimal timing of fertilizer application (i.e., when dense cover of seedlings has emerged) to avoid excessive leaching of nutrients. Fertilizer applications usually take place at the same time as sowing or just after the seedlings have emerged.

Nursery-Raised Plants

Using container-raised single species in restoration requires more effort than using seed. It is, however, an appropriate method for rare species or keystone species. Native plants can be raised in modern nurseries on a large scale where their growth is optimized. Such plants can be raised from seed or cuttings or even from cell propagation. Plants that require symbiotic association with soil microorganisms must be inoculated with an appropriate symbiotic partner (see Chapter 8). Plants grown in a nursery need a period of acclimation where they are grown for few weeks outside the nursery before transplanting in the field.

Proper timing for transplanting is vitally important. Container-raised plants must also be planted in appropriate microsites (planting holes) that are often amended with organic or inorganic fertilizers and cleared of all competing vegetation. Appropriate weed management is often needed. Use of herbicides or weed mats to prevent emergence of weeds around the plants is advantageous. If drought is a concern, installation of a temporary irrigation system must be evaluated. Direct planting of native species into restoration sites should mimic patterns in native ecosystems.

3.5 Ecosystem Diversity

Reduction in biodiversity, especially elimination of keystone species and functional groups, can affect the integrity and functioning of ecosystems. For example, the extinction of big predators that represent the highest trophic level can result in an increased population size of medium-size predators, which can have cascading impacts on other species. Recognizing the threats to biodiversity and attempting to restore biodiversity at all trophic levels therefore are important in restoring resilient and functional ecosystems.

At the ecosystem level a greater structural complexity supports a greater biodiversity. Increasing structural complexity of ecosystems increases available niches that then leads to higher biodiversity. Also, enhancing multispecies succession trajectories within an ecosystem may result in increased biodiversity. For this purpose it is critical to install natural disturbances such as fire regime or natural flood pattern of rivers.

Ecosystem degradation generally results in declining biodiversity. **Landscape conversion** has turned large areas of native ecosystems with high biodiversity into vast areas of low biodiversity. Examples include monospecific agricultural fields, industrial parks, and residential suburbs. Based on the level of anthropogenic disturbances, landscape conversion is classified in four main stages:

- The first stage is characterized by pristine wildlands that have not had any human impacts.
- The second stage involves low impacted areas such as free-range grazing of natural grasslands or a forest with scattered or low-density cultivation.
- The third stage concerns intensively used areas such as croplands, plantations, and urban and industrialized areas.
- The fourth stage is degraded land, which is useless for any ecological function or productivity.

Habitat Fragmentation

Habitat fragmentation can expose vulnerable populations to adverse environmental conditions that, in turn, may have detrimental effects on these populations. For example, small forest patches are highly susceptible to storm damage. Also, smaller patches can result in a drier microenvironment.

Habitat fragmentation results in an extensive **edge effect** of each patch. Natural forest edges typically have a complex environment and tend to support higher biodiversity than interior **core habitats**. A forest edge can harbor species typical of forest core, species characteristic of adjacent landscape, and opportunistic species (weeds). Habitat fragmentation results in exceptionally high proportion of edges that can result in a disproportionately high population size of species typically found on edges compared with those of species typical of intact core habitats. This can have a negative impact on the biodiversity of the core habitat because aggressive weeds or invasive plants can have easier access to the whole patch.

Edges typically facilitate invasion of non-native species. Restoration efforts for fragmented landscapes should aim at high biodiversity and complexity of the interior forests' cores. The ratio of edge-to-core habitat should be kept as low as possible. Edges that contain a high proportion of weeds and invasive species represent degraded landscapes and should be given a high priority in restoration. For example, controlling the shape of each patch can influence the intensity of the edge effect; long narrow patches have greater edges than circular patches. Active restoration around patches can buffer against negative environmental impacts, reduce edge habitat, and expand the size of the interior core habitat.

Connectivity can also improve as a result of restoring biodiversity in a fragmented landscape. This is achieved by building corridors or stepping stones to facilitate movements of organisms between isolated patches. Restoration strategies may also aim at the metapopulation level by restoring connections between adjacent populations. Naturally, local landscape heterogeneity must be considered during any of these restoration efforts. These topics are discussed further in Chapter 6.

To restore ecosystem diversity, large areas of diverse ecoregions must be restored to their full ecological potentials. In designing mega-restoration sites, it is essential to identify landscapes that have maximum restoration potential. Such work should aim at restoring viable populations of target organisms and functional ecosystems across the landscape. It is critical in this context to provide private landowners with incentives for maintaining high biodiversity on their properties. This issue is discussed further in Chapter 14.

Ecosystem Stability

Biodiversity plays a major role in ecosystem stability. The diversity-stability hypothesis assumes that ecosystem stability is achieved through high biodiversity. Increasing biodiversity through restoration therefore increases community stability. Moreover, ecosystem stability is directly related to diversity of functional groups (e.g., grasses, legumes, shrubs, trees). However, species that are eradicated from an ecosystem will most likely be replaced by another species of a similar functional group without any noticeable loss in ecosystem function.

Also, an ecosystem containing high biodiversity would most likely show high resistance and resilience to degradation.

One of the best examples of the influence of biodiversity on ecosystem stability is derived from a long-term field experiment of David Tilman at Cedar Creek Natural History Area, Minnesota. In the experiment diversity was controlled within plots so the number of savanna grassland species ranged from 0 to 32. This experiment involved 289 experimental plots (each one 169 m^2) and gathered information on productivity (biomass) of each plot through time. The result of this study showed that biodiversity within an ecosystem tends to be correlated positively with plant community stability.

Another factor controlling biodiversity is soil microorganism content. If symbiotic soil microorganisms are absent from a restoration site, establishment of their host plant species is simply hindered. (Impact of key soil microorganisms is discussed in Chapter 8.) Soil microorganisms also influence stability of ecosystems. For example, the stability of a grassland community in Switzerland was found to be tightly coupled to the diversity of symbiotic arbuscular mycorrhizal fungi. The study emphasized that the stability of a whole ecosystem depends on high diversity in plant–microbe interactions.

Summary

Restoration projects aim at building functional ecosystems containing high biodiversity. Ecosystem stability and resilience are enhanced by increasing biodiversity. Biodiversity is threatened worldwide, and species extinction is of serious concern. Extinction of species is on the rise, largely the result of human activities, mainly habitat fragmentation and destruction. Also, the incursion of invasive non-native species, chemical pollutants, and overhunting has greatly impacted species diversity. Overhunting has led to some of the most infamous extinctions and near extinctions in recent history. The great auk was hunted heavily for centuries, until the last pair was killed in 1844. The extinction of the great auk shows how direct hunting by humans can be a major causal factor in the extinction of species with limited available habitats for critical biological functions such as breeding. Biodiversity on restoration sites can be increased by directly transplanting or seeding a variety of native species. Restoration projects should aim at using native species that show high genetic diversity. Using locally adapted native plant species in restoration requires knowledge of their biology, especially seed germination and establishment. Seed germination of native species can be enhanced by various pretreatments, and seedling establishment can be ensured by judicious application of chemical fertilizers. Biodiversity can be restored at the landscape level by linking together fragmented patches and by increasing the size of these patches by building buffer zones.

3.1 Case Study

Ecological Genetics and Restoring the Tallgrass Prairie
Danny Gustafson, Department of Biology, The Citadel, Charleston, SC

Natural areas throughout the world are becoming increasingly degraded or lost due to anthropogenic activities. Habitat reduction, fragmentation, isolation, and disruption to movement of individuals and genes among populations can negatively influence the long-term viability of a population. A reduction in gene flow and an increase in mating among genetically related individuals also decrease genetic diversity. Although genetic diversity is rarely the first concern when restoring a population, restoration professionals acknowledge the importance of genetic diversity for the long-term success of their work. In this case study I focus on a significant component, purple prairie clover (*Dalea purpurea* Vent.), of an endangered ecosystem (North American Tallgrass Prairie) to illustrate the importance of ecological genetics (also known as conservation genetics) in restoration ecology.

North American Tallgrass Prairie has been reduced from 35 million ha to less than 63,000 ha in the less than 150 years since European settlement. Illinois once contained approximately 25% (8.9 million ha) of the North American Tallgrass Prairie but now has less than 0.01% of high-quality prairie remaining. The highest quality remnants are small pioneer cemeteries and linearly shaped railroad right-of-ways, often located within a matrix of agriculture and urban development. To restore a tallgrass prairie ecosystem, the restoration professional must often choose between local and nonlocal seed sources and single versus multiple seed donor populations. Selection of local versus nonlocal seed relates to the assumption that locally collected plant material is more likely to have evolved under similar environmental conditions. If there is a genetic basis for the locally adapted plants (ecotypes), then introduction of nonlocal maladapted plants reduces the probability of a successful restoration project as well as potentially disrupt the local gene pool with the introgression of nonadapted genes into remnant populations (outbreeding depression).

Founder effects, also of ecological and genetic concern in restoration ecology, occur when a small number of genetic individuals establish a population, resulting in low genetic diversity in the ensuing population. To reduce the likelihood of founder effects and inbreeding depression in a newly restored population, most restoration professionals advocate use of multiple local seed sources.

Purple prairie clover (*Dalea purpurea* Vent.) is a long-lived perennial member of the Fabaceae family and is a common member of high-quality remnant prairies. Members of Fabaceae are ecologically important because of their association with atmospheric nitrogen-fixing bacteria (*Rhizobium* sp.). These bacteria allow the plants to enrich the soil nitrogen levels and provide an important nitrogen source for herbivores. The showy purple inflorescences appear in early to mid-summer and are pollinated by common bee species (*Bombus* sp.). Purple prairie clover has a mixed mating system, meaning it can outcross and self-pollinate, although there is a substantial reduction in seed viability when flowers are self-pollinated relative to outcrossed flowers.

In this molecular ecology research, I focus on three basic questions: (1) are local (Illinois) populations genetically different from nonlocal (Kansas) population, (2) what are the levels of genetic diversity in remnant and restored prairies, and (3) are differences in plant performance related to seed source?

Gensburg-Markham prairie is one of the oldest and largest tallgrass prairies in Cook County, whereas the Morton Arboretum restored population included material collected from Gensburg-Markham prairie. The Mason County State Nursery contains remnant populations and seed increase plots used to provide plant material for roadside and wildlife plantings for the State of Illinois; therefore, I considered it a restored site for this research. The Black Hawk State Historical Site was established by a restoration professional from Knox County, Illinois, so genetic similarities between Black Hawk and Mason County is to be expected when local seed is used (**Figure 3.3**).

Genetic analysis showed differences between local Illinois remnants and the nonlocal Kansas tallgrass prairie. Genetic relationships among remnant and restored Illinois populations reflected geographical proximity, which is what one would predict if local seed was used to establish these restored sites. Genetic diversity decreased from the large Kansas prairie to smaller restored prairies to remnant Illinois prairies. Within Illinois the higher diversity estimates from restored versus remnant sites likely reflect the use of multiple, rather than one, local seed source. In all cases the persons who restored these purple prairie clover populations reported using seeds collected from at least two remnant sites within 50 to 100 miles of their restoration sites. These findings are important because this is one of the first studies to include restored populations, showing large-scale genetic structuring and that multiple local seed sources can help retain genetic identity regionally while increasing diversity locally.

In addition to the selectively neutral molecular markers, I conducted common garden and greenhouse experiments to test for differences in growth and competitive ability between plants from local and nonlocal seed sources. Nonlocal plants from Kansas were typically shorter than the local Illinois plants in the common garden study. In the competition experiment, plants from Kansas were shorter and had less total biomass and reproductive (inflorescence) biomass than plants from Illinois (**Figure 3.4**). If aboveground competition is an important mechanism for structuring the tallgrass prairie plant community, then Kansas purple prairie clover plants may be at a selective disadvantage when competing with taller local plants in a restored Illinois tallgrass prairie.

Based on the genetic and ecological differences between local and nonlocal purple prairie clover source populations used in this research, I would not recommend using plant material from Kansas in Illinois restoration projects. Since conducting this purple prairie clover molecular ecology research, I found similar patterns with two dominant

(continued)

Case Study (continued)

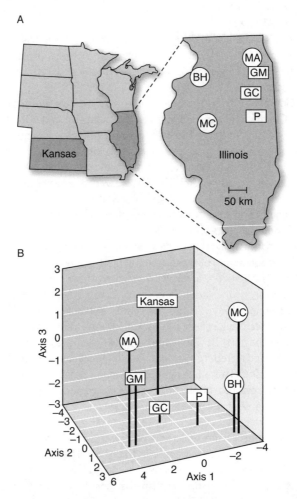

Figure 3.3 (A) Map of three remnant and three restored (circled) Illinois and one Kansas tallgrass prairies included in this *Dalea purpurea* study. (B) Principle components analysis (PCA), based on nine isozyme loci and 19 alleles, easily separated the Kansas and Illinois populations along the second axis, whereas relationships among Illinois populations largely reflect geographical proximity and "local" collection range. The first three axes accounted for 44%, 23%, and 15% of the variance in the data matrix, respectfully. (Courtesy of Dr. Danny J. Gustafson, The Citadel.)

grasses of the tallgrass prairie: big bluestem (*Andropogon gerardii*) and Indian grass (*Sorghastrum nutans*). Advocating use of multiple local seed sources for restoring local plant communities is consistent with the philosophical and practical approach of professional restoration ecologists, and my research supports this position. To what

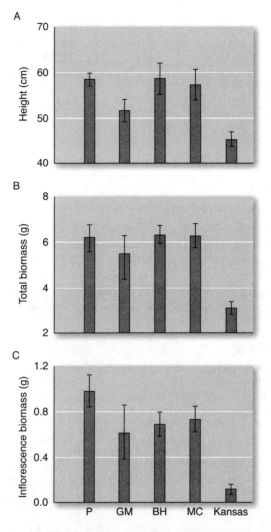

Figure 3.4 The results of a 3-month greenhouse competition experiment where mean plant height (A), total dry biomass (B), and inflorescence biomass (C) of plants from the nonlocal (Kansas) origin were significantly smaller than local (Illinois) plants. (Courtesy of Dr. Danny J. Gustafson, The Citadel.)

extent "local" is defined geographically and ecologically is likely to be ecosystem and species specific. What is clear, however, is that ecological genetic studies are a vital link among basic science (evolutionary biology, ecology, population genetics), applied science (restoration and conservation biology), and wise management of our natural resources.

Key Terms

Biodiversity 56
Chemical pollution 58
Core habitat 72
Edge effect 72
Extinction 60
Fragmentation 57
Hybridization 59

Landscape conversion 71
Minimum viable population 61
Native seed 67
Overhunting 59
Population viability analysis 61
Species diversity 56

Key Questions

1. Define the term biodiversity.
2. What are the main factors causing decline in biodiversity?
3. Describe the process of habitat fragmentation.
4. Describe restoration methods used to enhance biodiversity at the species level.
5. Describe methods used to restore genetic diversity of populations.
6. Describe the main factors to consider when using native seed in restoration.
7. Describe methods that aim at restoring ecosystem diversity.

Further Reading

1. Gaston, K. L. and Spicer, J. I. 2004. *Biodiversity: an introduction.* Malden, MA: Oxford, UK: Blackwell.
2. Novacek, M. J. (ed.). 2001. *The biodiversity crisis.* New York: The American Museum of Natural History and The New Press.
3. Wilson, E. O. (ed.). 1988. *Biodiversity.* Washington, DC: National Academy Press.
4. Wilson, E. O. 1992. *The diversity of life.* New York: Penguin Books.

4

SUCCESSION

Chapter Outline

4.1 Theories of Succession
4.2 Successional Processes and Restoration
   Primary Succession
   Secondary Succession
   Animals in Succession
4.3 Management of Succession
   Reducing or Inducing Disturbances
   Introduction or Removal of Species
   Nutrient Management
   Hydrological Manipulations
4.4 Monitoring Succession
4.5 Inferring Succession
Case Study 4.1: Restoration of Primary Sites

Succession is basically the recovery process of an ecosystem after a disturbance. Succession is driven by three basic factors: (1) availability of sites, (2) availability of different species giving rise to various succession trajectories, and (3) different capability of species to establish and compete. In addition, biomass and nutrient accumulation in soil increases with time and is an important factor in driving succession. Succession has been demonstrated worldwide in all biomes and at different trophic levels; however, the theory of succession is one of the most challenged in ecology.

4.1 Theories of Succession

Theories in community ecology and especially the theory of succession have traditionally been used as framework for restoration ecology. The theory of succession implies gradual changes of organismal communities with time on a particular site. The scale of disturbances dictates the type of succession that follows. Succession has therefore traditionally been divided into two distinct processes: primary succession, which takes place on barren lands, and secondary succession, which takes place after disturbances on sites that contain soil.

Although observations on vegetation changes have been documented for a long time, Eugene Warming (1895) was probably the first ecologist to emphasize that vegetation changes after disturbances were universal and unceasing. Successional changes were viewed as directional and predictable development of communities with time (**Figure 4.1**). It was assumed that a common succession pattern observed in various ecosystems must result from the same single underlying mechanisms. Succession patterns are also termed **succession trajectories** and incorporate the whole process from the initial state to the final state. Similar succession trajectories were, for instance, typically observed in forest regeneration. Recently, this view has been challenged, and succession is now thought to be a stochastic event and multidirectional where more than one final state is possible. This issue is discussed further in Chapter 5.

The theory of succession was developed in detail in the United States by Henry Cowles (1869–1939) and Frederic Clements (1874–1945). Clements defined succession as "a sequence of plant communities marked by the change from lower to higher life-forms." Moreover, Clements viewed succession as a directional and irreversible event. Furthermore, he assumed that succession proceeded deterministically to "climax vegetation," which represents the final state of succession. Also, succession was supposed to be driven by reaction (i.e., site modification). Clements described the climax community as a "superorganism." The climax community was the only possible final state of succession. In his analogy species were the organs and succession the ontogeny. He argued that each species played an important role in preparing the way for new species to enter the community in the succession. Clements' view on succession incorporates strong niche assembly where species fit within their physiological limits provided by the environment. However, often what appear to be the final state of succession may not necessarily be a stable community over a long period, and many communities are routinely disturbed and never reach a stable end state. In this context Clements did not recognize cyclic disturbances. Clements' view of succession is therefore based on potential regional trends in vegetation dynamics rather than on dynamics on much smaller patches.

Species interactions during succession have been the focus of many ecological studies. Succession models emphasize progressive changes in communities that replace each other with time in a deterministic way. In 1979 Joseph Connell and Ralph Slatyer proposed three models to describe these processes (**Figure 4.2**). The first model, *facilitation*, describes how pioneer species that colonize early in succession ameliorate the microenvironment and generally make primary sites more favorable for the establishment of mid- or late-successional species. Examples of facilitation are most common in a highly disturbed ecosystem. The facilitation model therefore has strong relevance in restoration of devastated ecosystems.

Theories of Succession 81

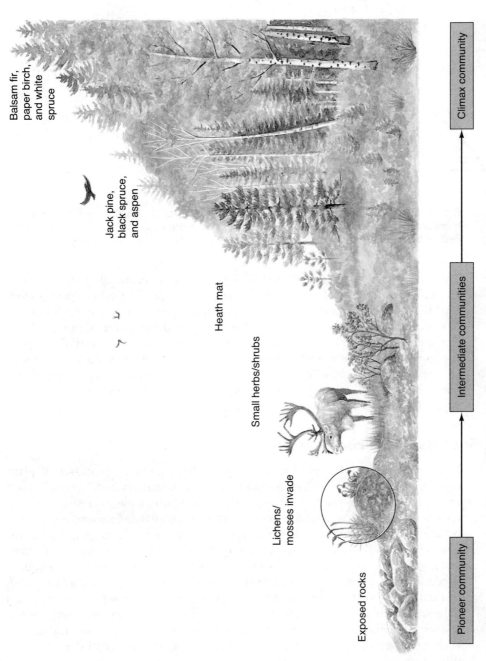

Figure 4.1 Primary succession of temperate forests in northern United States. Primary sites were formed as the Ice Age glacier retreated. Succession is demonstrated by distinct communities.

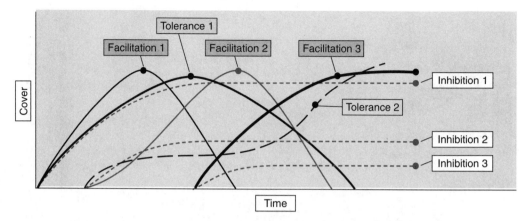

Figure 4.2 Succession models of Connell and Slatyer: facilitation, inhibition, and tolerance. Numbers (1, 2, 3) indicate different plant communities. (Adapted from D. C. Glenn-Lewin, P. K. Peet, and T. T. Veblen (eds.). *Plant Succession: Theory and Prediction.* New York: Springer, 1993.)

The second model, *inhibition*, describes how pioneer species alter the microenvironment on a site for their own benefit and to prevent other species from establishing. The inhibition model has relevance in restoration where succession is arrested at an unacceptable state. For example, dense cover of nurse species or plant litter can inhibit recruitment of native species in the initial stages of succession. The third model, *tolerance*, describes how plant species must tolerate certain adverse conditions, but due to their longevity and stature they eventually dominate the ecosystem. The tolerance model is therefore especially applicable in secondary succession of forests.

4.2 Successional Processes and Restoration

Restoration ecology has predominantly dealt with recovery processes of plant communities on devastated sites, and models in succession have certainly been useful in predicting changes in vegetation composition on such sites. In fact, restoration ecology is founded on successional models. Moreover, succession models have been widely used as blueprints in restoration of damaged ecosystems. Restoration effort is often marked by one initial event such as protection from grazing, seeding, or outplanting of nurse or native keystone species. Subsequent succession is expected to proceed to an acceptable final state in accordance with restoration goals. However, contrary to this prediction, succession is often found to be arrested at an unacceptable state. **Succession management** involving various restoration techniques therefore is needed to ensure that succession trajectories leads to an acceptable final states. It is also important in restoration to consider the timeframe of succession because the aim of restoration is often to accelerate succession. This may involve bypassing early successional phases and introducing mid- or late-successional species into restoration sites.

Primary Succession

Primary succession is characterized by the development of plant communities on "primary sites" where the substrate (sand, gravel, or just bedrock) has not yet been colonized by lichens, mosses, or vascular plants. The first photosynthetic organisms to colonize primary sites are commonly autotrophic bacteria (*Cyanobacteria*), lichens, and mosses. These species can easily disperse long distances. Primary sites have poorly developed soils that seriously lack macronutrients, soil organic matter (SOM), and seed bank. Because primary soils lack a seed bank, colonization is most frequently from seeds or propagules dispersed from adjacent communities. In addition, primary soils have typically harsh microclimates, including high irradiance and high fluctuation of temperature, and are subjected to drought.

Disturbances that create primary sites include natural processes, such as landslides, coastal erosion, catastrophic soil erosion, and volcanic eruptions, or human activities, including mining and industrial fields. Examples of primary sites include new volcanic islands, gravel and sand pits, coastal sands, mine tailings, and gravel moraines left by a retreating glacier. Primary sites can result in patchy landscape made up of different successional phases.

Soil processes are important in determining primary succession. During primary succession SOM develops gradually from the accumulated decomposed remains of the pioneer species. Through decomposition and mineralization (where organic matter is turned into inorganic nutrients) the nutrient content of soil of primary sites increases. Usually, once primary soils contain critical level of nutrients, pioneer vascular plants start to colonize primary sites. Primary succession is concomitant with the increase of SOM, especially with the increasing nitrogen (N) budget in the soil. The availability of N in soil is to a large extent controlled by the SOM through decomposition and mineralization. These processes are controlled by the activities of soil microbes (fungi and bacteria). An example of nutrient-driven succession is derived from studies on Glacier Bay, Alaska, where nitrogen- (N_2) fixing species were found to add significant amounts of N to the soil, which in turn was an important driving force of the succession. Increase in soil N content often occurs rapidly with succession. For example, the total soil N content increased from 1 to 15 g N m^{-2} during 16 years of succession on coastal dunes in the Netherlands.

Successional changes on primary sites are in agreement with the **Resource Ratio Hypothesis** of David Tilman where competition for resources is the driving factor. The Resource Ratio Hypothesis considers that resources (e.g., nutrients, water, and light) form a temporal gradient and that succession occurs mainly because plants compete over these resources and the outcome of competition drives the succession.

Ecosystem complexity increases during succession. As the ecosystem becomes more complex, ecosystem resistance and resilience typically increases (**Figure 4.3**). Also, the physiological environment changes during succession from an open sunny habitat to a more shaded one (Figure 4.3). These changes typically result in reduction in fluctuations of soil surface temperature and drought conditions.

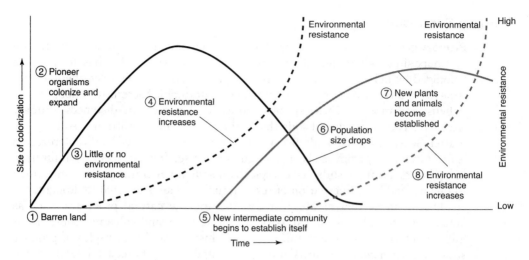

Figure 4.3 Changes in communities and environmental resistance with succession. The numbers indicate the progression of succession.

Models of primary succession have practical uses in the restoration of primary sites (e.g., mined lands, gravel and sand pits, industrial brownfields, and coastal sands) where soils are seriously degraded or missing altogether. Restoration on primary sites is typically initiated by seeding commercial grasses. This work is usually followed by application of chemical fertilizers for subsequent years. This work is relatively easy to implement because ordinary agricultural machinery can be used for this purpose. The initial aim of revegetation is usually to stabilize the ground and halt soil erosion but at the same time to initiate succession. The use of commercial grasses and chemical fertilizers may, however, result in productive plant communities, which may show low species richness (see **Case Study 4.1** on page 97). Successional trajectories after the initial restoration effort depends on seed supply from nearby sites (**Table 4.1**). If dispersal barriers exist for the native species, then primary succession will be slow and unpredictable. Restoration practices on primary sites can be improved by using mixtures of native seeds instead of commercial grasses to overcome dispersal barriers. Using mixtures of native seeds can determine successional trajectories. The ultimate choice of seed mixture for revegetation purposes, however, depends on restoration goals. It is important in restoration to consider the time scale of primary succession, which can vary tremendously. For instance, the transition from barren lands to temperate forests can take hundreds or thousands of years.

Restoration ecologists can learn about primary succession from long-term ecological studies. The ideal place to study primary succession is on newly emerged volcanic islands where the soil is initially made of sterilized pumice. An example of such volcano is Krakatau, an island located in the tropics between

Table 4.1 Factors Affecting Succession

Colonization depends on source of seeds
- From nearby sites
- From the seed bank
- From preadapted plants to the restoration site

Dispersal of seeds depends on
- "Seed rain" from nearby sites
- Secondary dispersal by wind, water, or animals

Establishment of plants depends on
- Availability of "safe site"
- Seed bank

Important factors in colonization of plants
- On primary sites, seed dispersal, germination, and vegetative propagation
- On secondary sites, recruitment from the seed bank or growth of vegetative underground parts of plants

Persistence of plants depends on
- Seed production of local plants for further population expansion

Java and Sumatra in Indonesia. Krakatau was devastated by a powerful volcanic eruption in 1883. All organisms on Krakatau were decimated and a barren new island, Rakata, emerged. The island was covered with a thick layer of sterilized pumice. Scientists were interested to know if a tropical rain forest similar to those found on nearby islands in Indonesia could develop on Rakata. Observations showed that colonization of plants and animals was stochastic, and a year after the eruption only a few grasses were found on the new island. Plants colonized the island gradually, however; 3 years after the eruption 15 plant species were found, 14 years after the eruption 49 plant species were found, and 46 years after the eruption 109 plant species were found, and 100 years after the eruption 148 plant species were found. The dispersal pattern was an important factor in the colonization process. The propagules of plants were brought predominantly by animals (47%), wind (32%), and to a lesser extent by ocean surface currents (22%). Dispersal of plant propagules by sea was, however, important during the first phase of colonization. Consequently, the coastal community was established rapidly. The second phase of plant colonization was promoted by animals. The colonization is still continuing, and today, Rakata is covered by a forest that resembles a tropical rain forest. There are, however, some distinct differences between the forest on Rakata and intact tropical rain forests on the nearby islands. For example, not a single tree species characterizing the forests on the nearby islands Java and Sumatra (44 km equidistance) has yet colonized Rakata. Also, the forest of Rakata is dominated by only few species. Another century of succession may be needed to attain a forest fully comparable with that of an old, undisturbed Indonesian tropical rain forest. Another small island (about 2.5 km^2), Anak Krakatau, emerged from Krakatau's submerged caldera near Rakata in 1930

(**Figure 4.4**). About 140 species of vascular plants were established on the island in 1992. The dispersal pattern of plants found on Anak Krakatau was distinctly different from the one of Rakata. Propagules of plants colonizing Anak Krakatau were brought predominantly by ocean surface currents (52%) and to a lesser extent by animals (24%) and wind (23%). Animals have since colonized the island, and in 1986, 24 species of resident land birds were found there.

Another island, Surtsey, emerged in 1963 after a submarine volcanic eruption (**Figure 4.5**). The island is about 2.5 km² and is located in the subarctic just 44 km south of Iceland. Bryophytes became established in 1967, and within 6 years 72 species of mosses and liverworts were found. Lichens were

Figure 4.4 Anak Krakatau eruption in 1930 (left) and a recent photo of Anak Krakatau showing recolonization of forest on the island (right). (Left courtesy of Dmitry Duden; right © Mark Clacy, 2007.)

Figure 4.5 Surtsey; the submarine volcanic eruption of 1963. (Courtesy of Howell Williams/ The National Geophysical Data Center, National Oceanic and Atmospheric Administration, U.S. Dept. of Commerce [http://www.ngdc.noaa.gov/].)

first detected in 1970, 25 species were recorded in 1990, and 87 species were recorded in 2008. Ten species of mushrooms were recorded in 2008. This demonstrates the importance of mosses, liverworts, fungi, and lichens (that can easily be dispersed long distance) in the initial phase of primary succession. The first vascular plant was found in 1965. In 2008, 63 vascular plant species were recorded in Surtsey. Not all plants that colonized the island established a viable population. About 51% of the colonizing vascular plants can be regarded as being successfully established (i.e., they reproduce and expand their population). The colonization pattern on Surstey has been distinctly different from the tropical islands mentioned above. Most propagules of plants were transported by birds (73%), by wind (16%), and by ocean surface currents (9%). Nearby small (but much older) islands have played an important role in the initial colonization and acted as "stepping stones" in this process. About 93% of plants found on Surtsey are also found on neighboring islands, and only 3% are from Iceland, and 4% are derived from more distant locations.

Plant species have gradually accumulated on the island, and it is assumed that about 100 vascular plants could potentially establish on Surtsey. The colonization of Surtsey provides insight into development of subarctic coastal ecosystem. Three main habitats were identified on Surtsey: (1) coastal sand dunes, (2) barren lava fields and, (3) sea-cliff gull colony. Niche assembly into each habitat showed different patterns. The coastal sand dunes were first colonized in 1965 when the keystone sand dune grass *Leymus arenarius* arrived. Coastal sand dune plants formed the first community on the island within a decade after the island was formed. A good example of facilitated succession was found during the strandline colonization on Surtsey where a coastal sand dune community is formed. In this case the strandline is colonized by a pioneer species (*Honkenya pepoloides*), and it acts as a nurse plant and facilitates colonization of the main dune building grass (*Leymus arenairus*), which in turn provides habitat for other species, such as *Martensia maritima*. Colonization in the coastal sand dune habitat has ceased but soil development (along with nitrogen accumulation) has been slow. Cover of vascular plants on the sand dunes was just above 20%. Colonization in the barren lava fields has been slow and erratic. Soil development has not taken place, and cover of vascular plants was very low (>1%). Gulls began to make nests on Surtsey in 1985 as sea-cliffs were formed by coastal erosion. In 2008, the gull colony exceeded 10 hectares and was characterized by very dense cover (100%) of perennial grasses. As the gull colony increased in size after 1985 more species of vascular plants were found to colonize the site. The gulls were the main vector of seed dispersal, and the gulls were responsible for soil nitrogen (N) and phosphorous (P) enrichment. The content of soil macronutrients has certainly been a factor driving the succession on Surtsey. The annual macronutrient input into soil mainly through gull feces has been estimated to be as high as 60 kg N per hectare and 12 kg P per hectare. A shift in plant species assemblage was recorded on the sea-cliff gull colony. The site was initially barren, but as gulls began to make their nests ruderal vascular plant species increased in cover. Perennial grasses (mainly *L. arenarius*, *Poa pratensis*, and *Festuca richardsonii*) increased rapidly in cover and dominate now the site.

On May 18, 1980 a major volcanic eruption of Mount St. Helens in Washington state in the United States resulted in a devastated landscape. Despite the general devastation around the volcano, isolated microsites of vegetation survived. Different types of substrate characterized the resulting landscape. The devastated land around the volcano was, however, analogous to a barren island surrounded by a forest. The nearby forest could potentially provide seed for the devastated landscape. The colonization and establishment of vegetation on these new soils were closely monitored. Primary succession was found to be slow, and little species turnover has been demonstrated. These studies showed that the seed bank contained in the microsites accelerated the colonization. Typical initial primary successional sequences involving cyanobacteria, algae, lichens, and mosses were found to be of limited importance for colonization of vascular plants on these sites. The pioneer plants established partially due to unpredictable events, and they still occur in scattered microsites. Also, successful colonizers need to be adapted to the harsh conditions, and seeds that are easily dispersed (from nearby forest) did not colonize effectively (due to unsuccessful germination) under these harsh conditions. Plant species accumulated steadily on each type of substrate, but plant cover is still very low, indicating severe growing conditions. These findings also stress that amelioration of the soil surface is an important precursor to successful establishment in harsh environmental conditions. Soil N was another limiting factor; therefore, symbiotic N_2 fixation by lupins was found to be important to facilitate establishment of other species.

Secondary Succession

Secondary succession is basically a recovery process of an ecosystem where a disturbance has impacted greatly or even decimated an existing community. Secondary succession takes place on "secondary sites" where disturbances leave soil in place. Secondary sites usually result from human activities such as logging and farming, including overgrazing. **Seed banks** exist on secondary sites and play an important role in the initial vegetation establishment. The determinant factor of secondary succession is especially existing seed bank of the intact soil and the proximity of seed sources determined by the distance to adjacent intact sites. Soils of secondary sites are often high in SOM and available macronutrients unlike soils of primary sites. Soil nutrients are often high in early secondary succession but reach lower levels toward mid-stages and then level off toward late succession.

The physiological environment changes typically from early to late successional stages, especially from sunny exposed habitat to a more shaded one. Consequently temperature fluctuations and drought conditions on the soil surface are reduced.

Examples of secondary succession can be found in studies on abandoned farmlands in northeast United States that have gradually returned to forest. A generalized model of this type of secondary succession can have five phases with corresponding characteristic species. The first phase is often dominated by annual weeds (e.g., *Cruciferae* and *Compositae*). The second phase is typically dominated by short-lived perennial herbs (e.g., *Aster pilosus*). The third phase is dominated by clonal herbs (e.g., *Solidago* sp.) and grasses (e.g., *Poa pratensis* and *Agropyron repens*). The fourth phase is dominated by shrubs (e.g., *Rhus* and *Rubus* sp.). The fifth phase (late successional) is dominated by several deciduous tree species, including eastern white pine (*Pinus strobes*) and red oak (*Quercus rubra*).

The aim of restoration on secondary sites is often to direct successional trajectories to a shrub- or tree-dominated state. Direct tree seeding is a common practice but requires control of seed-eating animals, preparing the site for sowing, and implementing weed-control management (see Chapter 11). Establishing native tree species from a seed mixture is commonly used to facilitate biodiversity on restoration sites. Transplanting of nursery-raised, container-based tree seedlings facilitates establishment of trees on secondary sites. Forest succession is, however, not always ensured by planting or seeding early-successional tree species. A long distance to an intact forest can form a dispersal barrier and seriously delay dispersal of late-successional trees. Strategic sequential planting of late successional tree species is therefore needed to ensure acceptable successional trajectories within an acceptable timeframe. This could, for instance, be accomplished by planting in multiple stages species of different successional phases, termed the "wave" approach. This approach probably mimics secondary succession, but it is labor intensive. An alternative approach is to plant at the same time species that belong to different succession stages. This approach aims at "stacking" secondary succession and is less labor intensive.

Animals in Succession

Succession models have predominantly focused on changes in plant communities. Animals are, however, common components of most ecosystems and have great influences on plant succession, and animal communities can also change along succession trajectories. Animals such as herbivores can regulate plant succession, and birds and bats can facilitate seed dispersal. Animals can facilitate or arrest specific succession trajectories and can therefore play an important role in succession management. For example, moderate grazing pressure may facilitate biodiversity, but overgrazing may lead to nutrient exhaustion, deterioration of plant communities and a loss of ecosystem integrity. Elimination of herbivores from degraded sites (by erecting fences) is often the first step in restoration projects. Exclusion of large herbivores can cause changes in succession trajectories, usually toward more shrub- and tree-dominating communities.

Another focus of restoration efforts can be on populations of endangered animals. For example, restoration of the spotted owl (*Strix occidentalis*) populations primarily involves protection from human-generated disturbances (mainly logging). The spotted owl is found in western United States and requires large areas of old (200–500 year) forest. A large old-growth forest (400–880 hectares) is necessary for a single pair of spotted owl. Spotted owl populations have declined as logging practices fragment the old forest. This work involves habitat restoration to ensure persistent population size of the target species. Also, restoration of endangered animals can involve reintroduction of animals into a restored habitat of different successional stages (discussed in Chapter 12).

Animals such as birds can be used as indicators of succession. Typically, during secondary succession of abandoned fields both bird density and diversity increase (**Figure 4.6**). In general, succession regulates bird species relative to their feeding habits. Early successional stages support granivorous birds, and mid-succession stages dominated by forest and shrubs support mainly fructivorous birds. Food availability alone does not adequately define suitable habitat, however. Another factor is structural complexity of forests that is often a good predictor of habitat quality of birds rather than successional status alone.

Animal communities can change as a result of plant succession and mainly due to changes in environmental factors during succession. These changes are regulated by the suitability of habitat for various animal species. Consequently, animals have traditionally been grouped into three successional types: (1) late successional species, (2) mid-succession species, and (3) species requiring a combination of successional stages. Management of successional phases to encourage colonization of specific animals is a challenging task.

Domestic animals can be used to manage successional trajectories. For example, grazing by domestic sheep can be used to change succession trajectory

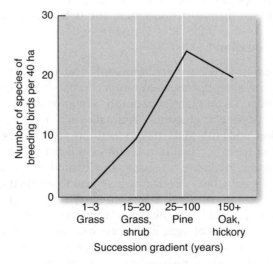

Figure 4.6 Changes in number of species of breeding birds along succession.

in grasslands by decreasing the dominance of a few dominant grasses and consequently increase biodiversity. In general, intensive grazing creates microsites for seed germination and allows some desirable species of plants to establish and expand their population. However, browsing can facilitate the invasion of aggressive weeds and invasive species. Periodic grazing followed by grazing exclusion is probably the preferred management practice to increase species diversity in grasslands. Furthermore, grazing and browsing can be implemented to discourage encroachment of shrubs and trees.

Animals not only affect succession by grazing and trampling but also by controlling insect pests, pollinating plants, and augmenting seed dispersal. Dispersal of seeds by animals is especially effective in enhancing the rate and trajectory of succession. Building perches or other structures to facilitate birds' movements from forests into degraded sites can augment the dispersal efficiency of birds (discussed further in Chapter 11).

4.3 Management of Succession

It is now assumed that succession does not necessarily have one stable endpoint but instead can result in several alternative stable states (discussed in Chapter 5). The aim of succession management is, therefore, to facilitate transition of communities from one state to another. Also, if succession trajectories deviate from desirable states, then succession management can possibly correct this problem. Varieties of restoration techniques are available that can be used to direct the succession on preferred successional trajectories. This can, for example, involve sequential introduction of native keystone species, removal of invasive non-natives, use of chemical fertilizers, and hydrological management. Initial site preparation is often needed on primary restoration sites. For example, if the soil surface is compact, then heavy machinery is needed to break (or rip) it up. Also, heavy machines can be used to make surface imprints. This can increase "**safe sites**" of seeds for various plant species. This is often followed by amelioration of the soil surface (using mulches, fertilizers, and/or liming). Such amelioration imitates the effect of natural plant litter.

Reducing or Inducing Disturbances

Succession can be modified by reducing disturbances such as fire, grazing, and deforestation. The clearing of forests creates various stages of succession. Annual plants and weeds typically increase their populations in cleared, sunny, forest patches that represent early successional stages. Also, disturbances that occur randomly (i.e., landslide, fire, or pest outbreak) can produce a patchy landscape made up of different successional phases. Landscape of different successional phases can also be the result of cyclic succession. A well-established example on such cyclic succession is demonstrated by the heather (*Calluna* sp.) community in northern Europe. Human impacts on successional processes have intensified in historical times. Today, various large-scale processes such as acid rain, nitrogen deposition, carbon dioxide emission, and global climate change are affecting successional trajectories worldwide. Removing sources of major disturbances is, however, essential in restoration practices.

Fire modifies many ecosystems worldwide. Fire can kill trees and shrubs and therefore suppress their dominance. Without fire, grasslands generally experience to tree- or shrub- encroachment. Growth rates of some plants can increase due to burning. Such plants are released from dominant species as the result of litter removal, increased light availability, soil warming, and greater availability of nutrients. Plants that are adapted to fire are termed pyrogenic. Fire can also act as a "switch" of community states, as in the case of pine- (pyrogenic) or oak-dominated forests in southeastern United States, as discussed further in Chapter 5.

Designed disturbance, on the other hand, is often required to create or eliminate site availability. Examples of designed disturbance include reinstalling natural flood pattern of a river or natural fire regime of a forest or grassland. Intentional use of fire in restoration practices is termed "prescribed burning." Prescribed burning is a powerful technique of succession management that has been used worldwide for a long time. Prescribed burning involves intentional use of fire to imitate natural fire regime and create a designed disturbance. Prescribed burning is effective in eliminating species that are not adapted to fire. Fires invariably modify successional trajectories by changing resources availability, species availability, and plant vigor or population structure. The use of fire is a cost-effective succession management technique. In addition, a variety of restoration goals can be met by implementing prescribed burning. For example, where the restoration goal is to control colonization, eradicate invasive species, or control species performances fire can be used effectively. Prescribed burning can have various effects on the ecosystem, such as increased seed germination, improved habitat for wildlife, thinned forests, and controlled pests. Using prescribed burning as a restoration tool to manage succession requires detailed planning, because not only fire but also smoke generated by prescribed fires is a factor that must be considered, especially in densely populated areas.

Introduction or Removal of Species

Careful selection of keystone species to be introduced in various successional stages is critical in the management of succession. The selection of such species is mainly determined by the physiological adaptation of the keystone species. Early successional species are generally short lived, and they are adapted to open sunny habitats. These plants typically germinate, grow, reproduce, and disperse their seeds quickly relative to mid- and late-successional plants. Light and fluctuating temperatures enhance seed germination of most plants in early succession. In addition, early-successional plants typically have high transpiration and photosynthetic rates. Controlled colonization of target keystone species involves active restoration techniques such as direct transplanting, seeding, and even adding topsoil with associated seed bank. On restoration sites only one or few keystone species are usually introduced at each time, but such species have great impact on succession and are expected to facilitate further colonization and establishment of native species. The performance of keystone species can change during succession, and series of keystone species are needed throughout succession. Such keystone species are termed "nexus species." It is essential to enhance growth and seed production of keystone plants. After keystone plants are established, restoration efforts, therefore, involve further enhancing their performance. This typically involves judicious application of fertilizers, removal of weeds or non-native species, and even temporary irrigation.

Nurse plants (commercial species) are commonly used in early succession to stabilize soil surfaces and ameliorate the harsh soil microenvironment. The nurse plants are often short-lived, but they may facilitate establishment of native plants. For example, in arid regions drought-tolerant shrubs can be used as nurse plants. These shrubs provide shade that facilitates further colonization of native plants.

Restoration practices aim at augmenting the performance of only few target keystone species. This can be achieved by eliminating competing weeds or non-native plants. Removal of such undesirable plants is necessary to enhance desired succession trajectory. Techniques of plant removal involve cutting, use of herbicides, prescribed burning, and cabling. Cutting (mechanical or manual) of vegetation is the most ubiquitous form of succession management. Cutting is especially useful in forests management where repeated cutting can eliminate altogether targeted unwanted tree species. Plants can be uprooted (manually or mechanically); however, this small-scale disturbance may facilitate further invasion of weeds or non-native species. Uprooting should be carefully monitored and should not be performed on a large scale due to the risk of soil erosion. Herbicides are commonly used to eliminate undesirable plants (weeds or non-natives). There is, however, a limitation to the use of herbicides because they usually need to be reapplied. Also, the negative environmental impacts of herbicide application need to be considered. Cabling is a technique used to modify the stature of shrubs or trees, for example, by linking two tractors or bulldozers with a chain at a certain height that is then pulled across a site. This method involves breaking down tall trees while leaving the understory vegetation intact. Cabling relieves understory shrubs from competition with tall trees. Cabling can strongly modify the population structure of shrubs and especially reduce the dominance of tall trees.

Nutrient Management

Nutrient management is commonly used to direct succession. Judicious use of chemical fertilizers can impact plant succession. Succession management can be achieved by manipulating the duration, type, level, and time of nutrient availability. A gradual change in the responsiveness of a community to nutrient addition occurs along succession trajectories. Early-successional species typically show great responses to nutrient application. Species performances can be manipulated by fertilizer application. For example, in revegetation practices the dominance of grasses is generally promoted by applying high levels of N fertilizers. Nitrogen is usually the most limiting nutrient in soil on primary sites. Primary sites are commonly revegetated by seeding commercial grasses followed by high doses of complete (N,P,K) fertilizers. However, less competitive species are consequently eliminated from primary sites. The use of commercial grasses and extended use of chemical fertilizers are effective in halting soil erosion and stabilize the ground. Grasses usually respond dramatically to fertilizer application, but often these responses last only for few years. Shifting competition in favor of grasses may therefore be accomplished by extending the duration of fertilizer application. To direct succession trajectories toward higher biodiversity, high rates of N fertilizers should be avoided.

A more agreeable approach is to use low rates of fertilizer application, which may increase biodiversity on sites with limited availability of nutrients without the dominance of only few grass species. Similarly, it is possible to direct successional trajectories by removing excessive soil nutrients. This can be accomplished by mowing and removing the biomass from the site. If soil nutrient availability is decreased, then succession trajectories can be arrested or even shifted to an earlier successional stage. This is often desirable because mid-successional stages can harbor high biodiversity.

Eutrophication (excessive nutrient enrichment) is a common environmental problem that affects succession where a few fast-growing species are favored. Nutrient exhaustion can be accomplished by various methods, such as intentional grazing of livestock, prescribed burning, and mowing and subsequent removal of the biomass. In extreme cases the nutrient-rich topsoil can be removed. Manipulation of soil pH can result in nutrient leaching and is often the first step in soil restoration (discussed in Chapter 8). Manipulating soil pH has tremendous effects on subsequent succession trajectories.

Hydrological Manipulations

Soil moisture is a critical factor in directing succession trajectories. Hydrological manipulations can affect successional trajectories. For example, manipulating the groundwater table acts as a designed disturbance for colonization. Drainage ditches can lower the groundwater table and subsequently favor more drought-tolerant plants. Similarly, flooded dikes and reservoirs can elevate the groundwater table and subsequently favor wetland species. The water level in wetlands, ponds, or lakes can be manipulated to control colonization and species performance of target plant species.

Irrigation is used to establish initial vegetation cover in restoration of arid lands. Trickle irrigation makes practical use of the water and is certainly a cost-effective method in restoration practices of arid regions. Irrigation ensures plant establishment on arid sites; however, the effects of irrigation are not long lasting.

Irrigation favors grasses but delays establishment of more drought-tolerant native species. Irrigation is usually an expensive operation and will probably not become a widespread method of succession management because of its temporary effects and difficulties in applying this method on a large scale in arid regions. Alternative restoration strategy is to use drought-tolerant native plants.

4.4 Monitoring Succession

Monitoring succession is a critical step in maintaining the effectiveness of implemented restoration actions. Monitoring provides early warning signs for adverse successional trajectories. The monitoring effort enables appropriate adjustments to changing environmental conditions resulting from unexpected disturbances. This in turn can adjust succession trajectories. Results of monitoring efforts also provide information on the cost-effectiveness of the restoration work. Moreover, results derived from long-term monitoring efforts provide vital information on restoration projects to the public, scientific community, and other stakeholders.

Varieties of methods are available for monitoring succession. Selection of monitoring methods depends mainly on the timeframe and budget of the restoration project. Monitoring succession requires initiating long-term ecological evaluation on restoration sites. For this purpose permanent plots are often used where measurements on vegetation dynamics can be repeated over several years. Data obtained from permanent plots include taxonomical survey of plants and measurements on vegetation cover. Permanent plots provide data for the interpretation of succession and also allows for testing of various management practices in restoration. For instance, effects of fertilizer application to accelerate succession can be tested on small permanent plots before being implemented on a large scale (see discussion on adaptive management in Chapter 14). Permanent plots should ideally be replicated across the landscape in various habitats within restoration sites. Plot size typically varies according to the community being examined. Plots can be relatively small (0.5–4 m^2) in communities dominated by grasses, herbs, or low shrubs. Permanent plots in tall shrub communities or forests usually should be much larger (100–1,000 m^2).

Vegetation cover can be measured by a variety of methods, such as the "point intercept" method that is practical for small plots, the "line intercept" method, or by using aerial photos for large plots. Canopy cover of trees can also be estimated by using aerial photos. Methods to estimate diversity of tree species involve plot and plot-less sampling methods. In the plots, species composition, litter, and soil cover should be recorded over time. This requires taxonomical knowledge on the local flora. Photos of permanent plots taken repeatedly through time can provide valuable information on succession. Such images provide rapid estimation on the cover of each species. The use of digital images of permanent plots is a new technique that allows rapid documentation of vegetation dynamics. Such images can be calibrated in the field before being analyzed. It is then possible to estimate rapidly the total cover of each species using computers. For large plots using aerial photos is the fastest method of documenting the vegetation. False color images are used to map vegetation cover over large areas and are particularly useful in sparsely vegetation environments.

4.5 Inferring Succession

Succession can be inferred by studying vegetation development (i.e., number of species and their cover) on restoration sites at different ages (chronosequences) provided that the sites share similar environmental conditions and that time is the main factor. For example, if restoration work is conducted in one particular area over many years, plots could be selected for certain time intervals (e.g., 2, 5, or 10 years) and vegetation development can be assessed within each plot. Using chronosequences relies on the assumption that all sites making up the time sequence have identical histories of disturbances, biotic influence, and environmental conditions. It is important to consider that certain weather patterns such as exceptional drought, temperature, or precipitation can affect these types of studies.

Succession can be inferred from an analysis of population age structures. This technique is limited to communities dominated by large shrubs or trees that can be accurately aged (using dendrochronology). It is assumed that saplings of long-living trees will eventually replace dominant trees of short-living species. It is therefore possible to predict successional trajectories from such estimations. Such an assumption may, however, not be valid if saplings of long-living trees do not persist long enough to fill gaps left by dominating trees.

Succession of plant communities over a long timeframe can be derived from preserved remains of plants (including pollens) buried in soil or sediment. Succession can be inferred by identification of these plant remnants in dated core samples. Carbon dating or use of teprachronology (dated volcanic ash layers) of the material can also establish a relative timeframe for succession. Past plant communities can now be constructed using molecular techniques. It is now possible to isolate DNA directly from plant remnants and soil and use polymerase chain reaction–based methods to amplify certain genetic markers (discussed in Chapter 8).

Models based on mathematical functions that simulate plant life history characteristics have been used to predict succession. These models are often general enough that they can be used under various conditions. These models can also incorporate effects of management activities on succession. In general, building a successional model requires extensive knowledge about the ecosystem in question. Models of succession can predict community-level or landscape-level changes through time as a result of management or natural disturbances. However, such predictions are probably not precise in dealing with environmental vagaries such as seed dispersal, invasive non-native species, or climate fluctuations. There are both advantages and disadvantages in using succession models. For example, the range of parameters in a model depends on the researcher's knowledge of the ecosystem being simulated. It is unlikely that a model can include all parameters affecting succession. Furthermore, it is impossible to predict certain factors that may dramatically alter successional trajectories in the future including invasion of non-native species, global climate change, and generally all stochastic disturbances.

Summary

The theory of succession deals with changes in plant communities with time following a disturbance. The succession theory has had strong influence on restoration ecology. Succession has traditionally been separated into primary succession and secondary succession. Primary succession deals with community dynamics on barren sites and is concomitant with soil development. Secondary succession deals with changes in communities on sites that have been disturbed but where soil is still intact. The outcome of competitive interaction of plants for resources (e.g., nutrients or light) is often the driving mechanisms behind succession. Interactions of plant species during succession have been described in three models: facilitation, inhibition, and tolerance. Animal interaction can determine succession trajectories. Also, animal communities are determined by successional stages of plant communities. Various restoration techniques are available to manipulate successional trajectories. These involve techniques that can lower the dominance of certain species and at the same time increase biodiversity. Monitoring succession by using replicated permanent plots is beneficial and can give early indication if further management action is nedded. Succession can be inferred by using methods such as vegetation chronosequences, population age structure, carbon dating, DNA profiles, and mathematical models.

4.1 Case Study

Restoration of Primary Sites
*Sigurdur Greipsson, Biology and Physics Department,
Kennesaw State University, Kennesaw, GA*

Restoration of primary sites is challenging because primary soils lack seed bank and plant colonization depends on the vagaries of seed dispersal from nearby native vegetation. Availability of "safe sites" on primary soils is critical for successful seed germination and establishment. Not only are primary sites limited by seed availability, but colonizing species must show physiological adaptations to such harsh microenvironments. High irradiance and fluctuating diurnal temperatures on the soil surface

(continued)

Case Study (*continued*)

is characteristic for these sites. Primary soils have low water-holding capacity and tend to dry quickly, which can limit seed germination on or close to the surface. Moreover, primary soils lack organic matter and have critically low amounts of available nitrogen (N) and low cation exchange capacity. Lack of N in the soil represents a critical threshold for restoration of primary sites. This can be overcome by introduction of N_2-fixing plants (lupins or alders) or by generous applications of chemical fertilizers. The absence of symbiotic soil microorganisms such as *Rhizobium* and mycorrhizae fungi is another limiting factor in restoration of primary sites. Pioneer plants usually establish in a heterogeneous pattern on primary sites. Further plant colonization is facilitated by these patches of pioneer plants and is termed *nucleated succession*. During the pioneer stage of primary succession colonizing species may accumulate without any species turnover. As soon as shrubs colonize primary sites, understory plants follow. Establishment of shrubs and trees increases available niches for animals on the primary sites. Consequently seed dispersal by animals is increased onto these sites.

Barren gravel lands in sub-arctic Iceland represent primary sites that are challenging for plant colonization and primary succession. The barren gravel soils typically contain very low amounts of N and soil carbon (C). Soils of barren gravel sites have N content in the range of 0.1 to 0.5 mg kg^{-1} and C in the range of 1 to 7 mg kg^{-1}. The soils of the target heathland community have N content ranging from 2.0 to 3.0 mg kg^{-1} and C content ranging from 3.5 to 5.5 mg kg^{-1}. Restoration strategies of such barren gravel sites involved spreading chemical fertilizers on the ground and seeding commercial grass species. The initial restoration effort was followed by fertilizers application for several years. Complete nitrogen-potassium-phosphorous (N, P, K) fertilizer was added annually in the amount of 400 kg per hectare. This practice resulted in increased N and C content of the restored soils through decomposition of plant litter. These restoration practices showed mixed results. The litter decomposition is a critical factor in the cold climate. Litter decomposition usually takes at least four to five decades. A few decades after the initial restoration effort, restored soils usually contain a low content of N (0.1–0.3 mg kg^{-1}) and C (3–4 mg kg^{-1}). Restored soils, however, can reach similar N and C contents as heathlands soils. The content of N in the restored soils depends on the rate of litter decomposition and eventual mineralization of organic matter. Before the restoration work, the barren gravel sites typically had very low cover (>5%) of vascular plants. The low plant cover is likely a result of soil nutrient deficiency, frost heaving, and abrasion from soil or sand particles that added to the difficulties plants have in colonizing such sites. About 20 pre-adapted and hardy vascular plant species are able to colonize and survive such harsh sites. Succession has been demonstrated on barren gravel sites where restoration has taken place. The target community of this work is a heathland community dominated by willows (*Salix* sp.), dwarf-birch (*Betual nana*), and/or birch (*Betula pubescens*). Such communities are found on reference sites with intact fertile soils. Following the initial restoration effort, the cover of commercial grass species reached more than 30%, but, as soon as application of chemical fertilizers stopped, their cover declined rapidly and may vanish within a few years. Shortly after the application of chemical fertilizers ceased native grasses typically increased their cover, which can reach more than 50%. The cover of native forbs also increased during the initial restoration effort and reached more than 60%. Most of the native forbs were already on the barren site but in low cover. The cover of native forbs declined as the cover of native grasses increased. Cover of mosses was very low on barren sites but increased shortly after fertilizer application stopped and reached more than 25%. Ten years after the initial restoration effort, plant litter dominated the cover (>55%) along with native grasses (>45%), but cover of native forbs was still low (>10%). Twenty-five years after the initial restoration effort, surface soil organic matter dominated the cover (>40%) along with mosses (>30%) and native forbs (>20%). Plant succession could be delayed by the dense cover of mosses, which can inhibit colonization of vascular plants. The dense cover of mosses could, therefore, represent another ecological constraint. The number of plant species colonizing sites under restoration increased with time and as many as 30 vascular plant species were found 25 years after the initial restoration effort.

The plant litter that dominated the sites 10 years after the initial restoration effort had decomposed into organic matter on the soil surface after 25 years. The rate of litter decomposition and mineralization of organic matter is critical for the buildup of the N content in the soil and for the progress of primary succession. Most restoration sites did not show clear turnover of plant communities. Plants that are part of the heathland community can colonize early in the succession, but their cover remains very low on restoration sites. It is still not known how much time it will take to reach the species assemblage of the heathland community with a dominant cover of *Salix* and *Betula* species.

Key Terms

Eutrophication   94
Monitoring succession   95
Nurse plants   93
Primary succession   83
Resource Ratio Hypothesis   83

Secondary succession   88
Seed bank   88
Succession management   82
Succession trajectories   80

Key Questions

1. Describe the three models of succession proposed by Connell and Slatyer.
2. What are the main differences between primary and secondary succession?
3. Give examples of primary succession.
4. Describe the role of animals in succession.
5. What methods can be used to manage succession?
6. What methods can be used to monitor succession?
7. What methods can be used to infer succession?

Further Reading

1. Bazzaz, F. A. 1996. *Plants in changing environments. Linking physiological, population, and community ecology.* Cambridge, UK: Cambridge University Press.
2. Gray, A. J., Crawley, M. J., and Edwards, P. J. 1987. *Colonization, succession and stability.* Malden, MA: Blackwell.
3. Miles, J. and Walton, D. W. H. 1993. *Primary succession on land.* Malden, MA: Blackwell.
4. Tilman, D. 1988. *Plant strategies and the dynamics and structure of plant communities.* Princeton, NJ: Princeton University Press.
5. Walker, L. R. and Morel, R. D. 2003. *Primary succession and ecosystem rehabilitation.* Cambridge, UK: Cambridge University Press.

5

ASSEMBLY

Chapter Outline

5.1 Equilibrium Theory of Island Biogeography
5.2 Ecosystem Resilience and Stability
   Resilience
   Resistance
   Stability
   Ecological Constraints
5.3 Alternative Stable States
   Regime Shift
   Restoring Alternative Stable States
5.4 Assembly Rules
   Sequence Introduction
   Species Compatibility
   Ecosystem Thresholds
   Ecosystem Filters
5.5 Unified Neutral Theory of Biodiversity and Biogeography
Case Study 5.1: Resilience and Ecosystem Restoration
Case Study 5.2: Phylogenetic Structure of Plant Communities Provides Guidelines for Restoration

Facilitating succession or assembly of a functional and integral ecosystem after disturbances is an important task of ecological restoration. Various ecological models have been used in these restoration processes. These models can predict changes in community composition with time and therefore aid in the restoration work. Some models have background in community ecology, whereas others are derived from biogeography. These include the deterministic model of succession (sensu Clements), the individualistic model of succession (sensu Gleason), and assembly models (sensu McArthur and Wilson and Diamond).

Succession models in particular have provided the theoretical framework for restoration of plant communities, as discussed in Chapter 4. These models have mainly been applied on terrestrial ecosystems. On the other hand, restoration of aquatic ecosystems has put much more emphasis on assembly models, which consider the possibility of alternative ecosystem states instead of only one final "climax" community. Recently, assembly models have been applied successfully in the restoration of terrestrial ecosystems. Many basic aspects underlying assembly models, such as the importance of arrival time of species to a new site in determining an alternative ecosystem state, are already incorporated in various succession models. The emphasis of succession models, however, is on the trajectory leading to a climax community, or final state whereas assembly models demonstrate the possibility of various different endpoints.

Restoration ecology benefits from using assembly models to consider the possibility of multiple stable states instead of one final state. Another practical aspect of assembly models in restoration work is to consider ecological constraints and especially what actions are needed to relie degraded states to move along an assembly trajectory toward a desirable ecological end state.

5.1 Equilibrium Theory of Island Biogeography

The equilibrium theory of island biogeography was proposed by Robert MacArthur and Edward O. Wilson in 1967 and has since had a profound impact on ecology, biogeography, and conservation biology. The equilibrium theory also has strong implications for restoration ecology. MacArthur and Wilson used the gradual colonization of offshore islands derived from a mainland source as a model system for community assembly. In essence, their work predicts that the number of species on offshore islands is based on a balance between the rate of random dispersal of colonizing species from the mainland and the rate of random local species extinction on the islands (**Figure 5.1**). The model, therefore, describes community assembly mainly as a result of random immigration and random local extinction. An important aspect of the model is that it assumes all species have the same probability of dispersal to an offshore island as they do to facing local extinction. The main parameters of the model are

1. Distance of the islands from the mainland determining the dispersal rate of new species
2. Size of an island determining the rate of species extinction

In practice, large islands close to the mainland are more likely to experience higher numbers of immigrating species and lower extinction of species than small islands. The opposite effect increases the farther the small islands are from the mainland. The size of these islands and distance from the mainland,

Equilibrium Theory of Island Biogeography 103

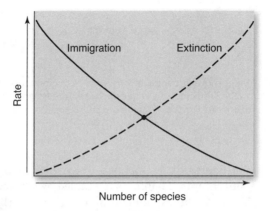

Figure 5.1 Model of equilibrium theory. Equilibrium in community assemblages is formed by the rate of immigration and extinction.

therefore, determine the number of species that exist there in a flux. Such random community assembly does not imply that these colonizing species are necessarily coadapted or preadapted to specific niches on the new site. The equilibrium theory does provide, however, a powerful insight into community assembly by emphasizing a few factors such as random dispersal and colonization resulting in random assemblages of species derived from the mainland source. It is assumed that random dispersal processes alone play an important role in structuring communities.

The prediction of the equilibrium theory is similar to Gleason's individualistic model of succession. The "individualistic model" of succession assumes that community composition is the result of a random process influenced by the availability of vacant niches and species dispersal. The individualistic model is analogous to a carousel where seats become vacant randomly and are filled by nearby passengers (that represent different species).

Practical uses of the equilibrium theory have been extended to conservation biology, especially to predict extinction rates of species in parks or nature reserves. At the same time the model has been used to predict the necessary size of reserves and need for restoration, buffer zones, and connectivity between fragments to facilitate species dispersal into parks and reserves where the goal is to avoid species extinctions. These factors (size and connectivity) are probably more important to avoid species extinction than protection alone. In fact, most nature reserves are too small to fully support biodiversity conservation in the long run. Large nature reserves are needed to inhabit large predatory animals that roam over huge territories. Such animals are often keystone species that influence functioning of large ecosystems. Restoration sites by themselves are often small and isolated away from pristine or larger intact habitats that can act as a source of colonizing species. Such sites must be effectively connected in the local landscape

Figure 5.2 Model showing community assembly as colonization by stepping-stone process. (Adapted from V. M. Temperton, et al. (eds.). *Assembly Rules and Restoration Ecology: Bridging the Gap Between Theory and Practice.* Washington, DC: Island Press, 2004.)

(discussed in Chapter 6). In practice, fragments of pristine landscape can be viewed as islands of intact habitat surrounded by a human-dominated landscape, such as agricultural lands, industrial parks, and urban and suburban areas. Restoring connectivity to facilitate dispersal rates of species can potentially restore functioning of such fragmented landscapes.

The equilibrium theory incorporates dynamic aspects of colonization. For example, in the mainland–island scenario offshore islands are colonized as stepping stones (i.e., species jump from one island to another) (**Figure 5.2**). The island that is closest to the mainland receives the highest number of immigrants from the mainland. However, colonization of islands that are farther away from the mainland depends on the species assemblage of the first island. The island farthest away from the mainland consequently has fewer immigrating species than the first island. Similarly, restoring connectivity between isolated fragments can be restricted or limited by dispersal distance from a source site and therefore create a dispersal barrier. In practice, such dispersal barriers can be alleviated by strategically restoring habitat patches to be "stepping stones" in the landscape.

5.2 Ecosystem Resilience and Stability

Resilience

Ecosystems can usually recover after mild disturbances. The time it takes an ecosystem to recover after a disturbance to its predisturbance ecological condition is termed resilience (see **Case Study 5.1** on page 116). Resilience is also defined as the amount of disturbance (frequency and duration) that an ecosystem can tolerate without causing a regime shift. Ecosystems vary in their resilience in that some recover rapidly after disturbances, while others need much longer time and may even require restoration efforts to do so.

Loss of resilience can trigger a shift to another ecological state. To avoid undesirable regime shifts in communities, the most pragmatic strategy is to restore and maintain resilience of the desired ecosystem conditions, in which various strategies can be used. Such efforts include increasing biodiversity (especially by increasing the number of species in diverse functional groups), controlling invasive species, and regulating disturbances that contribute to a shift in ecosystem state. Disturbances that provoke a shift in ecosystem state are usually stochastic and include climate extremes (strong storms, exceptional droughts), fires, eutrophication, and disease outbreaks. It is usually difficult to predict or control such disturbances. An alternative restoration strategy, therefore, is to put the emphasis on the resilience or the ability of the ecosystem to recover rapidly after any disturbances.

It is not unusual for a degraded ecosystem to show resilience to restoration efforts. To diminish this resilience, ordinary restoration efforts such as soil nutrient management or introduction of keystone native species can be used to facilitate a **regime shift** to a more desirable state. Restoration strategies should therefore focus on factors that induce a shift from a less desirable to a more desirable ecological state along assembly trajectory.

Resistance

Ecosystems can respond to disturbances by resisting ecological changes altogether. This is termed **ecological resistance**. In such a setting an ecosystem showing high resistance can be stable even while being frequently disturbed. An important task of restoration efforts is often to inhibit disturbances that may lead to ecosystem degradation. An alternative restoration strategy is to increase the resistance of the ecosystem instead of focusing only on inhibiting disturbances. This can be achieved, for instance, by restoring diversity of plants in diverse functional group. Doing so may increase the resistance of the ecosystem.

Stability

Ecological stability depends on various factors that change slowly, such as biodiversity, nutrient levels, soil properties, and existence of long-lived organisms. Many of these factors may be monitored and restored if needed to maintain stability. The importance of biodiversity for ecosystem stability was demonstrated by Tilman's field experiment on a prairie community in Minnesota. In his experiment biodiversity was manipulated and plots with high biodiversity were most stable under adverse climatic conditions.

Ecological Constraints

Disturbances and factors of degradation that are responsible for maintaining certain ecological conditions are termed **ecological constraints**. These include, for instance, invasion of non-native species, dispersal barriers, lack of available nutrients, eutrophication, habitat fragmentation, and loss of keystone species. Degraded ecosystems may have shifted to a new alternative state that cannot be easily restored to predisturbance conditions. An example of such a degraded ecosystem can be found in overgrazed semiarid grasslands that are invaded by shrubs. Here, decreasing grazing pressure of livestock may not be efficient in restoring the grass-dominated ecosystem. Restoration of such a degraded ecosystem often requires massive efforts, such as mechanical eradication of shrubs followed by ground stabilization and introduction of native grasses. Non-native species in a degraded ecosystem can represent another example. In this case the frequency and intensity of ecosystem processes such as fire regime, nutrient cycling, and hydrology might be permanently altered.

To restore degraded ecosystems it is important to identify ecological constraints and feedbacks that maintain the degraded conditions. Restoration efforts can release degraded systems from such constraints. This work might involve reducing grazing pressure; overcoming dispersal limitations by seeding native species, introducing keystone species; manipulating soil fertility; eradicating non-native species; and implementing other necessary restoration efforts.

Restoration of degraded ecological conditions to a more desirable state probably follows a different sequence than did the degradation process itself. For example, in ecosystems where eutrophication is a constraint, a perturbation program of nutrient exhaustion can be implemented. Mowing and biomass removal of alpine grasslands in Switzerland is, for instance, used effectively to exhaust nutrients to maintain high biodiversity of these grasslands.

Successional trajectories have often been unpredictable and controlled simultaneously by several different ecological constraints. Restoration efforts might, therefore, need to simultaneously manipulate multiple ecological constraints in a strategic way. Such manipulation is needed over a long period of time to direct the desirable assembly trajectory. These may include, for instance, the establishment of natural fire regime, nutrient cycling, and introduction of native plants. For example, restoration of heathlands in northern Europe involves the manipulation of abiotic and biotic constraints. In these cases the abiotic constraints were eutrophication and acidification, whereas biotic constraints were dispersal barriers and impoverished soil seed bank. Simultaneous manipulations of these constraints resulted in effective restoration of heathland ecosystems.

5.3 Alternative Stable States

Succession is often found to be unpredictable. Instead of ending in a single stable community, the climax, it may result in different endpoints, also termed "alternative communities." In fact, these different alternative communities have been considered as variations on successional trajectories (**Figure 5.3**). These variable outcomes of succession have been recognized and termed as arrested succession, polyclimax, metaclimax, and disclimax. **Alternative stable states** exist as a result of several ecological factors such as random species dispersal and colonization. This strong impact of early colonizing species on the community is termed "historical contingency." Once early arriving species are established, they can inhibit or delay colonization by other species and therefore maintain a characteristic ecological state. The stability of some alternative stable states can be explained by the fact that the dominating plants are clonal, form dense stands, and are long-lived. For instance, aspen (*Populus* sp.) and coastal redwoods (*Sequoia sempervirens*) can form such ecosystems.

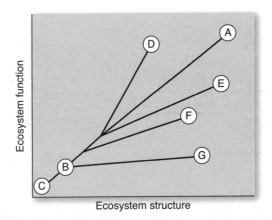

Figure 5.3 Model of alternative stable states with various endpoints as results of different assembly trajectories. (Reproduced from *Assembly Rules and Restoration Ecology* edited by Vicky J. Temperton, Richard J. Hobbs, Tim Nuttle and Stefan Halle. Copyright © 2004 by Island Press. Reproduced by permission of Island Press, Washington, D.C.)

Regime Shift

The existence of alternative stable states and regime shift between states is regulated by disturbances and has been identified in several ecosystems (**Figure 5.4**). Disturbances such as grazing pressure can regulate regime shift. For example, in rangelands a grass-dominated state can shift to a shrub-dominated state due to change in grazing pressure. Also, mountain birch (*Betula pubescens* sp. *czerepanovii*) forests in northern Scandinavia can exist in several states characterized by the dominance of the understory species. Even regime shifts between some of the most dominant states have been observed, for example, the shift between a lichen-dwarf, shrub-dominated forest to a more grass- and herb-dominated forest. Transition between these alternative states of the mountain birch forest was regulated by grazing and browsing of reindeer stocks. Such intensive reindeer grazing can further promote regime shift by transforming, for example, lichen–moss-rich heath tundra into grass–sedges-dominated tundra vegetation.

Regime shift between alternative ecological states can also be regulated by fire; for example, forests of northeast Florida can exist in two alternative states. The first state is characterized by pyrogenic longleaf pine (*Pinus palustris*) with open savanna. The second state is characterized by mesic oak (*Quercus* sp.) forest. Longleaf pine and oak show different responses to fire. Longleaf pines are adapted to frequent low-intensity ground fires. In fact, mature longleaf pines shed pyrogenic needles that promote such ground fire. Increased cover of pines leads to an increased number of pine needles on the ground that, in turn, promotes fires. The pyrogenic pine needles accumulate on the forest floor until

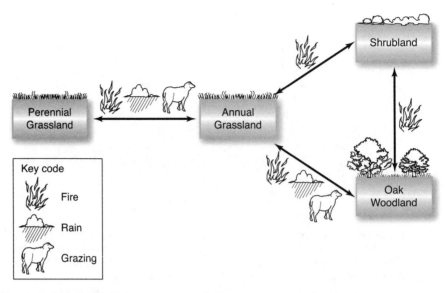

Figure 5.4 Model of alternative states in Californian grasslands. Switches between states are shown by arrows. (Adapted from V. M. Temperton, et al. (eds.). *Assembly Rules and Restoration Ecology: Bridging the Gap Between Theory and Practice*. Washington, DC: Island Press, 2004.)

a critical threshold is passed and a ground fire can break out (usually initiated by thunderstorm lightning). In contrast, young oaks are intolerant of fire, and mature oaks shed leaves that suppress ground fires. Both species, therefore, give positive feedback for further growth. In the absence of fire oaks eventually replace longleaf pine in the forest. However, low frequency of fire suppress oak growth or even eliminate them altogether, allowing longleaf pine to dominate the forest. In this ecosystem fire regulates the "switch" (also known as hysteresis) or regime shift between two alternative states. Here, the desired ecosystem state can be restored by using prescribed burning or fire suppression as a restoration tool. Another example of how fire can induce regime shift is the gradual conversion of heathlands in northern Europe into forests due to fire suppression. Restoration of heathland communities requires intensive fire that consumes the surface organic layer.

Eutrophication of terrestrial ecosystems has already caused regime shift in numerous places. Nitrogen deposition has done just that in wet heathlands in the Netherlands. These wet heathlands were previously dominated by bog heather (*Erica tetralix*) but have now been replaced by purple moor-grass (*Molina caerulea*). The restoration effort of these wet heathlands now focuses on removing top soil that has been overloaded with nitrogen. In addition, liming has improved the acidity of the soil.

In shallow lakes, regime shift between alternative states can be regulated by eutrophication and, especially, excess phosphorus loading. Such regime shifts are

Figure 5.5 Alternative states of shallow lakes; clear **(a)** and murky **(b)** waters. (Part a © Stanislav Komogorov/ShutterStock, Inc.; part b © Justforever/Dreamstime.com.)

common where clear water state dominated by aquatic plants is transferred to murky water state dominated by phytoplankton (algae) along with other ecological changes (**Figure 5.5**). Shallow lakes can rapidly shift from clear to murky states with each state being relatively stable. Curtailing eutrophication is most often not enough, however, to restore the clear water state of a lake. In the clear water state, sediments are stabilized by aquatic plants. The murky water state persists in the absence of aquatic plants due to waves that resuspend sediments. The murky waters decrease light penetration in the water column and, in turn, curtail the establishment of aquatic plants. Restoration of the clear water state involves biomanipulation. This involves implementing selective fishing that decreases the size of populations of planktivorous fish. When this is accomplished, an increase in populations of herbivorous zooplankton follows and, in turn, leads to a reduction in phytoplankton (algae) populations and increased light penetration through the water column. Increased light penetration facilitates establishment of aquatic plants, which stabilize sediments. Conversely, shifts from a clear to a murky water state can also result from overgrazing of aquatic plants by fish or waterfowl. It is noteworthy, then, that the disturbances that induce regime shift do not necessarily have similar impacts in the opposite direction. Considerable restoration efforts are usually needed to establish a clear water state of a shallow lake. In this context the nutrient levels of the lake need to be brought to a substantially lower level than the predisturbance clear water state of the lake. This is often accomplished by mechanically dredging nutrient-rich sediment and removing it from the lake.

In any restoration effort it is important to identify factors that regulate regime shift to alternative states. For this purpose a plan that outlines restoration efforts that can shift less desirable conditions to more desirable ones should be in place. Considering the possibility of multiple stable states along an assembly trajectory, restoration efforts need to be selected carefully to obtain the most desirable ecological endpoint. Using the passive restoration approach may return the degraded ecosystem to its predisturbance condition with little effort. Alternatively, the active restoration approach may be required to ensure agreeable results.

Models of alternative ecosystem states can be used to guide restoration of degraded conditions of an ecosystem. This is especially important where switches that regulate regime shifts have been identified and can be manipulated. The fact that a degraded ecosystem can possibly regenerate along an assembly trajectory to several different multiple stable states, any one of which has desired end points, has strong implications for restoration. The possibility of alternative stable states puts emphasis on tight monitoring and continuous active restoration efforts. This involves an initial restoration effort such as planting a variety of native species and should then be followed by intensive restoration activities as the ecosystem moves along the assembly trajectory.

Restoring Alternative Stable States

In planning restoration work the existence of multiple stable states requires the use of well-defined reference sites (see Chapter 14). Considering the possibility of multiple ecological states, the definition of reference sites may involve several choices, therefore, becoming a moving target along assembly trajectory.

Restoration efforts can be used to shift a degraded ecosystem to another more desirable one, for instance, by alleviating dispersal barriers of species into restoration sites. As outlined earlier, the tight control of species' arrival on site is important because it can be a determining factor in the community assembly. Additionally, dispersal barriers can result in a lack of native species that are critical for succession on restoration sites. In this regard dispersal barriers often serve only to maintain a certain ecological state. To overcome such dispersal barriers of keystone species, ordinary restoration efforts can be put into place, such as transplanting or direct seeding of these species on restoration sites. This is often followed by restoration efforts that aim at enhancing the establishment of these species on sites (i.e., nutrient additions or elimination of non-native species). Building perches on restoration sites for birds or bats, for example, increases seed dispersal. In practice, the restoration of prairie communities has been accomplished successfully by interseeding a mixture of native species (without any soil preparation) that are missing from the restoration sites. This step relieves dispersal barriers and also allows rare prairie species to establish themselves in the community. An alternative strategy might include the management of surrounding landscapes (i.e., building stepping stones or corridors) to alter seed pools and dispersal vectors, thus facilitating the establishment of desired species and therefore reducing dispersal barriers. Such efforts help in overcoming dispersal barriers of species that otherwise would not be able to colonize by natural means on isolated restoration sites.

5.4 Assembly Rules

Assembly rules were pioneered by Jared Diamond in 1975 in his study on bird communities of New Guinea, where characteristic assemblages of birds were observed in different habitats. Assembly rules are an alternative to the commonly used succession model in restoration (see Chapter 4). They predict that active restoration involving intensive control over species establishment within a restoration site is critical for attaining the final ecosystem state. On the other hand, restoration efforts that rely on succession models have traditionally focused on the final ecosystem state, determined by a selected reference site. Assembly rules are analogous to jigsaw puzzles and their intricately shaped, interlocking pieces; predicting that only certain species combinations are possible in each habitat. These species assemblages represent alternative stable states, as discussed previously.

Assembly rules make several important assumptions. First, they assume that communities are niche assembled. Second, unlike the deterministic model of succession with emphasis on the climax community, assembly rules assume that degraded ecosystems will not necessarily return to the original ecological state because the order of species colonization affects the final ecosystem state. Third, assembly rules predict that communities are relatively stable and not easily invaded by other species (native or non-native) (**Figure 5.6**)

Sequence Introduction

Assembly rules predict that appropriate species combinations must be introduced sequentially to a restoration site. Introducing many species simultaneously to a restoration site might lead to competitive exclusion and should be carefully examined. Because the order of species colonization affects the final ecosystem state, assembly rules can be used to predict the sequence of species

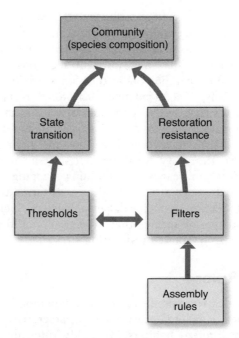

Figure 5.6 Interrelationships among the concepts of assembly rules. (Reproduced from *Assembly Rules and Restoration Ecology* edited by Vicky J. Temperton, Richard J. Hobbs, Tim Nuttle and Stefan Halle. Copyright © 2004 by Island Press. Reproduced by permission of Island Press, Washington, D.C.)

colonization on a restoration site toward the final ecosystem state. For instance, on a derelict mine site, tolerant species can be introduced to ameliorate the microenvironment. These are followed by introducing mid- and late-successional species, but only after appropriate microenvironmental conditions are reached. Similarly, on deflated coastal sand dunes plants are introduced strategically in a sequence to restore the dune ecosystem (see Chapter 9). Because of the emphasis of assembly rules on species arrival time for the final ecosystem state, intensive management of the critical sequence of species introduction is needed for restoration to be successful.

Assembly rules also assume that species characterizing the final (climax) community should not necessarily be introduced in the beginning of the restoration process. There may also be species essential in the initial restoration process other than those found in the desired final community. In fact, keystone species that are the driving force in determining the assembly trajectory often disappear from the community before the final state is reached; these species are termed **nexus species**. Nexus species are important in determining which alternative state the community will reach. Such species play a very important role in the ecosystem. For instance, they can stabilize soil conditions, fix nitrogen (legumes), or provide habitat for birds that increase seed dispersal on sites. An important task of restoration efforts should therefore be to strategically introduce nexus species into degraded ecosystems along the assembly trajectory.

Species Compatibility

The assembly process requires "adaptive management" (see Chapter 14) where the compatibility of different species combinations is carefully examined. Assembly rules assume that only certain combinations of species can coexist in a particular community (the jigsaw puzzle analogy). The use of assembly rules in restoration efforts should help in predicting which combination of species coexist in a given habitat on restoration site. Species combinations should therefore be selected from a regional pool for certain environmental conditions, habitats, or successional phases (i.e., early, mid, or late). Assembly rules also suggest that large prey and predators be introduced together, as should symbiotic mycorrhizal fungi and mycorrhizal-dependent species, appropriate nitrogen-fixing symbiotic bacteria and legumes, and so on.

Ecosystem Thresholds

Important components of assembly rules are **ecosystem thresholds**. These represent certain conditions that prevent regime shift from degraded to less degraded ecological conditions. A restoration effort is therefore needed to manipulate thresholds and transform the ecosystem to a more desirable state. This usually includes both ordinary restoration efforts and long-term aftercare. Biotic and abiotic factors can be responsible for the ecosystem thresholds. Biotic thresholds include heavy grazing pressure and lack of native species in the local seed pool. Abiotic thresholds include soil degradation and changed hydrology. In the restoration process itself, it is important to identify thresholds and methods to manipulate their effects. Of course, ordinary restoration methods can still be used to overcome restoration thresholds. For instance, if overgrazed and degraded rangelands cannot shift to less degraded ecological conditions due to soil compaction, massive restoration efforts, such as effective soil ripping to increase water infiltration, are needed. This example demonstrates how manipulation of ecological thresholds can be used effectively to restore ecosystem functioning.

Ecosystem Filters

The environmental conditions of a restoration site dictate which species can occupy a particular habitat; these are termed **ecological filters**. Only species preadapted to the environmental conditions of a site can establish successfully (see **Case Study 5.2** on page 119). Ecosystem filters are responsible for selecting species out of a regional pool that are appropriate to occupy a site. For instance, a landslide within a forest creates a barren site and will only provide a niche for a limited number of forest species that are preadapted to disturbed sites. This landslide therefore acts as a filter. Restoration efforts should focus on manipulating filters to facilitate and fasten a desirable species combination toward the final state. This work should ideally come at the same time as thresholds are being manipulated. Both abiotic and biotic filters are critical in this process.

Abiotic filters might include climate, soil type, and landscape structure (patch size and isolation). Biotic filters might include competition, predation, trophic interactions, propagule availability, mutualisms, and order of species colonization. Restoration efforts should include modification of abiotic filters such as remediation of toxic soil and improving nutrient status of soil. Concurrently, modification of biotic filters might include introducing native species on restoration sites or controlling invasive species (weeds and non-native species). Such modifications should enhance a regime shift from a degraded to a more agreeable ecological state.

5.5 Unified Neutral Theory of Biodiversity and Biogeography

The unified neutral theory of biodiversity and biogeography (UNTB) was recently proposed by Stephen Hubbell. In essence, the UNTB is an extension on the equilibrium theory of MacArthur and Wilson. According to the UNTB, community composition is determined by regional biogeographic processes such as random species dispersal, ecological drift, random speciation, and extinction without any role of special niche assembly. **Ecological drift** refers to random fluctuations in species abundances. The UNTB predicts that species are equal in their ability to colonize sites, and random processes structure communities that exist in a flux. Stochastic processes affecting communities are therefore at central importance in the UNTB.

The UNTB is based on a zero-sum game that is neutral and played by individuals that are identical in their functioning in the community (**Figure 5.7**). In a zero-sum game species can only increase in abundance if other species in the community decrease in abundance. The UNTB assumes that as vacant niches become available in the community they are filled up randomly without any special pre-adaptation or niche requirement. This is in contrast to the conventional view of community assembly where species are assumed to be preadapted to a specific niche.

One of the best ways to test the predictability of UNTP is to use the mainland and nearby island model system in community assembly. It is assumed that species in such a scenario would experience ecological drift, random colonization, local extinction, and adaptive radiation where new species arise.

A study on the forest assemblages of three small islands with similar environmental conditions found in the Bay of Panama, Central America, has collaborated predictions of the UNTB. These islands were once connected to the mainland when sea levels were much lower than what they are today. It is assumed that these islands harbored similar communities when they were connected. As the islands became isolated, ecological drift, local extinction, and

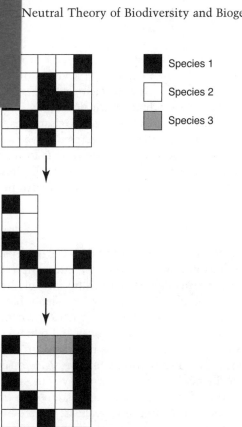

Figure 5.7 Model of a zero-sum ecological drift. New species can only enter in the community if abundance of existing species decreases.

random colonization would probably have affected the community composition of each island. Interestingly, each island now has different assemblages of dominant tree species despite similar environmental conditions. These findings are consistent with prediction of the UNTB.

Additional support for the UNTB comes from a transplant experiment within Canadian forests. Transplanted trees that were not preadapted to local conditions did as well as local ones, therefore, supporting the prediction of the UNTB.

Another practical aspect of the UNTB relates to the prediction that most communities are open and easily invaded by non-native species. This prediction has strong implications for restoration ecology. Most communities appear to be easily invaded by non-native species. Considering the widespread occurrence of non-native species, it is questionable if restoration efforts should particularly focus on controlling or eradicating non-native invasive plants. Further considering the usual lack of funding, more passive restoration approaches could be adapted for any large-scale restoration programs.

Although there is a strong need for resto‌‌‌‌‌‌‌‌‌‌‌‌‌ommunities that have been nearly decimated, it is not known ho‌‌‌‌‌‌‌‌‌‌‌‌‌ities will respond to the ongoing global climate change. One a‌‌‌‌‌‌‌‌‌‌‌‌‌ the assembly of future communities, which will probably be affected by global climate change and tremendous pressure from non-native species, could be to allow the assembly of species to randomly establish on restoration sites. This is not, however, to say that some aggressive invasive species that form monopopulations or alter ecosystem functioning should not be eradicated. These are all important issues that restoration practitioners will have to deal with in the near future.

Summary

Models in community ecology and biogeography have been used successfully in ecological restoration. Traditionally, succession models have been used on terrestrial ecosystems, whereas assembly models have been used on aquatic ecosystems. The equilibrium theory of island biogeography puts emphasis on random immigration and random extinction in community assembly. Ecosystems can resist changes from disturbances and remain stable. Ecosystem resilience is determined by the time it takes to recover from a disturbance back to the predisturbance state. Restoration strategies using assembly models focus on increasing the resilience of a desirable ecosystem state and reducing resilience of a degraded state of an ecosystem. Ecosystems can exist in alternative stable states where regime shift (between states) is regulated by disturbances. Assembly rules predict that only certain species combinations can exist in a community and puts strong emphasis on niche-assembly. Assembly rules also assume that appropriate species combinations must be introduced on restoration sites. The unified neutral theory of biodiversity and biogeography describes community assembly through ecological drift, random dispersal, and random speciation. It does not assume any role for niche assembly in this process.

5.1 Case Study

Resilience and Ecosystem Restoration
Lance Gunderson, Department of Environmental Studies, Emory University, Atlanta, GA

The idea of resilience was originally proposed as a way of explaining unexpected and dramatic change in ecosystems. The concept describes how ecosystems change in ways that are not just linear and predictable, but ones that are sudden and unpredictable. This model of change involves a transformation, whereby systems change structure and processes into qualitatively different regimes.

Ecosystems can be described as self-organized systems that are defined within a given spatial and temporal domain. Yet many, if not all, of these systems are affected to some degree by external events or processes that occur at broader or larger scales. For example, storms such as cyclones influence structure and function of coastal ecosystems, or fires occur in many forest and grassland terrestrial ecosystems. Biotic interventions occur in the form of invasion of non-native organisms, land use transformations, and harvests of renewable resources, among others. These broader scale interventions can be described as disturbances or **perturbations** to ecosystems.

In ecology, the term resilience has at least two different meanings and contexts. Both meanings apply to the interaction between internal system structures and processes and external perturbations. The word resilience is derived from Latin roots meaning "to jump or leap back." Hence, the general meaning of resilience is the ability to recover from external disturbances or adjust easily to change. Resilience is used by many ecologists to describe how quickly a system returns to its previous state after a perturbation. In 1973, however, the theoretical ecologist C. S. Holling argued that many systems, population systems to ecological systems, change into fundamentally different systems and resilience is the property that mediates the transitions among different system configurations, rather than just postdisturbance recovery. Each of these two definitions is elaborated in the next paragraphs.

Some ecologists (and engineers) define resilience as the time required for a system to return to an equilibrium or steady-state after a perturbation. Implicit in this definition is that the system exists near a single or global equilibrium condition. Hence, the measure of resilience is how far the system has moved from that equilibrium (in time) and how quickly it returns to that state. There is an implicit assumption of global stability; that is, there is only one equilibrium or steady state and hence resilience is the ability of the system to return to that prior state. This type of resilience has been called engineering resilience.

The second definition of resilience characterizes the systems with multiple equilibria, pathways, or configurations. **Ecological resilience** in this case examines the dynamics that can transform a system into another regime of behavior (i.e., to another stability domain). In this case resilience is measured by the magnitude of disturbance that can be absorbed before the system changes into another system or regime, as a result of shifting controls in key variables and processes.

Regime Shifts

An ecological regime shift occurs when characteristic or defining features of an ecosystem change. The change fundamentally alters the way the system looks (its structure) and functions (processes), thus creating a new regime. The ecological components of coupled systems undergo dramatic transformations, or regime shifts, as a result of human interventions.

(continued)

Case Study (*continued*)

All over the planet humans directly and indirectly modify ecosystems to secure a supply of goods and services. In many systems this can result in ecological regime shifts, for example,
- Forests change and habitat is lost as humans remove trees for fuel, timber, and pulp.
- Lakes, rivers, and estuaries become eutrophic from nonpoint pollutants.
- Overgrazed rangelands become woodlands.
- Excessive water use leads to soil salinization.
- Over-fished coral reefs become covered with algae.

Each example describes an ecological regime shift, whereby the structure and processes that characterize one regime are replaced by others. Some of the changes are brought about by direct manipulation, such as agriculture and forestry practices. Others, such as water pollution and algae-covered reefs, are the indirect result of other activities. In rangeland ecosystems a shift in state can occur because of a loss in biodiversity due to overgrazing. If grazing pressure is high, grazers remove many drought-tolerant plant species. When the system is subjected to a drought, few (if any) tolerant species survive, leaving the system vulnerable to colonization by shrubs and other woody species. Overgrazed rangeland systems often shift from a grass-dominated system to one dominated by woody plants. In other systems nutrient enrichment leads to regime shifts. Many freshwater systems, such as lakes or wetlands, receive inputs of nitrogen or phosphorus (or both) from surrounding areas of intensive agriculture or urbanization. The Everglades wetland is one example where small increases in soil phosphorus concentrations make the system vulnerable to a regime shift. Historically, much of the Everglades wetland was covered by monotypic stands of the sedge *Cladium jamaicensis*. In areas of elevated soil phosphorus concentrations, perturbations such as fire, frost, or drought can lead to a marsh system dominated by *Typha* sp. Other mechanisms that lead to a loss of resilience include the removal of keystone species (top predators, key grazers) or the modification of the physical environment. In many cases the alternative regimes are less productive, less desirable, and become the focus of ecosystem restoration.

Ecosystem Restoration and Resilience

Around the United States and the world, managers and governments are attempting restoration of resource systems. Much effort is placed to actively manipulate systems to reach a desired or restored condition. In many cases restoration can be considered as regime management, that is, restoration involves changing the system from a degraded or undesirable state to a restored state.

In attempting to restore degraded regimes to more desired ones, managers are faced with issues of reversibility and hysteresis. In some systems it may be physically impossible to restore the system to a desired state. In these cases regime shifts are unidirectional and cannot be recovered. The extinction of a keystone species is one example of irreversible regime shift. Costs of restoration may determine the degree to which regimes can be restored. The nature of alterations may also determine the degree to which a system can be restored. For example, restoration of anadromous fish populations may not be possible without the removal of human-constructed dams. Another consideration for managers is the issue of hysteresis, which suggests that the path to a restored regime may be very different from the one that led to the degraded state. In the overgrazed rangeland example, simply lowering grazing pressure will not restore the system once it has shifted to a woody state.

Ecological restoration in many cases involves active management that seeks to shift from an undesired regime to a desired one. Many environmental issues, such as cattail stands or the decline in the number of wading birds in the Everglades, can be described as undesired regimes. Restoration of desired regimes requires a careful assessment of ecosystem dynamics and exploration of feasible policy options. In many cases restoration policies are numerous and depend on how ecosystems are thought to respond to various actions. A growing set of experiences indicate that many large-scale restoration projects can only proceed through an adaptive management process, because of the inherent uncertainty of system responses. In an adaptive management framework, policies are acknowledged as guesses about system response, and actions are designed to help better understand how the system responds. Generally, ecosystem-scale experimentation is needed to understand how to shift regimes for restoration purposes.

5.2 Case Study

Phylogenetic Structure of Plant Communities Provides Guidelines for Restoration

Jeannine Cavender-Bares, University of Minnesota, Saint Paul, MN;
Nicole Cavender, Chief Programmatic Officer/Director of Restoration Ecology,
The Wilds, Cumberland, OH

One of the fundamental goals of restoration ecology is to understand the factors that influence assembly and establishment of colonizing species after disturbance. A long tradition of research and theory in community ecology provides a useful framework for the newer, more applied discipline of restoration ecology. One of the central differences between restoration and community ecology is that in restoration ecology the endpoint of the assembly process is defined by agreed upon restoration goals rather than by ecological conditions and dynamics alone (Temperton et al., 2004).

(continued)

Case Study (*continued*)

The linkage between ecological structure (e.g., species diversity, habitat complexity) and ecological function (e.g., biogeochemical processes, disturbance regime) has the potential to advance the practice of restoration. Theoretical and empirical work focused on this linkage is critical to advancing the science of restoration ecology and the practice of restoration (Palmer et al., 2006).

Current issues in community ecology relate directly to decisions about the restoration goals themselves. There is currently significant debate about the extent to which community assembly is influenced by deterministic processes including niche differentiation and matching of organismal traits to the environment or by stochasticity and neutral processes, in which species are essentially equivalent (Hubbell, 1979; Goldberg and Werner, 1983; Hubbell, 2001; Tilman, 2004; Hubbell, 2006; Leibold and McPeek, 2006). To use an analogy of the late Stephen J. Gould (1989), if nature's tape were replayed again and again, would the same communities result? If community structure is random and merely the result of historical contingency, perhaps attempting to restore them to specific endpoints is misguided. On the other hand, if community assembly follows specific rules leading to predictable outcomes, these may serve as guidelines for reassembling communities after disturbance. Therefore, understanding the extent to which communities are randomly or deterministically assembled influences how the goals of ecological restoration are set.

According to one deterministic perspective on community assembly, the assembly process can be understood in terms of a series of filters that includes both the physical environment and the interactions of species (Lambers et al., 1998). Together, these filters determine the composition and structure of local communities. Early on in the development of community ecology, Schimper (1898) described the physical environment as a filter that eliminates species that have arrived but lack the physiological traits to grow and survive under those conditions. Species interact with one another and can form a biotic filter that determines whether species can persist in the presence of other species. In the first half of the twentieth century, it was theoretically and empirically demonstrated that multiple species that compete for the same resources cannot coexist (Gause, 1934). The principle of competitive exclusion is considered a central component of the biotic filter. Lack (1944) was one of the first to point out that closely related species living together in nature might coexist by partitioning resources between them, and MacArthur and Levin (1967) demonstrated mathematically that competition could set a limit to the similarity of coexisting species. Hutchinson (1957, 1959), an animal ecologist, extended the idea of resource partitioning in his conceptualization of the N-dimensional niche, which was later applied to plant communities by Bazzaz and collaborators (Bazzaz, 1996). The axes of species occurring in multidimensional niche space were the various biotic and abiotic factors in the environment along which species could partition resources. Distributions of species, therefore, were thought to reflect their relationship with both the physical environment and other species, including predators, prey, pathogens, hosts, pollinators, dispersal agents, and other mutualists. This deterministic view of community assembly holds that niche differentiation allows for the coexistence of species, particularly those in the same trophic level.

Ricklefs (1987) highlighted the importance of historical processes in influencing local diversity and urged incorporation of historical, systematic, and biogeographical information into community ecology. He reminded ecologists that the equilibrium theory of island biogeography (MacArthur and Wilson, 1967) was based on a balance of regional processes (those that increase colonization) and local processes (those that cause local extinction). He argued that limiting similarity was in most cases a weaker force than regional processes in community assembly and that local diversity, rather than being determined solely by local environmental factors and limiting similarity, was consistently dependent on regional species diversity (Schluter and Ricklefs, 1993). The roles of dispersal, disturbance, and stochastic processes in community assembly, which played a central role in the theory of island biogeography (MacArthur and Wilson, 1967), were given new prominence by Hubbell (2001) in his unified neutral theory of biodiversity. Hubbell challenged the perspective that deterministic niche processes influence community assembly, asserting that ecological communities are open, continuously changing, non-equilibrial assemblages of species whose presence, absence, and relative abundance are governed by random speciation and extinction, dispersal limitation, and ecological drift. According to this view, species differences do not predict outcomes of competition, species do not specialize for specific habitats, and interactions between species and with the environment are not relevant to community assembly.

More recently, niche theory has been merged to varying degrees with neutral theory (Tilman, 2004; Leibold and McPeek, 2006), acknowledging the importance of both niche-based and neutral processes in community assembly. Where any given community falls along the spectrum between these two extremes depends perhaps on community age, the extent to which current species interactions have influenced the evolutionary process, and the heterogeneity of the environment. Proponents of both perspectives generally agree that large-scale processes such as speciation, migration, and dispersal determine how many and which species form the regional species pool from which local communities are established (Ricklefs, 1987; Lambers et al., 1998; Ricklefs, 2004). In the face of human dispersal of organisms around the globe, a changing regional species pool sooner or later will alter local community composition, even if other factors remain constant.

Florida Plant Communities

Empirical evidence from oak-dominated forest communities in north central Florida, at the confluence of northern temperate and subtropical ecotones, provides support for a largely deterministic, niche-based model of community assembly. A filtering process is apparent in these plant communities because species distributions are not random with respect to the environment or with respect to each other. Most strikingly, in this system, 17 species of oaks (genus *Quercus*) occur in close proximity, begging the question of how so many closely related species can co-occur. Closely related species have much of their evolutionary history in common, and, therefore, are presumed to share many phenotypic attributes and to have similar niche preferences. There are limits

(continued)

Case Study (*continued*)

to how similar coexisting species can be (MacArthur and Levins, 1967), making the sympatry of such a large number of congeners challenging to explain.

The oak species occur in three broadly defined communities in north central Florida: scrub, sandhill, and hammock. Hammocks themselves have been subdivided by other authors into hydric, mesic, and xeric hammocks based on the hydroperiod and hydrology of the soils. These communities differ significantly in soil moisture availability and fire regime (Cavender-Bares, et al., 2004b). The first indication that a filtering process is at work in the assembly of these communities is that the distribution of oak species across the major environmental gradients is not random. Rather, their distributions are predictable based on the functional traits they possess (Cavender-Bares and Holbrook, 2001; Cavender-Bares, et al., 2004b). A matching of trait to the environment is found both under field conditions where traits may vary plastically with environment, as well as in a common garden, where environmental variation is minimized. Furthermore, species show evolved trade-offs, indicating that they specialized for one set of environmental conditions at the expense of another. The matching of phenotypic traits to the environment has long been recognized in other systems and has been demonstrated in plants across the Earth's major biomes (Reich et al., 1999; Wright et al., 2004). Functional traits of species, therefore, can serve as guidelines for where they should be planted across environmental gradients.

Perhaps more interesting is the evidence for a biotic filter that emerges when the phylogenetic structure of these communities is examined. Across a fertility gradient, the number of woody species occurring in a 0.10-hectare (ha) plot increases with soil fertility until a saturation point is reached while the number of oak species is capped at three, regardless of habitat (Cavender-Bares et al., 2004b). This indicates that there is a limit to the number of oak species that can co-occur irrespective of the physical environment, and this number may be linked to the phylogenetic diversity in the group. There are three major clades that occur in this region: red, white, and live oaks. The white oaks and live oaks together form a clade that is sister to the red oaks. Comparing observed co-occurrence patterns of species to null models, in which species distributions were randomized, we found that the oak species were phylogenetically overdispersed (Cavender-Bares et al., 2004a). Statistically, this means that closely related oak species (those within the same clade) are unlikely to co-occur within the same 0.10-ha plot while oak species from different clades are more likely to co-occur than expected by chance. In other words, only one member from each of the major clades was likely to occur in any given plot. The pattern of phylogenetic overdispersion is a result of the evolutionary history of the group in which the oaks appear to have adaptively radiated into contrasting soil moisture and fire regimes (Cavender-Bares et al., 2004a). As a result, there is considerable functional diversity among species within the same clade, and functional traits important for habitat specialization show convergence among distantly related oaks. The overdispersion of close relatives may prevent competitive exclusion or reduce density-dependent mortality because of clade-specific pathogens (Webb et al., 2006; Gilbert and Webb, 2007). It may also reduce introgression of close relatives (Cavender-Bares et al., 2009).

The so-called "phylogenetic repulsion" (Webb et al., 2002) of close relatives has important implications for community restoration. Density dependent processes, such as disease and competition, may prevent the long-term coexistence of close relatives, and highest-diversity, oak-dominated communities may be realized when communities are drawn from distantly related oaks. Maximizing phylogenetic diversity, even within a single lineage such as the oaks, may, therefore, be an important restoration goal. With the availability of tools and data for phylogenetic analysis, simple metrics to determine phylogenetic structure and diversity of communities are readily accessible (e.g., Webb et al., 2005; Webb et al., 2008, reviewed in Vamosi et al., 2009; Cavender-Bares et al., 2009).

In a subsequent study, we asked whether only the oaks were structured in this manner or whether all plant species showed non-random distributions. We found that when all plant taxa were included in the analysis, species showed phylogenetic clustering (Cavender-Bares et al., 2006). In other words, closely related species that shared many functional traits in common were more likely to occur together than expected by chance. This pattern resulted from a matching of functional traits to the environment and conservatism of traits through evolutionary history. Within communities, species' traits were more similar than expected by chance. We did not find definitive evidence that other groups of close relatives (such as pines or hollies) were overdispersed. Several other research teams, however, have found evidence for phylogenetic overdispersion among speciose clades in which the member species occur in the same region (Slingsby and Verboom, 2006). These results suggest that both environmental filtering and species interactions are important in structuring communities but at different scales. At small spatial scales and among close relatives, evidence for species interactions emerge. At large spatial scales and among diverse taxa, evidence for matching of phenotypes to the environment is apparent. These results do not preclude the importance of stochastic processes and historical contingency in influencing community assembly. They do show, however, that many plant species in Florida have specialized for particular environments and that environmental filtering plays an important role in community assembly.

Phylogenetic Diversity and Ecosystem Function

The phylogenetic structure of communities shows promise for predicting ecosystem processes and properties that may be targets of ecological restoration. There is increasing evidence that phylogenetic diversity is linked to ecosystem function in plants (Maherali and Klironomos, 2007; Cadotte et al., 2008; Cadotte et al., 2009). In both plant and plant-mycorrhizal communities, studies have demonstrated that phylogenetic diversity can predict community productivity better than species richness or functional group diversity. These studies provide support for the hypothesis that phylogenetically diverse communities can maximize resource partitioning and hence use greater total resources. This is based on evidence that the more differentiated species are the greater their resource exploitation (Finke and Snyder, 2008). If phylogenetic relatedness predicts ecological similarity, phylogenetic diversity should

(continued)

Case Study (*continued*)

enhance complementarity and increase ecosystem productivity by maximizing total resource uptake. By the same logic, high phylogenetic diversity may be predicted to increase ecosystem stability by ensuring that sufficient ecological strategies are represented in an assemblage to ensure persistence of the ecosystem in the face of changing conditions. Similarly, phylogenetic diversity may be linked to nutrient cycling, resistance to invasion, soil carbon accumulation, and other ecosystem processes, goods, and services (Cavender-Bares et al., 2009). Such links, if they continue to be substantiated, support the argument that phylogenetic diversity has higher utility than species richness as a conservation criterion for management decisions (Faith, 1992; Gerhold et al., 2008).

Restoration of Vascular Plant Sommunities on Degraded Land

Restoration efforts are often directed at areas that have been severely degraded or affected by mining activity. Such is the case at The Wilds, a 3,700-hectare center for conservation research and education located on reclaimed strip-mined land in Muskingum County, Ohio. Mining for coal began on these lands in the 1940s and was completed by 1984. The process of coal extraction requires the complete removal of vegetation, topsoil, and rock, so that the coal seam can be exposed and extracted on the surface.

A majority of what is now The Wilds was coal mined by the Big Muskie, the world's largest coal mining dragline. Following coal extraction, much of the reclamation included replacement of rocky overburden and topsoil, grading, and shaping to the approximate original contour of the land followed by re-vegetation.

Although the land at The Wilds before European settlement was deciduous hardwood forest, re-vegetation efforts for reclamation included planting cool-season, non-native grasses and legumes. The area of The Wilds has now been recovering from this disturbance for more than two decades, but it remains extremely altered from its original state and has associated environmental problems. Loss of the native seed bank and microflora, severe soil compaction, low nutrients, and presence of invasive species all must be addressed while attempting restoration.

In ecosystems that have been dramatically altered and have crossed the threshold of irreversibility, it becomes important to consider the landscape context. The site, such as the one described, holds restrictions to what can actually be achieved, and it is extremely difficult to target historical references. It, therefore, becomes necessary to set ecosystem functional goals that can be achieved in a shorter time frame. Historical references can be used to set long-term goals, but more short-term goals such as increasing biodiversity, phylogenetic diversity, improving soil structure, and enhancing wildlife become more realistic targets in early restoration. Computation tools that allow analysis of phylogenetic diversity (methods reviewed in Vamosi et al., 2009 and Cavender-Bares et al., 2009) may provide a useful approach for measuring and monitoring indicators linked to functional goals.

As an example, in 2003, a large-scale restoration effort began at The Wilds with the intention of improving components of ecosystem function, habitat quality, and

Table 5.1 Changes in Species, Family, and Phylogenetic Diversity Before and After Restoration in Reclaimed Mine Land in Southern Ohio

Time Period	n Species	n Vascular Plant Families	Sum of Branchlengths (my)	Phylogenetic Diversity (Faith's PD Index)	Mean Phylogenetic Distance Between Species Pairs
Before	18	11	581	0.218	84.17
After	93	30	2,621	0.984	87.08

biodiversity. Two major goals included improving vascular plant diversity and increasing the butterfly populations. Invertebrates are essential to self-sustaining ecosystems and can be useful to measure restoration success (Webb, 1996; Majer, 1997; Wheater and Cullen, 1997; Halle and Fattorini, 2004). Butterflies, and pollinators in general, have shown major declines in recent years, and increasing their numbers and richness was an essential goal to the project. With their dependence on a wide variety of plants for various stages of their life cycle, increasing plant diversity became an important driver for this restoration project.

Before restoration activities began, the site comprised mostly cool-season, non-native grasses with few high nectar-generating plants (**Table 5.1**). From 2003 to 2007, a variety of herbaceous plant species, mostly native to Ohio, were introduced by seed using a no-till drilling technique and hand broadcasting. Many of the herbaceous species chosen for augmentation included those adapted to tallgrass prairie ecosystems. Because prairie species develop deep and fibrous root systems, they may be better adapted to poor and compacted soils and improved soil organic matter structure (Burke et al., 1995; McLauchlan et al., 2006; Matamala et al., 2008).

A long-term monitoring transect was established simultaneously with vegetation augmentation to monitor butterfly activity. An 870-m transect was established throughout 6 hectares of habitat following the methodology used by The Ohio Lepidoptera Society's Long-term Monitoring Program (adapted from Pollard and Yates, 1993). This fixed transect was mowed regularly and divided the site into sections according to habitat changes, so that observations could be made according to location and habitat. The transect was surveyed for butterflies weekly from early spring through late summer (2004–2007), and observations were made including the presence of vascular plants.

During the initial four years of restoration, perennial vascular plants increased from 11 species to 93, and vascular plant families increased from 11 to 30. A hypothesis of the phylogenetic relationships of species was generated using Phylocom (Webb et al., 2004). Faith's phylogenetic diversity index increased from 0.218 to 0.984 demonstrating a dramatic increase in vascular plants across the tree of life, indicating that not only

(continued)

Case Study (continued)

Table 5.2 Total Butterfly Species Richness and Average Individual Butterfly Counts Surveyed Over a 23-Week Period Between 2004 and 2007

Monitoring Year	2004	2005	2006	2007
Species Richness	26	23	37	33
Average # of Individuals	653	787	1,403	2,138

were more species represented in the system but also more evolutionary innovations must be represented. The phylogenetic distance between any two species in the system also increased (largely due to the colonization of a conifer, *Taxodium distichum*) although not significantly post-restoration. This indicates that species accumulated (both through management and from the regional pool) and occurred in a consistent and random manner from across the vascular plant phylogeny. Thus, while species were selected for management and are likely to have persisted in the system based on adaptive functional traits, the new species were not highly concentrated in any particular evolutionary clade. This highlights the diversity of functional strategies that persist in the same environment. In studies of the economic spectrum of plant traits (Wright et al., 2004), for example, a high proportion of the total variance in the functional attributes of plants is found at the same site.

At the beginning of the restoration project an average of 653 butterflies were butterflies were recorded, a 227% increase, and butterfly species richness had increased by 42% (**Table 5.2**).

Restoration activities are ongoing at the site with goals of restoring more than 80 hectares primarily for enhancing pollinator habitat. Although the restoration project is still developing, the integration of monitoring tools such as vascular plant and butterfly diversity and phylogenetic diversity provide guidance in meeting restoration goals in both the short and long term.

Conclusion

The non-random structure of oak-dominated communities, both in terms of the phylogenetic relatedness of species within communities and in the degree to which traits match the environment, indicates that deterministic processes are at play in assembly of these communities. Understanding the filters that are operating in a community and the traits that are critical for establishment can serve the goals of restoration ecology. The matching of functional traits to the environment, particularly hydraulic architecture of plants, indicates that specific microsites should be selected for planting individual species. In north central Florida, attention should be paid particularly to the fire regime and the hydrology, a notion well understood by the Florida Park Service. Ongoing restoration efforts rely heavily on prescribed burning to maintain fire-dependent communities. Floridian plant communities also provide an example of how community structure can be understood in an evolutionary context

(Cavender-Bares and Wilczek, 2003). The repeated pattern of phylogenetic overdispersion among the oaks indicates that local coexistence among members of different clades is more likely than among members of the same clade. As a result, phylogenetic overdispersion of close relatives has specific implications for restoration of oak-dominated communities. High diversity should be expected to persist in the long term only when distantly related oak species, rather than closely related species, are planted together.

In degraded lands where novel communities must be created *de novo* to improve land and habitat quality, the emphasis is less on restoration of specific community types. Here, as well, phylogenetic structure can provide guidance in meeting ecological goals. In restoration efforts at The Wilds, a dramatic enhancement of phylogenetic diversity in vascular plants, incorporating species with a diversity of rooting depths and resource use strategies, has improved soil structure and ecosystem productivity. It has also lead to a significant increase in the diversity and population sizes of other trophic levels including butterflies. Together, these case studies highlight the importance of matching functional attributes of plants to the environment and the restoration successes achieved in maximizing functional diversity in degraded land. Links between functional and phylogenetic diversity are complex (e.g., Prinzing et al., 2008; Cavender-Bares et al., 2009; Cadotte et al., 2009), but they reinforce their importance as conservation criterion. Inclusion of phylogenetic structure and diversity as an indicator for monitoring progress in ecosystem restoration is an emerging and promising approach.

Case Study References

Bazzaz, F. A. 1996. Plants in changing environments: linking physiological, population, and community ecology. Cambridge, UK: Cambridge University Press.

Burke, I. C., W. K. Lauenroth, and D. P. Coffin. 1995. Soil organic matter recovery in semiarid grasslands: Implications for the conservation reserve program. *Ecological Applications* 5:793–801.

Cadotte, M., J. Cavender-Bares, T. Oakley, and D. Tilman. 2009. Using phylogenetic, functional and trait diversity to understand patterns of plant community productivity. *PloS ONE* 4:e5695.

Cadotte, M. W., B. J. Cardinale, and T. H. Oakley. 2008. Evolutionary history and the effect of biodiversity on plant productivity. *Proceedings of the National Academy of Sciences* 105:17012–17017.

Cavender-Bares, J., D. D. Ackerly, D. A. Baum, and F. A. Bazzaz. 2004a. Phylogenetic overdispersion in Floridian oak communities. *American Naturalist* 163:823–843.

Cavender-Bares, J. and N. M. Holbrook. 2001. Hydraulic properties and freezing-induced cavitation in sympatric evergreen and deciduous oaks with, contrasting habitats. *Plant Cell and Environment* 24:1243–1256.

Cavender-Bares, J., A. Keen, and B. Miles. 2006. Phylogenetic structure of Floridian plant communities depends on taxonomic and spatial scale. *Ecology* 87:S109–S122.

Cavender-Bares, J., K. Kitajima, and F. A. Bazzaz. 2004b. Multiple trait associations in relation to habitat differentiation among 17 Florida oak species. *Ecological Monographs* 74:635–662.

Cavender-Bares, J., K. Kozak, P. Fine, and S. Kembel. 2009. The merging of community ecology and phylogenetic biology. *Ecology Letters* 12:693–715.

Cavender-Bares, J. and A. Wilczek. 2003. Integrating micro- and macroevolutionary processes in community ecology. *Ecology* 84:592–597.

Faith, D. P. 1992. Conservation evaluation and phylogenetic diversity. *Biological Conservation* 61:1–10.

(continued)

Case Study (continued)

Finke, D. L. and W. E. Snyder. 2008. Niche partitioning increases resource exploitation by diverse communities. *Science* 321:1488–1490.

Gause, G. F. 1934. *The Struggle for Existence*. New York: MacMillan.

Gerhold, P., M. Partel, J. Liira, K. Zobel, and A. Prinzing. 2008. Phylogenetic structure of local communities predicts the size of the regional species pool. *Journal of Ecology* 96:709–712.

Gilbert, G. S. and C. O. Webb. 2007. Phylogenetic signal in plant pathogen-host range. *Proceedings of the National Academy of Sciences* 104:4979–4983.

Goldberg, D. E. and P. A. Werner. 1983. Equivalence of competitors in plant communities: A null hypothesis and a field experimental test. *American Journal of Botany* 70:1098–1104.

Gould, S. J. 1989. *Wonderful Life: the Burgess Shale and the Nature of History*. New York: W. W. Norton.

Halle, S. and M. Fattorini. 2004. Advances in restoration ecology: insights from aquatic and terrestrial ecosystems. Pages 10–33 in V. M. Temperton, R. J. Hobbs, T. Nuttle, and S. Halle, eds. *Assembly Rules and Restoration Ecology*. Washington, DC: Island Press.

Hubbell, S. 2001. *The Unified Neutral Theory of Biodiversity and Biogeography*. Princeton, NJ: Princeton University Press.

Hubbell, S. P. 1979. Tree dispersion, abundance and diversity in a tropical dry forest. *Science* 203:1299–1309.

Hubbell, S. P. 2006. Neutral theory and the evolution of ecological equivalence. *Ecology* 87:1387–1398.

Hutchinson, G. 1957. Concluding remarks. *Cold Spring Harbor Symposium in Quantitative Biology* 22:415–427.

Hutchinson, G. E. 1959. Homage to Santa Rosalia, or why are there so many kinds of animals? *American Naturalist* 93:145–159.

Lack, D. 1944. Ecological aspects of species-formation in passerine birds. *Ibis* 86:260–286.

Lambers, H., F. S. Chapin, III, and T. L. Pons. 1998. *Plant Physiological Ecology*. Berlin: Springer.

Leibold, M. and M. A. McPeek. 2006. Coexistence of the niche and neutral perspectives in community ecology. *Ecology* 87:1399–1410.

MacArthur, R. and R. Levins. 1967. The limiting similarity, convergence and divergence of coexisting species. *American Naturalist* 101:377–385.

Maherali, H. and J. N. Klironomos. 2007. Influence of phylogeny on fungal community assembly and ecosystem functioning. *Science* 316:1746–1748.

Majer, J. D. 1997. Invertebrates assist the restoration process: an Australian perspective. Pages 212–237 in K. M. Urbanska, N. R. Webb, and P. J. Edwards, eds. *Restoration Ecology and Sustainable Development*. Cambridge, UK: Cambridge University Press.

Matamala, R., J. D. Jastrow, R. M. Miller, and C. T. Garten. 2008. Temporal changes in C and N stocks of restored prairie: implications for C sequestration strategies. *Ecological Applications* 18:1470–1488.

McLauchlan, K. K., S. E. Hobbie, and W. M. Post. 2006. Conversion from agriculture to grassland builds soil organic matter on decadel timescales. *Ecological Applications* 16:143–153.

Palmer, M. A., D. A. Falk, and J. B. Zedler. 2006. Ecological theory and restoration ecology. Pages 1–10 in M. A. Palmer, D. A. Falk, and J. B. Zedler, eds. *Foundations of Restoration Ecology*. Washington, DC: Island Press.

Prinzing, A., R. Reiffers, W. G. Braakhekke, S. M. Hennekens, O. Tackenberg, W. A. Ozinga, J. H. J. Schaminee, and J. M. van Groenendael. 2008. Less lineages—more trait variation: phylogenetically clustered plant communities are functionally more diverse. *Ecology Letters* 11:809–819.

Reich, P. B., D. S. Ellsworth, M. B. Walters, J. M. Vose, C. Gresham, J. C. Volin, and W. D. Bowman. 1999. Generality of leaf trait relationships: A test across six biomes. *Ecology* 80:1955–1969.

Ricklefs, R. E. 1987. Community diversity: relative roles of local and regional processes. *Science* 235:167–171.

Ricklefs, R. E. 2004. A comprehensive framework for global patterns in biodiversity. *Ecology Letters* 7:1–15.

Schimper, A. F. W. 1898. *Pflanzen Geographie und Physiologische Grundlagen*. Jena, Germany: Verlag von Gustav Fischer.

Schluter, D. and R. E. Ricklefs. 1993. Species diversity: An introduction to the problem. Pages 1–9 in R. E. Ricklefs and D. Schluter, eds. *Species Diversity in Ecological Communities.* Chicago: University of Chicago Press.

Slingsby, J. A. and G. A. Verboom. 2006. Phylogenetic relatedness limits co-occurrence at fine spatial scales: evidence from the schoenoid sedges (Cyperaceae: Schoeneae) of the Cape Floristic Region, South Africa. *American Naturalist* 168:14–27.

Temperton, V., R. J. Hobbs, T. Nuttle, and S. Halle. 2004. *Assembly Rules and Restoration Ecology: Bridging the Gap Between Theory and Practice.* Washington, DC: Island Press.

Tilman, D. 2004. Niche tradeoffs, neutrality, and community structure: A stochastic theory of resource competition, invasion, and community assembly. *Proceedings of the National Academy of Sciences of the United States of America* 101:10854–10861.

Vamosi, S. M., S. B. Heard, J. C. Vamosi, and C. O. Webb. 2009. Emerging patterns in the comparative analysis of phylogenetic community structure. *Molecular Ecology* 18:572–592.

Webb, C. O., D. D. Ackerly, M. A. McPeek, and M. J. Donoghue. 2002. Phylogenies and community ecology. *Annual Review of Ecology and Systematics* 33:475–505.

Webb C.O. and Donoghue M. J. 2005. Phylomatic: tree assembly for applied phylogenetics. *Molecular Ecology Notes* 5:181.

Webb, C. O., G. S. Gilbert, and M. J. Donoghue. 2006. Phylodiversity-dependent seedling mortality, size structure, and disease in a bornean rain forest. *Ecology* 87:S123–S131.

Webb C. O., D. D. Ackerly, and S. W. Kembel 2008. Phylocom: Software for the analysis of phylogenetic community structure and character evolution. Version 4.0

Webb, N. R. 1996. Restoration ecology: science, technology and society. *Trends in Ecology and Evolution* 11:396–397.

Wheater, C. P. and W. R. Cullen. 1997. Invertebrate communities of disused and restoration blasted limestone quarries in Derbyshire. *Restoration Ecology* 5:77–84.

Wright, I. J., P. B. Reich, M. Westoby, et al. 2004. The worldwide leaf economics spectrum. *Nature* 428:821–827.

Key Terms

Alternative stable states 106
Ecological constraints 105
Ecological drift 114
Ecological filters 113
Ecological resilience 117

Ecological resistance 105
Ecosystem threshold 113
Nexus species 112
Perturbation 117
Regime shift 105

Key Questions

1. Describe the main predictions of the equilibrium theory of island biogeography.
2. Define the terms *ecological resilience* and *resistance*.
3. How can ecological constraints curtail restoration efforts?
4. Define factors that regulate regime shifts.
5. What implications do assembly rules have on restoration ecology?
6. What are the main differences between assembly rules and the UNTB?

Further Reading

1. Cody, M. L. and Diamond, J. M. (eds.). 1975. *Ecology and evolution of communities.* Cambridge, MA: Belknap Press of Harvard University Press.
2. Hubbell, S. P. 2001. *The unified neutral theory of biodiversity and biogeography.* Princeton, NJ: Princeton University Press.
3. MacArthur, R. H. and Wilson, E. O. 1967. *The theory of island biogeography.* Princeton, NJ: Princeton University Press.
4. Roughgarden, J., May, R. M., and Levin, S. A. (eds.). 1989. *Perspectives in ecological theory.* Princeton, NJ: Princeton University Press.
5. Temperton, V. M., Hobbs, R. J., Nuttle, T., and Halle, S. (eds.). 2004. *Assembly rules and restoration ecology.* Washington, DC: Island Press.

6

LANDSCAPE

Chapter Outline

6.1 Connectivity
   Matrix Restoration
   Corridors
   Stepping Stones
6.2 Metapopulation
   Metapopulation Networks
   Metapopulation Dynamics
   Metapopulation Restoration Projects
6.3 Landscape Restoration
   Passive Restoration
Case Study 6.1: A Metapopulation Approach to Restoration of Pitcher's Thistle in Southern Lake Michigan Dunes

Landscape ecology focuses on processes that take place across heterogenous landscapes composed of interacting ecosystems. It is founded on the equilibrium theory of island biogeography outlined in Chapter 5. Processes in landscape ecology apply particularly well to fragmented habitats where the aim of restoration efforts is to increase connectivity by improving migration between isolated habitat patches (fragments).

Approaches in landscape ecology are beneficial in restoration ecology. For example, information on spatial heterogeneity of landscapes can improve outplanting strategies that aim at mimicking natural patches. Basically, this approach attempts to replicate natural spatial patterns of native species according to landscape features on restoration sites. This type of outplanting strategy is relatively easy to implement. Because spacing of outplanted material according to landscape features can increase complexity, this, in turn, enhances biodiversity.

In planning restoration projects it is important to consider features of the landscape, especially when selecting reference sites (see Chapter 14). In this context it is important that the restored site fits into the surrounding landscape. In practice, restoration of severely degraded sites such as mined lands has placed an importance on selecting native plants for restoration purposes that fit the surrounding landscape. Finally, assessment of restoration projects should ideally incorporate landscape perspectives in the evaluation (see Chapter 14).

A landscape is a heterogeneous area composed of various interconnected ecosystems. It can be, for example, a forest that is intersected by streams, wetlands, ponds, and meadows. The landscape harbors patches or critical habitats that provide niches for populations of different organisms. The dominating ecosystem of the landscape is called the **matrix**. The matrix can, for instance, be an intact forest or an arable field.

Within the matrix, habitat **patches** can be found. Patches often contain remnant (original) habitats that differ substantially from the matrix. Such patches may form critical habitats for species that are habitat specialists. Patches are often randomly distributed within the landscape and are distinguishable by various features such as density, isolation, size, and shape. The density of patches represents the porosity of the matrix. Sharp boundaries may exist between patches and the matrix. For example, a sharp edge is found where a forest abruptly meets an agricultural field. In this scenario discrete patches within the landscape can effectively represent islands (sink), and the recruitment site (source) can represent the mainland. Immigration and extinction of populations that live in such patches are pivotal processes in the dynamics within the matrix.

The movement of organisms between patches is affected by the **connectivity** within the matrix. This connectivity is facilitated by corridors and stepping stones. Corridors connecting many patches together can form a network. Restoration efforts that aim at increasing connectivity within the landscape focus on restoration of the matrix, corridors, and stepping stones.

The impacts of landscape fragmentation on the survival of many species are often modeled using the metapopulation approach. A **metapopulation** is an interconnected network of distinctive populations inhabiting patches that are embedded in a matrix. In fact, many species live in ordinarily isolated habitat patches such as ponds or wetlands and forest fragments embedded in an agricultural matrix. Species living in such patches are often highly specific in their habitat requirements. The metapopulation approach has been used successfully in the restoration of many species living in fragmented environments.

Landscape restoration involves ecological processes that operate over large areas, and it is particularly practical in the passive restoration of these large areas. In this chapter we discuss landscape restoration at length, providing relevant examples as needed.

6.1 Connectivity

Fragmentation along with large-scale habitat loss can result in reduced or curtailed connectivity between populations inhabiting remaining habitat patches (fragments). The movement of organisms between isolated patches is reduced or prevented by landscape barriers, also called **connectivity thresholds**. Such barriers include human-made structures (roads, power lines, dikes) or natural features (rivers, mountains, meadows) that cut through the landscape and curtail the movement of organisms. Fragmented populations that are isolated are potentially more prone to genetic drift and extinction than large populations or metapopulations. Also, the functioning of fragmented populations is influenced by the quality of the patches and the surrounding matrix. Isolated populations may exist for some time in a highly fragmented landscape. The extinction process initiated by fragmentation, however, could be difficult to halt in the short term. Isolated populations that are embedded in a degraded matrix usually have a low chance of long-term survival.

Fragmentation and habitat loss generally increase the extinction risk of local populations. Without any restoration effort, isolated populations inevitably become extinct. The extinction risk of different species is mainly determined by the size and quality of the intact or critical habitat that remains as well as the connectivity between patches.

The minimum size of a critical habitat required for long-term persistence of a species is called the extinction threshold. It is critical to determine the extinction threshold during the restoration of the fragmented landscapes of endangered species. This **extinction threshold** is especially important to calculate when patches of the critical habitat are being lost randomly.

Connectivity involves any features in the landscape that facilitate the movements of organisms between patches. Of course, improving connectivity between isolated patches decreases the probability of extinction. This "rescue effect" can come in many forms of restoration efforts (i.e., matrix restoration, corridors, stepping stones) and has had beneficial impacts on rare and endangered species.

Matrix Restoration

Matrix restoration focuses on large swatches of existing habitat known to support movements of organisms between populations. Matrix restoration increases the movement of organisms through the landscape and facilitates the colonization of native species. It can also constrain non-native invasive species. Such efforts reduce the extinction risk of populations that live in fragments and may enhance the connectivity of corridors and stepping stones that already exist within the matrix. Such improvement in connectivity is important for the existence of isolated populations, especially when considering that many ecosystems are already fragmented.

One prominent example of matrix restoration can be found in the large-scale restoration efforts on bushlands across southwestern Australia, one of the world's hotspots of biodiversity. "Gondwana Link," the name given to this massive project, involves efforts to increase connectivity through 6 to 7 million hectares of public land. The project involves working groups at different levels (local, regional, and national). The active restoration component involves replanting bushland over more than 1,000 km to reconnect the woodlands of the drier interior to the wet coastal forests.

Corridors

Corridors are linear features in the landscape that link together isolated patches and increase the movement of organisms between patches. Corridors harbor ecosystems with similar features to the original ecosystem. Corridors, therefore, usually differ substantially from the matrix. They have various shapes ranging from narrow linear hedgerows to large swatches of habitat that connect existing patches, and they are found in both terrestrial and aquatic habitats (e.g., stream corridors can effectively connect different water bodies).

Planning the restoration of corridors over large landscapes is facilitated by Gap analysis using geographical information system, or GIS, technology. This work involves overlaying a series of maps showing different landscape features such as land use, species abundance, and intensity of land degradation. Such information can, for example, indicate on spatial scale clusters of habitat patches that could potentially be connected by restoration efforts. It is practical to focus restoration efforts on such clusters.

Connectivity between isolated patches can be restored by constructing corridors between them. Active restoration, along with succession management, is effective in the construction of these corridors. Such efforts result in increased movements of organisms between isolated patches. Corridors should be designed to be large enough to provide substantial core habitats that encourage the movement of target species. Elongated corridors, conversely, may provide insufficient core habitat space, which may have a negative effect on the movement of target organisms. Some large animals are, in fact, very reluctant to move along elongated, narrow corridors.

An example of the effective use of corridors is found in a study on the rare mountain pygmy possum (*Burramys parvus*). (Two subpopulations of this species were studied.) The first subpopulation inhabited an intact landscape, but the second one was found in a fragmented landscape. The subpopulation found in the fragmented landscape exhibited a more skewed sex ratio and lower survival rates than did the intact subpopulation. After construction of corridors that allowed migration between patches, the genetic structure of the fragmented population was restored and the survival rate became equal to that of the intact subpopulation.

Although the advantages to their implementation and use are many, corridors carry risks that have the potential of outweighing their benefits. For instance, they can actually facilitate stochastic hazardous events in isolated patches. An example of such an event is the movement of predators toward target populations inhabiting isolated fragments. Also, corridors can increase the risk of fire, facilitate the invasion of non-native species, or bring diseases and pathogens to the target population living in isolated patches. In these cases the negative impact on rare and endangered species could be tremendous.

A real-life example of the negative effects of corridors is found in the studies on the wild turkey (*Meleagris gallopavo*) in the southeastern United States. Predation on wild turkey was facilitated along narrow riparian habitats. Such narrow corridors allowed predators to search for nests of wild turkey. To mitigate this situation, corridors were increased in width to make it more difficult for predators to search the critical habitat.

Stepping Stones

Small patches of critical habitat that are embedded in the matrix can form a network of **stepping stones**. Such interconnected patches can facilitate the movement of target organisms through the matrix. Although not directly connected, stepping stones can improve connectivity within the landscape and are an alternative to corridors.

Restoration of stepping stones involves reestablishing new patches in strategic locations to improve connectivity and enlarging the size of already existing patches. These isolated patches must be large enough to provide the minimum size of the critical habitat. They can effectively be increased in size by restoring buffer zones around the perimeter of already existing patches. Active restoration efforts on **buffer zones** could, for instance, provide protection by building enclosures (fencing), outplanting or seeding native plants, eradicating invasive species, or managing secondary succession. Such buffer zone restoration efforts can, in turn, reduce the potentially negative effects of sharp edges surrounding the patches.

Among the benefits of buffer zones is their ability to protect the core of the patch against storms and potentially invasive species. The restoration of buffer zones also reduces the direct impact on the existing core habitat found within the surrounding patches. Such core habitats typically harbor remnants of native species. Buffer zone expansion, another restoration approach, can actually increase intact core habitats while protecting them from hazardous stochastic events.

Priority in buffer zone restoration is usually given to large patches. It is generally cost effective to restore buffer zones around larger patches rather than smaller ones. In this context the shape of the patch is also important for the survival of the target organisms. For instance, a patch in a circular shape maximizes the core habitat, whereas a thin rectangular patch is affected by its edge and only contains a narrow core habitat.

The ongoing effort to save Oregon's endangered Fender's blue butterfly (*Icaricia icarioides fenderi*) is a study in the use of buffer zones and stepping stones for restoration purposes. The survival of this species depends on patches of a critical habitat harboring remnants of a prairie ecosystem that contain Kincaid's lupin (*Lupinus sulphureus* sp. *kincaidii*). Today, only about 0.5% of the original habitat remains. Lack of natural disturbances and invasion by non-native species has already further degraded many existing prairie fragments. It is estimated that only about 5,000 butterflies exist in about 12 fragments. However, several patches contain Kincaid's lupin but not the Fender's blue butterfly. These patches do, however, represent potential habitat for the Fender's blue butterfly. In such cases, it is beneficial to focus restoration efforts on large patches that meet the minimum size required for survival of the species and those that are located within the connective threshold (1 km) of the species. This study also revealed a cost effectiveness in restoring degraded patches to a useful habitat for the blue butterfly as well as a success rate favoring the restoration of small (>5 ha), effectively connected (within 1 km from each other) habitat patches over large, isolated ones. However, it is critical for the success of this restoration program to identify stepping-stone patches within the matrix. Most of the existing fragments of the critical habitat were less than 5 hectares and were, therefore, too small to maintain a viable population in the long term. In addition, these fragments were located at a distance greater than the connectivity threshold for the Fender's blue butterfly. In this case, habitat restoration that aims to increase connectivity by building stepping stones made of patches of the critical habitat is, therefore, an effective restoration strategy.

6.2 Metapopulation

Metapopulation Networks

Metapopulation networks usually consist of source and sink populations. A "source population" is typically large and persistent and contributes individuals that colonize vacant patches in the matrix. "Sink populations" are much smaller, more isolated populations that are prone to extinction. Effectively connected metapopulation networks may lower the risk of extinction and increase genetic variation within small and isolated populations. When movements (gene flow) between populations are curtailed, serious genetic problems may appear. In such cases even small movements between populations improve the genetic composition of the metapopulation. Conversely, without any movements between populations, isolated and small populations will likely become extinct within a short time.

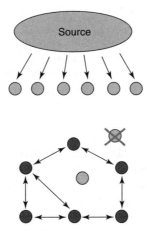

Figure 6.1 General features of metapopulation. Source and sink populations are represented in the drawing at the top. Movements of individuals are shown by arrows. The metapopulation network is represented (below) with extinct population (crossed out) and a vacant patch (gray).

Metapopulation Dynamics

Metapopulation dynamics include processes at the local population level such as extinction, immigration, and colonization (**Figure 6.1**). A metapopulation exists in a balance between the extinction of local populations and the establishment of new populations in vacant patches. Extinction of local populations is common in metapopulation dynamics. It is even possible that the whole metapopulation network becomes extinct due to emigration or stochastic disturbance. In this context disconnected metapopulations are more threatened than are effectively connected metapopulations. Effective network connections within metapopulations facilitate immigration and may reduce extinction risks of isolated populations.

Metapopulation dynamics are affected by the size and number of habitat patches within the landscape as well as the distance between them. The size of the patches often determines the population size of the species in question. Also, the integrity of the matrix influences connectivity within the metapopulation. The isolation of patches is yet another factor influencing immigration and colonization rates. Isolation is predominantly determined by the distance between patches.

Populations of endangered or rare species that are facing serious decline or threat of extinction and are not effectively connected in a metapopulation network have been termed the "living dead" because their long-term survival is questionable. The extent of such populations is not known, but with increased fragmentation of native habitat it is assumed that this problem is on the rise. Species vary, however, in their tolerance to fragmentation and isolation.

Metapopulation Restoration Projects

Metapopulation ecology has direct application to the planning and management of restoration projects. In restoration, special attention is placed on the integrity of isolated populations and the functioning of metapopulation networks (**Figure 6.2**). Restoration efforts that focus on the metapopulation network can reduce the probability of the permanent extinction of isolated populations. Local populations that might appear isolated may, in fact, persist for some time due to equilibrium between extinction and colonization. Restoration efforts that aim at improving metapopulation networks involve locating vacant habitat patches to establish new populations. Such populations should be located strategically to improve the functioning of the metapopulation network (**Figure 6.3**).

Unoccupied patches may play an important role in the long-term persistence of metapopulations. Restoration of the metapopulation network involves, for instance, strategic establishment of new populations that improve movements within the metapopulation. Because of connectivity thresholds, species may be absent from patches that represent critical habitats. On the other hand, species may be present where environmental conditions are not necessarily favorable. Such populations are often maintained by effective immigration from a large source population.

Restoration of metapopulation functioning can also involve integration of large patches (that act as source populations) into the network. Such restoration is not an alternative to large-scale landscape restoration. However, if rare or endangered species are contained in small habitat fragments, it is essential to reduce connectivity thresholds. Although it can be challenging to predict the optimal spacing between patches, restoration efforts must consider the importance of this spacing to reduce connectivity thresholds. At the same time patches should be far enough from each other to reduce the chance of stochastic events

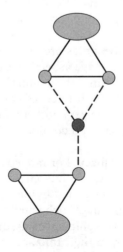

Figure 6.2 Hypothetical arrangement of a restored population in a metapopulation network.

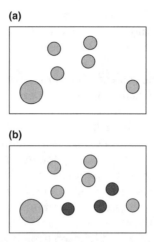

Figure 6.3 **(a)** Hypothetical restoration of an isolated population within a metapopulation network. Restoration of three strategically located patches (dark gray) improves the functioning of the metapopulation network **(b)**.

that can lead to local extinction. Preserving dissimilar patches (e.g., wet and xeric habitat) that can still function as a critical habitat for the population in question is another extinction-prevention strategy.

The metapopulation approach is beneficial in the restoration of endangered or rare species in highly fragmented and degraded landscapes. Long-term survival of isolated populations can occur only via effective restoration of metapopulation networks. Landscape restoration aiming at ecosystem and habitat diversity mitigates the effects of fragmentation. Restoration therefore should not aim only at small habitat fragments but also at the landscape level where fragments can be effectively connected together. It is, therefore, pragmatic in restoration to target large areas still containing fragments of the original habitats rather than focusing only on conserving relatively small isolated fragments containing intact habitat. Investing heavily on the restoration of small isolated fragments might be a mistake. Also, expecting long-term survival of populations in such small fragments is in most cases unrealistic.

Karner Blue Butterfly An example of the use of metapopulation dynamic in restoration is found in studies on the endangered Karner blue butterfly (*Lycaeides melissa samuelis*) that lives in fragmented prairies in central Wisconsin. The Karner blue butterflies can migrate over 3 km (which set the connectivity threshold between available habitat patches). For colonization purposes, Karner blue butterfly restoration sites required the presence of lupine species. Also, because it was critical for the success of the restoration program that Karner blue butterflies have access to heterogeneous habitats, restoration sites occupied both xeric and wetlands habitats. In the event of drought Karner blue butterflies would need access to wetlands, and in wet years they would need access to xeric habitats. Their ability to colonize both wet and xeric habitats was critical to ensure long-term survival of the metapopulation.

Bighorn Sheep Another example of metapopulation approaches in restoration can be found in the studies involving translocations of bighorn sheep (*Ovis canadensis*) in the western United States. The bighorn sheep are confined to mountainous areas from Texas to British Columbia. They are an important prey for top predators such as wolves, bears, and mountain lions. It is estimated that before European settlement the total population of the bighorn sheep was about 2 million. Today, the total population size of bighorn sheep is just 28,000, and only a few small isolated populations exist. This sharp population decline is mainly due to diseases (such as pneumonia) brought on by domesticated sheep, competition from livestock, and direct hunting.

The critical habitat of bighorn sheep is among mountain meadows adjacent to cliffs that provide an escape route from predation. Long-term survival of the bighorn sheep depends on the availability of this critical habitat. This habitat, however, has declined in the western United States due to encroaching forests whose expansion is mainly due to fire suppression.

A habitat assessment using GIS technology was completed for more than 3.8 million hectares to identify potential patches that could serve as a critical habitat for the bighorn sheep. Through habitat assessment, potential restoration sites were identified. Patches that were small or had short distance (<16 km) to domesticated sheep were not used in the restoration program. Generally, habitat patches with potential for restoration should be clustered so they can eventually form metapopulation networks. Also, restoration sites should not have natural barriers to animal movement and should be close to accessible water sources.

Active restoration has involved the removal of domesticated sheep, prescribed burning to increase the size of the critical habitat, and translocation of bighorn sheep into available patches. Populations of bighorn sheep were linked in a metapopulation network before human disturbances. Restoration of metapopulation networks was essential to increase the chance of survival for translocated animals. Also, restoring metapopulation networks increases the chances of recolonizing vacant habitat patches as they become available in the near future.

A strict criterion was made for the selection of bighorn sheep to be translocated. The selection was based on several criteria, including genetic information and disease history. Bighorn sheep that were nonrelated were targeted for translocation. Then, translocations were conducted into prioritized habitat patches. The restoration effort next focused on establishing large populations (125 animals) connected to a metapopulation structure. Successful colonization of translocated bighorn sheep into mountain meadows depended on many landscape features, including size of habitat patches and distance between patches. In this respect the initial population size determined by the size of habitat patches was critical for the success of the translocation. It was discovered that populations that were translocated into large patches were more likely to colonize nearby smaller patches.

To track the movement of bighorn sheep that were translocated, a collar that included a global positioning system (GPS) unit was used. The GPS unit recorded the location of the animals every 5 hours for a year. The collar was programmed to drop off the animal on a certain day to then be collected. The stored information in the GPS unit was then used to map the movements of the animal. The use of GPS units to track movements of the bighorn sheep has helped in identifying connectivity thresholds to their movements.

Connectivity thresholds such as large rivers, dikes, high fences, continuous conifer forest, and flat terrain between existing populations and potential patches reduced colonization rates. Connectivity thresholds were more serious obstacles in reducing the colonization of unoccupied patches than the absolute distance between the patches.

The restoration of bighorn sheep populations through translocation was successful, and in 5 years about 22% of the potential habitats were inhabited. Also, in 7 years the translocated animals increased 25% in number. Previous efforts in translocating bighorn sheep that did not consider the importance of metapopulation structure were not as successful.

Florida Scrub Jay One last example of a restoration effort using the metapopulation approach can be seen in studies on the endemic and endangered Florida scrub jay (*Aphelocoma coerulescens*). It is estimated that the total population of Florida scrub jays has declined by about 90%. In 1993 the total number of Florida scrub jays was just about 4,000 pairs. The distribution of this species has also declined rapidly in Florida. Population size and adequate distribution are both important for the long-term survival of the species. Population viability analysis indicated that a population with fewer than 10 breeding pairs has about a 50% chance of extinction within one century. On the other hand, a population with 100 breeding pairs has only about a 3% chance of extinction within one century.

The Florida scrub jay requires a critical habitat confined to early-successional xeric scrub oak (*Quercus* sp.) communities that are found on infertile sandy soils throughout Florida. This critical habitat is characterized by small oak trees and low shrubs that have high canopy cover. The understory is typically non-vegetated and sandy. Natural fire regime in the past maintained this ecological state. However, most of the critical habitat of the Florida scrub jay is now highly fragmented within a matrix of agricultural fields and residential areas. Fire suppression, land conversion, and invasive plants have even further degraded the scrub communities in Florida. Fragmentation and isolation of habitat patches has prevented the spread of natural fire. Consequently, sand pine has colonized those patches and will eventually dominate this habitat. However, a dense forest of sand pine is an unsuitable habitat for the Florida scrub jay.

During the restoration planning, about 191 subpopulations of Florida scrub jay were outlined. It was alarming that more than 80% of these subpopulations were smaller than 10 breeding pairs. Only 3% of the subpopulations contained more than 100 breeding pairs. It was, therefore, predicted that most of these subpopulations would go extinct in the near future if no action were taken. The maximum dispersal distance was estimated to be about 3.5 km between subpopulations. The Florida scrub jay avoids crossing over large matrixes between habitat patches, especially if the distance exceeds 3.5 km. Also, because connectivity between subpopulations is reduced, the need for translocation and reintroduction increases (see discussion in Chapter 12).

The restoration effort has, to date, focused on the largest subpopulations of the Florida scrub jay that can effectively act as source populations. Priority has been put on maintaining habitat patches that can support at least 10 breeding pairs of scrub jay. Habitat patches that have less than 10 breeding pairs or contain seriously degraded habitat are not targeted for restoration unless such patches play an important role as stepping stones or harbor unique genotypes. Metapopulations were separated by a distance of 12 km between habitat patches. Such a distance represents a serious connectivity threshold. Isolated populations were connected by building stepping stones to increase connectivity within metapopulations.

Restoration of the critical habitat is, however, a necessary precursor for increasing population size of the Florida scrub jay. Restored habitat patches should preferably be within 3.5 km distance to source populations. Active restoration of the critical habitat involves logging large trees (sand pine and oaks) and implementing prescribed burning. Apart from logging and prescribed burning, mowing overgrown vegetation has been implemented to reduce the size of small oak scrubs. In addition, mowing facilitates low intensity fires. Typically, mechanical manipulation is conducted only once, and after that prescribed burning is used to maintain the critical habitat. Also, prescribed burning can eliminate and control non-native species. High frequency of burns also prevents the accumulation of dense thickets.

Prescribed burning is generally conducted in "management blocks." This strategy provides a protected area for the Florida scrub jays while another adjacent area receives prescribed burning. Since 1999 prescribed burning of overgrown habitats has restored more than 8,000 hectares of the critical habitat in central Florida. Restoration of the critical habitat for the Florida scrub jay has been used as a protocol for degraded sites elsewhere in Florida. An example of restoration of metapopulation networks of Pitcher's thistle in Southern Lake Michigan dunes is provided in **Case Study 6.1** on page 146.

Alternatives to metapopulation models are the spatially explicit, individual-based landscape models. These models incorporate species behavior in complex landscapes and are particularly useful for suggesting conservation strategies in large landscapes.

6.3 Landscape Restoration

Landscape restoration considers ecological processes that operate on large spatial scales. Such ecological processes include natural disturbances (i.e., fire, invasive species, and even global climate change). Also included in landscape restoration are human perturbations such as habitat fragmentation and land conversions. At the population level, processes of interest include immigration (gene flow), emigration, and extinction. In particular, dispersal of organisms and eventual colonization are of importance.

Passive Restoration

An example of a large-scale landscape restoration can be found in the work of Dan Janzen on Area de Conservación Guanacaste (ACG) in Costa Rica (**Figure 6.4**). The size of the restoration site is about 110,000 hectares. The ACG is estimated to harbor an exceptionally high biodiversity of more than 235,000 species.

Anthropogenic destruction of native forests in Costa Rica has been going on for four centuries. Intentional fires have been used through the dry season to convert primary forests into rangelands and maintain the grassland state. Widespread fires, however, are not considered to be a natural component of the native ecosystem. It is possible that small ground fires are started by lightning on these rangelands during the rainy season, but such fires do not usually extend into the shady and moist understory of the adjacent forest.

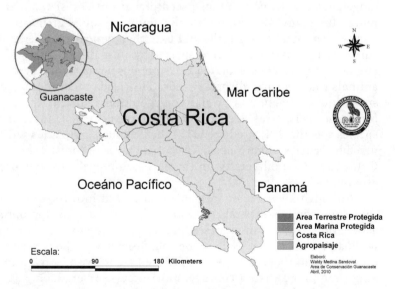

Figure 6.4 Map of Area de Conservación Guanacaste (ACG) in Costa Rica. (Courtesy of Waldy Medina, Area de Conservación Guanacaste.)

The landscape-level restoration of the ACG forest started in 1985. Passive restoration was used in this ambitious project. It was challenging to locate the reference site (containing old-growth Meso-American dry-tropical forest ecosystem) in the degraded landscape. The final ecosystem is, therefore, considered to be a moving target (see Chapter 14). The initial aim of the restoration was to halt degrading factors such as fire, hunting, logging, and farming. Also, it was realized that it would take a long time (five centuries) for restoration sites to reach the high biodiversity and structural complexity of old-growth, dry-tropical forest. The final ecosystem, therefore, will probably be an approximation to an old-growth, dry-tropical forest but not an exact replica.

The main focus of the landscape restoration of the ACG was on forest fragments harboring old-growth forests. The degraded rangelands typically harbor fragments of the original forest. Such fragments are found in the degraded landscape and represent different stages of secondary succession. Multiple succession trajectories are therefore found within the ACG. Forest fragments containing old-growth forests were especially important in providing seed that can colonize the surrounding degraded rangelands. Because of this, buffer zones were established around existing forest fragments. In fact, the restoration potential of the ACG depends on the biodiversity of the forest fragments. Effective seed dispersal into degraded rangelands is critical in the recolonization process. A great variety of seeds is dispersed effectively into the degraded rangeland by wind during the dry season. Seed rain from existing forest fragments is critical because soil seed bank is absent in the degraded rangelands due to very intense seed predation. Forest fragments also act as refuge for many vertebrates in the rainy season. Furthermore, these fragments form perches for birds and bats that typically disperse seeds into the surrounding grasslands. Seed dispersed by animals typically takes place under and near tree perches. Isolated trees on grasslands act as perches and attract animals that, in turn, disperse seeds. This leads to an increased perimeter of such isolated fragments by native vegetation that can act as a nucleus for further seed dispersal. Seed dispersal into the degraded rangelands is, however, species specific. It is predicted that the degraded grasslands will become colonized by species of various successional stages of forest over several decades. Consequently, biodiversity on the degraded rangelands is expected to gradually increase.

Although a variety of species are dispersed into the degraded rangelands, the establishment depends on various factors such as appropriate symbiotic association with soil microbes, appropriate microenvironment, and appropriate pollinators for further seed production. Species with large seeds take longer to establish than those with wind-dispersed seeds. Also, some forest trees take a long time (more than a century) until they start producing their own seeds. Exceptionally large seeds that are typically dispersed by forest-dwelling agoutis or tree-dwelling primates are largely missing from the seed rain. In this case establishing natural patterns of the larger-seeded species can be implemented by active restoration. Strategic outplanting of large-seeded trees can certainly accelerate this process.

Passive restoration of the ACG forest is a cost-effective option compared with a large-scale, active restoration involving a more intensive approach such as the outplanting of nursery-raised trees. In this context limited funds are better spent in purchasing degraded rangelands that contain mosaic of forest fragments rather than initiating intensive active restoration over much smaller area. Purchasing degraded rangeland for restoration purposes is an acceptable approach. This approach could probably serve as a model for nearly all remaining dry-tropical forest in Costa Rica. Such an approach is especially more feasible if high numbers of forest fragments are already embedded into the landscape. Consequently, large areas of marginal farmlands were purchased and connected to the already existing ACG forest. It appears to be difficult to purchase land containing large swaths of intact, old-growth, dry-tropical forest compared with degraded rangelands. Continuous purchase of degraded lands will increase the total area under restoration.

An important aspect in planning the large-scale restoration of ACG forest using landscape restoration was to gain political support at local, national, and international levels. Without political support it would have been difficult to implement this project. Moreover, this project can be used as a model for much larger restoration effort of tropical rain forests, as discussed in Chapter 11.

Summary

The landscape is made up of the dominating ecosystem, the matrix, embedded with habitat patches harboring populations of different species. Fragmentation and habitat loss of ordinary landscape has forced many populations to exist in isolated fragments. The movement of organisms between patches is determined by landscape connectivity. Such movements are facilitated by linear corridors and stepping stones that connect discrete patches. Restoring the connectivity between isolated patches can rescue isolated populations from extinction. Restoration of buffer zones around patches is also important in enhancing their functioning. Clusters of patches can form stepping stones that facilitate movements across the landscape. A metapopulation network is made up of interconnected populations. Metapopulations exist in equilibrium between population extinction and recolonization of vacant patches. Metapopulation dynamics is affected by various landscape features, some of which act as connectivity thresholds. Restoration efforts that focus on metapopulation functioning put emphasis on increasing connectivity within metapopulation network. Large-scale landscape restoration has been successfully implemented in drastically degraded dry-tropical forest. This work was accomplished by halting anthropogenic degrading factors and implementing a large-scale passive restoration approach.

6.1 Case Study

A Metapopulation Approach to Restoration of Pitcher's Thistle in Southern Lake Michigan Dunes

Noel B. Pavlovic, USGS Great Lakes Science Center, Lake Michigan Ecological Research Station, Porter, IN and A. Kathryn McEachern, USGS Channel Islands Field Station, Ventura, CA

The southern Lake Michigan shoreline has changed greatly since the arrival of Europeans. Today, within the context of isolated landscape remnants, conservation of species can be challenging because of the disruption of processes that are important for their survival. Disrupted processes can affect the species at the population and/or landscape level, necessitating a metapopulation perspective in planning their conservation. One such species is Pitcher's thistle (*Cirsium pitcheri*), an endemic to the narrow belt of sand dunes in the western Great Lakes that is threatened in the United States and Canada (**Figure 6.5**). It reaches the southern limit of its range at the Indiana Dunes.

Pitcher's thistle grows in the nonforested sand dunes. Suitable habitats encompass a successional gradient from the beach to foredunes stabilized by beach grass (*Ammophila breviligulata*) to secondary dunes and blowouts dominated by little bluestem (*Schizachyrium scoparius*). Pitcher's thistle also persists in blowouts that

Figure 6.5 Seedling, juvenile, and flowering adult of Pitcher's thistle (*Cirsium pitcheri*) (from upper left to right). (Courtesy of Noel B. Pavlovic/USGS.)

extend inland through otherwise forested dunes. Although these habitats represent a gradient of early- to mid-successional stages, all types can occur in close proximity to each other because of the spatially variable dynamics of the dune system. Plants and populations thrive with the presence of bare sand and active sand movement, but smaller plants are less able to withstand deeper sand burial than larger plants. Pitcher's thistle populations decline as litter and vegetation cover increase above 50%.

Unlike many perennial plants, Pitcher's thistle only blooms and sets seed once after a juvenile life span from 5 to 8 years. To survive from seedling to flowering adult, Pitcher's thistle requires a relatively stable habitat for juvenile growth. Too frequent disturbance or rapid successional change may cause local populations to decline. The life span and fates of individuals are influenced by the fluctuations in the abiotic environment, habitat changes, frequency of herbivory and trampling, and individual genetic variation.

Pitcher's thistle survival is intimately tied to processes that create sand dune habitats and arrest dune succession. Sand is supplied by currents that move sand down both sides of Lake Michigan. Winter storms and periodic 32- and 160-year lake-level fluctuations influence sand movement and dune formation. Foredunes form when lake levels fall, causing more sand to blow landward, thus creating suitable habitat for Pitcher's thistle. When lake levels rise and erosion destroys dunes, vegetation, and Pitcher's thistle plants, the sand bluff created is the source for lesser quantities of sand that blow landward, thus rejuvenating populations inland. Paleohistorical data suggest that after the 160-year high lake levels, when lake levels are beginning to decline, great quantities of sand often blow inland to create the blowouts, thus creating new habitat. Through centuries of succession, blowouts may eventually be invaded by trees and become forested, unless catastrophic storms or volumes of sand again blow inland, burying trees. This rejuvenation of populations is illustrated by the effects of varying lake levels on lakeside Pitcher's thistle populations further north. High water levels in 1986–1987 eroded foredunes, causing Pitcher's thistle to be restricted at low density to a narrow secondary dune. In the late 1990s low lake levels created foredunes and a broad beach that allowed the expansion of Pitcher's thistle populations lakeward, often attaining higher densities than on the narrow secondary dune.

These processes create a dynamic habitat mosaic where Pitcher's thistle populations wax and wane and long-term survival depends on the persistence of the metapopulation rather than populations. Key processes for metapopulation dynamics are seed dispersal within and among populations, environmental fluctuations, and disturbances that occur independently among populations. Spatial landscape variation can create refuges for species during extreme environmental events. For instance, the drought of 2005 caused high seedling mortality in typical habitats, but a population that was on a more northerly slope had seedlings that survived into the summer, a pattern opposite what is normally observed. The metapopulation concept is relevant to frame conservation strategies for Pitcher's thistle because this model links population dynamics with landscape processes

(continued)

Case Study (*continued*)

across scales and levels of biological organization: individuals, populations, and metapopulations.

Disruption of interpopulation dispersal and destruction of populations in the southern Lake Michigan dunes occurs because of the construction of jetties, harbors, industrial complexes, and shoreline towns that jut into Lake Michigan. These changes have altered lakeshore sand supply and near-shore movement of sand, causing shoreline erosion to the west of obstructions and sand buildup on the east side of obstructions. Habitat destruction and disruption of shoreline processes have isolated populations of Pitcher's thistle and have possibly reduced the probability of future blowout creation. Although the species did occur on the beach in the early 1900s, over a century of human recreation and shoreline erosion, interacting with lake-level fluctuations, has eliminated Pitcher's thistle along the foredunes that provide a corridor for population and genetic dispersal.

Populations can decline because of successional change, loss of genetic variability, herbivory, catastrophes, trampling, and environmental extremes such as drought. If a population appears to be declining, then mitigation actions occur at the population level. Declines from succession may have to be arrested by removal of woody species, litter, and herbaceous dominants. If reproduction is low, introducing seedlings may bolster the population and increase genetic diversity. Some populations may be experiencing low seed production because of increasing herbivory, necessitating the reduction of herbivores or actions to increase plant vigor. If patches of populations are being impacted by recreation, this activity may have to be eliminated or reduced, as has been done successfully with one population. Construction of a boardwalk trail in one area of the Indiana Dunes National Lakeshore has allowed populations to persist and attain high density compared with an adjacent area where recreational trampling remains unchecked.

At the metapopulation level, colonization between distantly isolated habitats may not be occurring; thus, management seeks to promote dispersal processes that allow genetic interchange among populations and allow the creation of new populations as suitable habitat forms. Also, the more populations present, the greater the chance for metapopulation persistence. Because normal seed dispersal is mostly within 5 m, human-assisted dispersal may be required to allow invasion to new suitable habitats, although rare but important long-distance dispersal may occur through the action of birds. Isolated populations may decline as they lose genetic variability and experience lower seed production. Human-assisted dispersal by introducing seed or seedlings from nearby populations may increase local genetic variability and increase population size. In some instances introducing new populations between existing populations may restore colonization to suitable habitats outside of dispersal range. At one site, experimental reintroduction of seed from multiple populations in 1994 has resulted in a viable population that bridges patches that were formerly isolated and declining. Some options to increase the connectivity between populations may not be feasible at present, such as removal of structures, industrial complexes, and harbors to restore shoreline and sand flow, but they should not be rejected entirely if situations change.

> Restoration of Pitcher's thistle in the southern Lake Michigan region has involved protection of populations and creation of new ones. In the past 13 years Pitcher's thistle has not been actively introduced into foredune habitats, but it has expanded into two sites from blowouts without human intervention. Another population is declining because jack pine (*Pinus banksiana*) growth is blocking winds and oaks are invading. The decline will be monitored, while Pitcher's thistle will be reintroduced nearer to the lake. Rare species, such as Pitcher's thistle, would thrive and prosper in a dynamic intact landscape, but with disrupted shoreline and dune processes in a fragmented landscape, Pitcher's thistle may likely become extirpated without metapopulation management. The long life span of Pitcher's thistle requires a long-term commitment to document and understand the forces that influence its metapopulation dynamics and viability.

Key Terms

Buffer zone 135
Connectivity 132
Connectivity thresholds 133
Corridors 134
Extinction threshold 133
Matrix 132
Metapopulation 132
Patches 132
Stepping stones 135

Key Questions

1. Define the term matrix.
2. How is Gap analysis used in restoration?
3. Give examples of connectivity thresholds.
4. Describe the main features of a metapopulation.
5. How can connectivity be increased between populations?
6. Describe landscape restoration.

Further Reading

1. Burel, F. and Baudry, F. 2004. *Landscape ecology*. Enfield, NH: Science Publishers.
2. Hanski, I. 1999. *Metapopulation ecology*. Oxford, UK: Oxford University Press.
3. Hanski, I. and Gaggiotti, O. E. 2004. *Ecology, genetics and evolution of metapopulations*. Amsterdam: Elsevier Academic Press.
4. Hansson, L., Fahrig, L., and Merriam G. (eds.). 1995. *Mosaic landscapes and ecological processes*. London: Chapman and Hall.
5. McCullough, D. R. (ed.). 1996. *Metapopulations and wildlife conservation*. Washington, DC: Island Press.

7

INVASIVE SPECIES

Chapter Outline

7.1 Process of Invasion
7.2 Effects of Invasion on Ecosystems
7.3 Methods of Control
    Prevention
    Eradication
    Containment
    Chemical and Biological Control
7.4 Restoration to Constrain Invasion
    Niche Preemption
    Fire Management
    Increasing Biotic and Abiotic Resistance
Case Study 7.1: Kudzu—the Notorious Invader of Southern United States
Case Study 7.2: Pale Swallow-Wort: An Emerging Threat to Natural and Seminatural Habitats in the Lower Great Lakes Basin of North America

An invasive **non-native species** is of concern because it can devastate an ecosystem's structure and function and can cause environmental and economic harm. During invasion **native species** are often outcompeted from their own habitat and even decimated. Invasion of non-native species is considered to be the second greatest cause of serious species decline and extinction only after habitat destruction. Non-native species are also known as aliens, exotics, introduced species, nonindigenous species, immigrants, and exogenous.

A non-native species does not necessarily have to be invasive, but a small proportion of non-native species (about 10%) usually become invasive. A species is called **invasive** if it establishes a rapidly growing population and proliferates in an ecosystem where it was previously not present.

Non-native species are continuously being introduced for different reasons to an ecosystem where they have not lived before. Some are introduced deliberately for horticultural, agricultural, and recreational purposes. In fact, agriculture is based on the practice of introducing non-native crops and livestock into new habitats. This includes non-native species that are introduced for land reclamation and soil conservation (see **Case Studies 7.1 and 7.2** on pages 165 and 169, respectively). Other non-native species (such as rats, mice, dogs, cats, and goats) are brought accidentally or intentionally during expanded human settlements. These introduced species are all notorious in decimating local species. Increased international travel and trade has also resulted in accidental introduction of many species. Today, the increase in international trade is probably the main cause for the unprecedented rate of introduction of non-native species worldwide. One effective transportation of aquatic species worldwide is the release of ballast water from cargo ships. The use of seawater as ballast in cargo ships began in the 1880s. Today, the scale of the transported ballast water is enormous. Big cargo ships can carry more than 60 million liters of ballast water, and more than 45,000 cargo ships move up to 12 billion tons of ballast water around the world each year.

The economical impact of invasive species is enormous and is on the rise. Direct annual economic loss from invasive species is about $1.4 trillion worldwide and $200 billion in the United States. Florida alone spends about $50 million each year just to control invasive plants. Some species are particularly troublesome; for instance, the zebra mussel (*Dreissena polymorpha*) is responsible for fouling vessel hulls and clogging industrial pipelines with an annual control cost of $140 million for the United States and Canada. The invasion of the star thistle (*Centaurea solstitialis*) in the Sacramento Valley, California, has caused annual economical loss of $56 million as a result of a decline in the water supply. It is estimated that the invasion of the salt cedar (*Tamarix* sp.) in the United States has an annual cost of up to $180 million in lost water supplies. In the United States at least 4,500 non-native species are now established, of which 675 species cause severe economic harm. The adverse economic effects of invasive species include direct damage to native commercial species, adverse effects on ecosystem services, and cost of control measures. The total economic impact of invasive species is probably underestimated because this does not include the cost of total eradication or restoration of affected ecosystems. Also, potential cost of species extinction, which could include loss of economic opportunities from native species, is not included. Invasive species have also disrupted critical ecological functions in many ecosystems, and in turn this affects potential economic output provided by these ecosystems. Invasion can also interfere with aesthetic values by creating excessive shade or blocked vision by tall growing non-native vegetation. This in turn can reduce recreational or other values of the ecosystem. Another example of the negative aesthetic impact of invasive species is loud noise from non-native coqui frogs (*Eleutherodactylus coqui*) in Hawaii.

Invasive species pose a challenge for ecosystem restoration. First, control methods are needed to halt and eventually eradicate invasive species, and this must be followed by restoration of native ecosystems. Second, restoration of degraded ecosystems can be implemented to constrain invasion of non-native species. Examples of these strategies (control, eradication, restoration) are provided in this chapter.

7.1 Process of Invasion

Invasion is a complex process that includes several steps, starting with translocation and introduction of a non-native species to a site outside its native range (**Figure 7.1**). The introduction to a new site is followed by a slow initial spread (lag growth) of the non-native species. The lag growth could involve acclimatization of the non-native species. Absence of appropriate symbiotic organisms—such as pollinators, nitrogen-fixing bacteria, or mycorrhizal fungi—can often be a limiting factor for initial population growth. Invasive species are often generalist

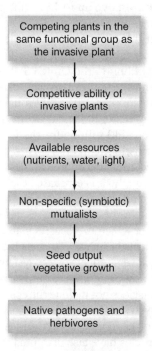

Figure 7.1 Causal relationship between factors and processes may affect invasion of non-native species into plant communities. (Reproduced from *Invasive Alien Species: A New Synthesis (SCOPE 63)*, edited by Harold A. Mooney, Richard N. Mack, Jeffrey A. McNeely, Laurie E. Neville, Peter Johan Schei, and Jeffery K. Waage, copyright SCOPE, Island Press, 2005.)

in their symbiotic relationship and can therefore partner with local symbiotic organisms. Colonization of appropriate pollinators and increased vectors of seed dispersal by local animals of course facilitate further population expansion in the new range. The formation of a population of non-native species that maintains itself (i.e., is not dependent on immigration) is termed naturalization. The establishment of **naturalized** populations is followed by the formation of dense monopopulation and rapid population expansion through invasion. This typically involves competition between native and non-native species. Invasive plant species can outcompete native vegetation, which could result in local extinction of native plants. Eventually, the invasive population may stabilize when it reaches its physiological–geographical boundaries.

Why do invasive species proliferate in their adventive range? The main mechanism behind invasion involves release from natural enemies from their native habitat including predators, herbivores, and pathogens. This has also been called the "release from natural enemies hypothesis." The success of invasive non-native species might depend on the extent of release from pathogens. In support of this hypothesis, plant species brought from Europe to North America have much less fungal and viral infections in their naturalized range than in their native range. Although invasive species have escaped from pathogens in their home range, they typically accumulate fewer pathogens in their adventive range. **Invasiveness** of non-native plants is therefore a function of both escape from enemies in their native range and low pathogen load in their adventive range. Invasiveness of ecosystems is usually measured as the proportion of non-native species to native species in an ecosystem.

Traits of non-native species that promote invasiveness include rapid growth, early maturation, self-pollination, large seed production, effective good (seed) dispersal ability, high seed germination rates, and ability to survive under a wide range of environmental conditions. Strong competitive ability is probably the most important feature of invasive species. The extent of invasion is a function of both the aggressiveness of the non-native species and the vulnerability or invasiveness of the ecosystem in question. Invasiveness of ecosystem depends on degradation factors of such as habitat fragmentation, habitat conversion, eutrophication, and generally increased frequency of natural and/or anthropogenic disturbances.

Ecosystems vary in their susceptibility to invasion. A few ecosystems show abiotic resistance to invasion due to unsuitable environmental conditions. These include exceptionally dry or cold environments found in deserts, tropical dry forests, and arctic ecosystems. On the other hand, islands, lakes, and rivers are particularly susceptible for invasion. Rivers and lakes provide suitable conditions for many species and islands may have high proportion of endemic species.

Ecosystems can also show biotic resistance for invasion. Biotic resistance is expressed as high competitiveness of native species and high resilience of the native community. Biodiversity is another determining factor of biotic resistance that can influence the invasiveness of non-native species. Communities harboring low species richness are usually vulnerable to invasions, especially if no or few species exist in the same functional group as the invasive species. Reducing the number of species in any ecosystem will, therefore, increase the likelihood of invasion.

The magnitude of invasion is determined by the competitiveness of the invading species and rate of invasion. Several factors can influence the rate of plant invasion. The number of introduction points (foci) into the new range and the number of invasive individuals is probably the greatest determining factor of invasion rate. Continuous invasion of the non-native species depends on availability and size of appropriate habitat adjacent to the introduction point.

The rate of invasion is often astounding. A good example of this is the well-documented invasion of the European starling (*Sturnus vulgaris*) into North America. The initial introduction of this bird took place in 1890 when about 120 starlings were released in New York's Central Park during a Shakespearian festival. Starlings are omnivorous, aggressive, and tenacious, which probably makes them successful invaders. They spread quickly to many urban areas and formed large colonies. Today, the European starling is found across America from Mexico to Alaska. The total population size is estimated to be at least 100 million individuals. Some of the largest populations contain a few million individuals.

Starlings often outcompete native bird species for their habitats and are ferocious in food competition. Attempts to eradicate starlings or halt their invasion have so far been unsuccessful. Another example of an aggressive invasive species is the introduction of fire ants (*Solenopsis invicta*) in the southern United States. Fire ants typically outcompete native ants where they invade. Fire ants were accidentally introduced in the 1930s from South America to the port in Mobile, Alabama. They have invaded the southeast United States and are found from North Carolina to Texas.

7.2 Effects of Invasion on Ecosystems

Invasion of non-native species into native communities can affect species composition, especially if a dominant monopopulation is formed. The ecosystem functioning is consequently changed. The impact of the invasion is best demonstrated by examples from islands and peninsulas that harbor high proportions of **endemic** species, but the principle also applies to land-locked ecosystems.

Invasion of non-native species has particularly affected ecosystems on islands and peninsulas. Particularly hard hit in the United States are Hawaii and Florida, where many ecosystems are now dominated by non-native species. For instance, on Hawaii about 80% of plants are non-native and 14% are invasive. Approximately 12% of the native flora of Hawaii is now extinct since European settlement. On Hawaii, about 25% of insects, 40% of birds, and almost all freshwater fishes are invasive. Similarly, in Florida invasive species comprise high proportions of the flora and fauna.

Another good example of the damage of invasive non-native plants on an island ecosystem is that of miconia (*Miconia calvescens*) in French Polynesia. This species not only has changed the structure of forests by dominating the forest canopy but also eliminates native understory species. Miconia is native to the tropical rain forests of Central America, where its natural habitat is typically small clearings in the forest. It colonizes rapidly such clearings that are caused by natural disturbances in the tropical forest. This plant was introduced as an ornamental plant to Tahiti in 1937. Since then it has invaded rapidly and formed dense, monospecific stands, currently covering more than 70% of the island. The thick canopy of miconia prevents the growth of an otherwise diverse understory community.

Introduction of non-native animals can intensify grazing pressure, which consequently affects plant communities. Grazing by non-native animals can have devastating effects on island ecosystems. Grazing degrades vegetation, causes extinctions, and facilitates invasion of non-native plants. For instance, goats were introduced to the tiny island of St. Helena, in the Atlantic Ocean in 1513. These goats have not only changed plant communities but also are responsible for the extinction of more than 50 endemic plants on the island.

Invasive non-native species not only modify structure of plant communities but also affect ecosystem functioning. Invasion of non-native plants can alter the soil chemistry. For example, invasion of the non-native legume *Myrica faya* in Hawaii has increased N inputs 500% to the ecosystem. This high input of N into the native ecosystem has consequently lead to drastic changes in the community composition. Invasive species can induce soil erosion with devastating consequences. The invasion of miconia on Tahiti has increased soil erosion on the island. Miconia has a very shallow root system, which does not anchor the soil on steep hills that are typical for the island. Because of the shallow roots of miconia landslides have increased drastically on Tahiti. Invasive plants can also alter fire regime either by suppressing or increasing frequency of natural fire. Several non-native grass species introduced to Hawaii have increased by 300% the extent of fire in woodlands. Another example is the invasion of cheatgrass (*Bromus tectorum*) into western North America where it dominates more than 40 million hectares and has increased fire frequency by more than 1,000%. Furthermore, grazing pressure by introduced animals has resulted in an alternative ecosystem dominated by pyrogenic grasses. Invasive plants can alter soil's hydrology by increasing drastically evapotranspiration. This can eventually lead to salinization of soils that has adverse effects on the ecosystem. The invasion of the star thistle (*Centaurea solstitialis*) and salt cedar (*Tamarix ramosissima*) in the United States has adversely affected hydrology of local areas.

Invasion can have a direct effect on the genetic make-up (genome) of resident native populations. Hybridization can now be more easily detected by using molecular techniques. Hybridization between native and non-native species is common (see discussion in Chapter 12). This results in changed genetic diversity of the resident population. For example, the native Pecos pupfish (*Cyprinodon pecosensis*) in Texas has hybridized with the invasive sheepshead minnow

(*Cyprinodon variegatus*) to the extent that the original genotype of Pecos pupfish is probably extinct. Hybridization of non-native and native species has even resulted in new invasive species. The American cordgrass (*Spartina foliosa*) formed an aggressive hybrid with the native cordgrass in England (*S. anglica*) that is now invading salt marshes in Europe and the United States.

Invasive non-native species can have devastating effects on native communities by inducing diseases. For instance, in Hawaii non-native birds from Asia are host to avian pox and avian malaria. The diseases are transmitted to native birds and have brought some endemic bird species to the brink of extinction. Another example of such an interaction is the introduced Asian chestnut blight fungus (*Cryphonectria parasitica*) that has virtually devastated every native chestnut tree (*Castanea dentata*) in the eastern United States (**Figure 7.2**). The chestnut used to be the dominant tree of the forest canopy ranging from Georgia to Maine. The chestnut blight was first observed in New York City in 1904 on plants imported from Asia. The spread of this fungal disease was phenomenal, and in less than half a century it spread over more than 91 million hectares of the eastern United States. Consequently, the chestnut was almost brought to extinction. Today, the American chestnut is very rare and is only found in a few isolated places. The fungus had devastating effects on the entire forest ecosystem.

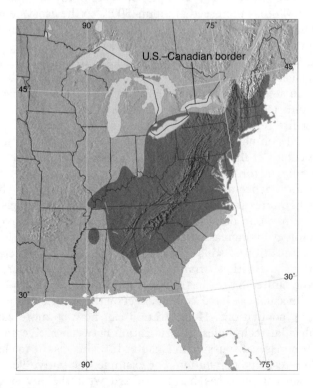

Figure 7.2 Historical distribution of the American chestnut.

Not only were the chestnut trees decimated, but several insect species that live only on the chestnut are now endangered or extinct. Most of its natural habitat is still infected with the chestnut blight, therefore, preventing any restoration efforts by replanting American chestnut back into these infested forests.

Restoration of the American chestnut is in progress using genetic resources of the few populations of the American chestnut that still exist. The crux of this work has involved developing blight-resistant hybrids using traditional breeding techniques in which the disease-resistant Asian chestnut is crossed (hybridized) with the American chestnut. The hybrid is resistant to the chestnut blight. Backcrossing is then completed when Asian-American hybrids are crossed with pure American trees to increase the genome of the American chestnut in the hybrid. The main aim of this work is to transfer blight resistance to the hybrid and then phase out most of the Asian trait of the hybrid. The first chestnut blight-resistant trees saplings are already growing. Seed will be available for large-scale restoration after 5 to 15 years. The hybrid must then be transplanted strategically into forests of its former range. Eventually, mass production in nurseries of the hybrid tree will follow. This project has good potential, but enormous work lies ahead.

Invasive non-native species can also prey directly on native species with disastrous consequences. For example, the introduced Nile perch (*Lates niloticus*) in Lake Victoria is responsible for the extinction of many endemic cichlid fish species (family Cichlidae). Another example is that of the brown tree snake (*Boiga irregularis*), a native to New Guinea and Australia, which was introduced accidentally to the island of Guam. Without natural enemies on Guam the brown snake population multiplied exponentially and caused extensive preying on the island's native birds. Almost 73% of native forest bird species on the island are now extinct.

7.3 Methods of Control

Prevention

Prevention measures against invasive non-native species is the first step to stop them from spreading (**Figure 7.3**). This includes education and public awareness about the damaging effects of invasive species. Also, it is critical to gain public support for a prevention program. Prevention is reinforced by legislation on introduction of invasive non-native species (discussed further in Chapter 14). Preventive measures can focus on the source of introduction. For instance, preventive measures have been implemented to avoid accidental introduction of aquatic organisms. These include ballast-water treatment and on-ship treatment targeting all living organisms in the ballast water. It is important to ensure political support, especially from the public and stakeholders before a preventive program is initiated. Sufficient funding must be secured for preventive programs and also for subsequent monitoring and restoration of native ecosystem.

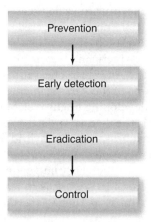

Figure 7.3 Factors to consider in the invasion process of non-native species. (Reproduced from *Invasive Alien Species: A New Synthesis (SCOPE 63)*, edited by Harold A. Mooney, Richard N. Mack, Jeffrey A. McNeely, Laurie E. Neville, Peter Johan Schei, and Jeffery K. Waage, copyright SCOPE, Island Press, 2005.)

It is much more difficult to deal with invasive species once they have established than it is to prevent their arrival. In this respect it is critical to obtain basic biological information on potential invasive species. This includes species introduced for agricultural, horticultural, or recreational purposes. Such information should include details on life cycle and potential vectors of dispersal. Early detection of potentially invasive species is critical for preventing its spread and facilitating its eradication before it becomes unmanageable.

If prevention fails invasive species may represent a challenge for restoration ecologists, especially when eradication is required. Three main strategies are available to deal with invasive species that have already become established (**Figure 7.4**): eradication, containment, and control. Selection of appropriate strategies depends on the extent of the problem and the aggressiveness of the species in question.

Eradication

Eradication involves the elimination of all individuals of an invasive species. Methods of eradication are fairly well established and are used successfully on a large scale in different ecosystems. Monitoring sites that have been eradicated of invasive non-native species should be considered as a part of the eradication program. This should be followed by restoration of native ecosystems.

Flora Methods for the eradication of plants include mechanical methods, which involve removing individuals of the non-native species by using tools (shovel, cutters, or machete) or by using specifically designed machines (mowers,

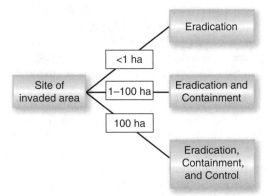

Figure 7.4 Management options to control non-native invasive plants. (Reproduced from *Invasive Alien Species: A New Synthesis (SCOPE 63)*, edited by Harold A. Mooney, Richard N. Mack, Jeffrey A. McNeely, Laurie E. Neville, Peter Johan Schei, and Jeffery K. Waage, copyright SCOPE, Island Press, 2005.)

plows, or harrows). All sites are not easily accessible by machinery and therefore require more labor-intensive methods. Eradicating plants by using hand tools is only practical when dealing with small isolated populations but is impractical if such populations are dispersed within a large area of native plants. Also, pulling of deep-rooted or rhizomatous plants is not practical, and minor soil disturbances could promote further invasion. Notorious invaders that form monospecific stands are typically eradicated by mowing, frequent harrowing, deep plowing, or bulldozing the entire site. Mowing is most effective when applied at a sensitive growth stage, such as just before flowering. Grazing management is another method that can effectively eradicate populations of invasive species. For instance, goats and native bison have been used to reduce kudzu infestation on a small scale in the southern United States.

Valuable experience has been gained in eradication of many agricultural pests. This effort has typically involved different combination of methods such as quarantine, chemical controls, and education. Such strategies can certainly be used in eradication of invasive species. It is critical for successful restoration to select effective combination of methods of eradication for non-native invasive species. Such methods should ideally have minimum impact on native ecosystems and should facilitate restoration of native plants. Successful eradication of invasive plants has been achieved in various ecosystems by using a combination of methods. For example, about 260 hectares of the noxious invasive shrub camelthorn (*Alhagi pseudalhagi*) was eradicated in California by combining eradication methods (herbicides and grazing). Eradication of the notorious French and Scotch broom (*Cytisus monspessulanus* and *C. scoparius*) provide a good example of using combinations of methods. Both broom species are N_2-fixing shrubs that are native to Europe but have invaded North America and formed dense, monospecific stands. The brooms have a strong impact on the structure and function of native ecosystem. Herbicides are typically used to eradicate brooms as they invade forests. Other methods of eradication include more labor-intensive hand pulling, cutting, and prescribed burning. Brooms are also cut using heavy machinery (a tractor or bulldozer); however, plants may resprout and therefore repeated cutting is required. Cutting is typically applied on extensive stands where burning is not possible due to close vicinity to residences. Hand pulling is labor intensive and cutting is not as effective. All these methods are, however, effective at reducing cover of broom. It is important to repeat these methods, however, because the brooms have large seed bank in the soil and can resprout after minor soil disturbances. Combining methods was especially effective in eradicating the brooms. Cutting broom in summer and burning it in the fall is a promising method of control. After initial eradication, sites are typically burned repeatedly to remove new broom seedlings. Frequent fires are, however, counter-effective because they can promote invasion by other non-native species. Close monitoring for the emergence of the brooms is, therefore, essential. Repeated burning and pulling of the brooms are, however, effective methods in promoting passive restoration of native plants.

Another approach is "integrated pest management," which involves a combination of control methods selected for a particular invasive problem. This approach often provides the most effective and acceptable control measures. Combining methods of eradication must be carefully selected and tested in small scale trials for each target invasive species before being implemented on a large scale.

Fauna Eradication of invasive vertebrates involves direct hunting and use of bait stations with toxic chemicals. Eradication of invasive freshwater fish species involves the use of toxins specific to fish, for instance, lampricide was successfully used against the sea lamprey in the Great Lakes. Invasive insects are usually eradicated by the use of insecticides or biopesticides. In small contained areas, baits or traps are used. However, widespread application is needed if the target insect is invading rapidly. The mass release of sterile male flies referred to as the "sterile insect technique" is effectively used against insect outbreaks affecting agricultural production such as screwworm fly (*Cochliomyia hominivorax*) and Mediterranean fruit fly (*Ceratitis capitata*). Use of the sterile insect technique may, however, be prohibitively expensive in small-scale restoration projects.

Not all eradication programs have been successful. For example, the attempt to eradicate the introduced fire ant from the southern United States has so far been unsuccessful. If the fire ants could be totally eliminated, the native ants would recolonize affected sites. The initial chemical control (using heptachlor) in 1957 resulted in deaths of wild animals and livestock. This was followed by a large-scale application of another chemical in baits (mirex), but the fire ant rapidly reinvaded areas from which it had once been eliminated. Moreover, mirex residues were discovered in other nontarget organisms and its use was soon terminated. Meanwhile, the fire ants have advanced their invasion dramatically.

Containment

Containment measures aim at restricting the invasion of non-native species to certain well-defined areas. The invasion of non-native species is restricted by using methods of eradication described above around populations of the non-native species. Any non-native species invading out of this defined area should ideally be eradicated. Monitoring the invasive species is important in this respect because the target invasive species might exist in small interconnected populations. In such cases an effective strategy is to disrupt small interconnected populations of invasive species.

Chemical and Biological Control

Control measures are implemented if eradication does not work. The invasive species is controlled to contain a certain level of economic and/or ecological damage. Methods of control are often the same as those previously described for eradication programs. Chemical control, especially insecticides and herbicides or toxic baits for animals, are widely used for this purpose. Chemical control (herbicide) on plants is often used in conjunction with other mechanical methods (mowing or cutting). The success of control measures such as herbicide application is much greater if followed by restoration efforts. Applications of herbicides are most effective when chemicals are translocated to underground tissues. Therefore, timing of herbicide application is critical to coincide with movement of carbohydrates to plants' underground storage tissues. This usually takes place during late summer to early fall. Herbicides are used directly (spot spraying) on targeted non-native species or by spraying it over large monopopulations of invasive species. The former method is more agreeable as the latter might affect other species than invasive organisms. Spot-spraying plants is ideal for small infestation and is accomplished by trained labor. Large-scale spraying is accomplished, for instance, by using crop dusters (small aircraft). One drawback of relying on chemical control is that repeated application is usually required. Also, the high cost of chemical control might be prohibitive. The environmental impact of chemical control is discussed in Chapter 10.

Biological control (biocontrol) is traditionally achieved by the introduction of natural enemies into invaded sites from the original range of the invasive species in question. Benefits from biological control programs in the United States are estimated to exceed $180 million annually. Some environmental risks (release of pests), however, are associated with this method. Leafy spurge (*Euphorbia esula*) is a non-native species that has invaded more than 2 million hectares in United States and Canada. Programs aiming at biological control of leafy spurge in the United States were initiated in the 1960s, and today

15 non-native insect species are used for this purpose. These insects limit growth of leafy spurge by consuming foliage, roots, and seed. Biological control can effectively reduce leafy spurge stem densities by as much as 90% over large areas.

Control of invasive non-native animal species can potentially be achieved by building up populations of native predators. Nevertheless, controlling invasion of non-native animals can be viewed as a step toward total eradication and eventual ecosystem restoration.

7.4 Restoration to Constrain Invasion

Restoring ecosystems to their full potential can be used as a strategy to control invasive species.

Niche Preemption

Niche preemption is the main principle behind restoration efforts that aim at reducing the impact of invasion. This approach establishes native species in the same functional group as the invasive species in question. This approach is also called "integrated restoration management." A good example is derived from studies on a notorious perennial tussock grass (*Agropyron cristatum*) that was introduced to the northern Great Plains of the United States after the 1930s. Today, this grass dominates about 17 million hectares of semiarid lands. Invaded stands of *A. cristatum* typically show reduced biodiversity, and the soil chemistry is also altered. Soil below *A. cristatum* stands has lower levels of available nitrogen and about 25% less total carbon than native prairie soil. This grass forms monospecific stands, some of which have lasted for more than a half century and therefore represent an alternative ecosystem state. The potential of using ecological restoration as a strategy to control this invasive grass was tested in an old field in the Grasslands National Park, Saskatchewan, Canada. The intensity of invasion of *A. cristatum* was compared between non-restored abandoned grassfield and restored prairie grasslands. Restoration was accomplished by broadcasting seed of native prairie grasses in the same functional group as *A. cristatum*. This effort reduced the invasion of *A. cristatum* by one-third. Furthermore, restoration practices could perhaps be improved by determining the optimal density of species in native functional groups and implementing eradication of the invasive species before large-scale restoration.

Niche preemption also includes restoration efforts after eradication of non-native species. After the invasive species is eradicated, restoration of the site by native species is essential to avoid reinvasion. Eradication of the invasive salt cedar in New Mexico by mechanical and chemical methods is most successful when this effort is followed by restoration of native trees such as cottonwood and black willows.

When eradication of invasive species is followed by restoration of native species, the prevalence of invasive species is usually reduced. For instance, in the Great Smoky Mountains National Parks, populations of non-native pasture plants are gradually being eradicated. For this purpose an integrated restoration management involving herbicide treatment, prescribed burning, and seeding native plants has effectively decreased the abundance of several non-native species. At the same time abundance of several native species that were previously not found in these pastures has increased. Eradication of invasive plants must therefore be followed by restoration of native species as quickly as possible to fill the empty niche and prevent the re-invasion of any invasive species.

In another example, the effect of eradication on invasive species for 1 year without any restoration was compared with repeated eradication efforts over 5 years followed by active restoration. This study took place in invaded forests in Ontario, Canada, dominated by sugar maple (*Acer saccharum*) and American beech (*Fagus grandifolia*). Different methods of eradication such as application of glyphosate (herbicide), hand pulling, cutting inflorescence, and adding a thick layer of mulch were tested on monopopulations of invasive understory species. These methods of eradication were tested once or repeated over 5 years followed by restoration efforts. The invasive species were garlic mustard (*Alliaria petiolata*), dame's rocket (*Hesperis matronalis*), and celandine (*Chelidonium majus*). Restoration efforts involved planting mature plants of the following native species: trout lily (*Erythronium americanum*), sharp-lobed hepatica (*Hepatica acutiloba*), mayapple (*Podophyllum peltatum*), and bloodroot (*Sanguinaria canadensis*) at a density of two shoots per square meter. The one-time use of glyphosate and hand pulling had adverse effects; the population of the invasive plants increased. The cutting and adding a mulch layer was also ineffective in eradicating the invasive species, but at least their populations did not increase. The repeated control efforts over 5 years followed by restoration effort were, however, effective in reducing the population of the invasive species and also by increasing the abundance of native plants. This example demonstrates that incomplete eradication without any restoration efforts can make matters worse. Also, long-term commitment is necessary in eradicating invasive species that should be followed by restoring native ecosystems.

Fire Management

Fire management, including fire suppression or prescribed burning, is used as a tool in restoration of invaded sites. Increasing the frequency of fire can intensify invasion. Curbing induced disturbances such as fire is often necessary to eliminate invasion of non-native species. For instance, an African grass jaragua (*Hyparrhenia rufa*) was deliberately introduced during the past century into pastures in Central America to improve cattle pastures. This grass can build up dense and tall (up to 2 m) stands. The grass litter is pyrogenic during the dry season, and frequent outbreaks of fire maintained this grass after cattle grazing was abandoned in what is now part of the national park system in Costa Rica. Fire is not a natural component of these forests and fire constrained restoration of native forest. It was, however, noticed that by suppressing fire outbreaks forest species began to recolonize the pasture and gradually outcompeted the shade-sensitive jaragua grass. It may, however, take the next half-century before an abandoned pasture is recolonized by the intact tropical dry forest species. On the other hand, prescribed burning is used effectively to eradicate invasive plants. Prescribed burning is most effective just before flowering of the invasive plant. Prescribed burning is commonly used in the United States to manage invasive species such as the Australian pine (*Casuarina equisetifolia*).

Increasing Biotic and Abiotic Resistance

It is assumed that restoring degraded ecosystems will reduce the risk of invasion. A degraded ecosystem is generally more prone to invasion because invasive species are usually preadapted to disturbed habitats. It follows then that degraded ecosystems have generally higher proportion of invasive species than less degraded or pristine ecosystems. Studies in the United Kingdom have shown that the proportion of non-native plants increases with degradation of the ecosystem. Restoration of invaded sites should focus on increasing biotic and abiotic resistance of the degraded ecosystem. Biotic factors such as the thickness of a forest canopy usually affect composition and density of understory species and is a strong determining factor for the forest's vulnerability to invasive species. In this case the shade of the canopy represents biotic resistance of the ecosystem. Invasive non-native understory species that are adapted to disturbed, open habitats are typically shade intolerant and can therefore be curtailed by restoring a thick canopy of native trees. The density of the canopy of native trees is one of the most important factors in determining an ecosystem's invasibility. An alternative method of eliminating the jaragua grass mentioned above from abandoned tropical pastures is, therefore, to out-shade it by planting dense stands of gmelina (*Gmelina arborea*), a fast growing non-native tree. A dense canopy is formed in about a year and the jaragua grass is eliminated (out-shaded) from the pastures in 3 to 5 years. The gmelina promotes recolonization of native understory species. The gmelina is typically cut down after 6 to 8 years, and this facilitates establishment of the rain forest. This method of using rapidly growing commercial trees as nurse plants is a powerful restoration tool that is commonly used for restoration of abandoned tropical pastures.

Ecological restoration places emphasis on **biotic resistance** as a strategy to slow down or inhibit invasion. Non-native invasive species interact with native species, pathogens, and herbivores that can potentially limit their impact on the ecosystem. Native pathogens may sometimes reduce the invasiveness of non-native species. Communities containing high species diversity will most likely harbor high diversity of pathogens. Such assemblages of native pathogens will probably with time increase the pathogen load of the invasive species, which will in turn possibly reduce their invasiveness. Increasing biotic resistance therefore reduces an ecosystem's invasibility.

Summary

Problems associated with invasive non-native species are on the rise. This is in part due to increased long-distance dispersal vectors that did not exist before and also due to increased degradation of native ecosystems. Notorious invading species can permanently alter the structure of an ecosystem by forming monospecific stands. Invasive species also alter ecosystem function such as nutrient cycling, hydrology, and fire regime. It is important to identify potential invasive species before they spread into native ecosystems because eradication is in some cases prohibitively expensive. Eradication of invasive species is possible using a variety of mechanical and chemical methods. Integrated restoration management involves using combination of methods to eradicate invasive species. This work must be followed by restoration of native ecosystem to prevent re-invasion. Native ecosystems can show resistance (biotic or abiotic) to invasive species. Biotic resistance might not only be affected by species richness of the native ecosystem but also by the presence of native species belonging to the same functional group as the invasive species. Restoration of degraded ecosystems can also be used as a strategy to halt invasion of non-native species.

7.1 Case Study

Kudzu—The Notorious Invader of Southern United States
Patricia Kinney, Georgia Power, Savannah, GA

In a less-than-affectionate term, kudzu (*Pueraria montana*) is often referred to as "the vine that ate the South." In its native Asia, kudzu was used to make food, medicine, and paper for thousands of years. Kudzu was introduced to the United States in 1876 at the Philadelphia Centennial Exposition and appeared at the New Orleans Exposition in 1883. It quickly became a popular ornamental among southern gardeners, gracing porches with its sweet-smelling purple flowers. During the early 1900s kudzu seeds and root crowns were commonly available in mail-order catalogs. Farmer C.E. Pleas of Florida operated one such mail-order catalog, marketing kudzu as nutritious, high-protein forage for livestock. The U.S. government also played a significant role in the spread of kudzu. Historically, the South had an agrarian economy, and many years of intensive crop production and inappropriate agricultural techniques left poor, eroded soils. To combat soil erosion the U.S. government created the Soil Erosion Service, which subsequently encouraged farmers to plant kudzu to stabilize the soil. From the 1930s to 1940s the Soil Erosion Service distributed approximately 85 million kudzu seedlings to farmers, paying them $19.75 per hectare planted. Concurrently, the Civilian Conservation Corps planted kudzu to control erosion on slopes, roadsides, and public lands. Altogether, more than 1.2 million hectares of kudzu was planted through government programs.

(continued)

Case Study (continued)

As farming became less profitable, people moved from rural to urban areas, leaving fields of kudzu unattended. Kudzu rapidly became widespread in the Southeast, and by 1953 the government stopped advocating its use. The U.S. Department of Agriculture (USDA) declared kudzu a weed in 1972, and it was placed on the Federal Noxious Weed List in 1997. Although most prevalent in Georgia, Alabama, and Mississippi, kudzu's current range extends from Florida to New York and west to Texas. Kudzu now blankets more than 3 million hectares in the Southeast and is spreading at an alarming rate (**Figure 7.5**).

Because of kudzu's rapid vegetative growth and ability to outcompete native plants, it is classified as an invasive species in the Southeast. Invasive plant species display certain growth characteristics that facilitate their spread, and kudzu is no exception. Taking advantage of warm temperatures and plentiful rainfall, kudzu can grow up to 0.1 m per day or 20 m per year. Kudzu spreads primarily through runners,

Figure 7.5 Kudzu distribution in North America. (Reproduced from USDA/NRCS. *The PLANTS Database*. Baton Rouge: National Plant Data Center, LA, March 30, 2010 [http://plants.usda.gov].)

rhizomes, and new plants that root at vine nodes. Kudzu also produces a limited amount of seed, with flowers forming mainly on hanging vines in sunlit areas. However, anecdotal evidence suggests that kudzu may be producing more seeds as it adapts to its new environment. Kudzu's rapid vine growth is supported by carbohydrate stores in massive taproots that can weigh up to 180 kg and reach depths of 3 m. Up to 30 vines can stem from one root crown, and individual vines can extend for 33 m. These vines can wreak havoc on native ecosystems, shading the forest floor and reducing biodiversity.

Kudzu damages natural ecosystems by literally smothering native vegetation (**Figure 7.6**). It limits the space, water, sunlight, and nutrients available to native species. Young trees and shrubs are killed by girdling as vines wrap tightly around them. Kudzu easily tops tall pine trees, where its intertwining vines form thick canopies that prevent sunlight from reaching the forest floor. Mature trees eventually die from shading or are uprooted by the weight of the vines, while understory vegetation struggles unsuccessfully to survive in shady conditions. Many native animals and insects also are negatively affected as kudzu displaces native forage and habitat. A common hypothesis to explain the success of invasive plants suggests that exotic species flourish in their new homes because of the absence of their natural enemies, including herbivores, insects, parasites, and bacterial and fungal diseases. But like other invasive non-native plant species, kudzu also may take advantage of mutualistic relationships to enhance its invasive ability.

Kudzu's prolific vegetative growth requires large amounts of macronutrients (nitrogen and phosphorus), and its ability to acquire these nutrients, even in poor environments, makes kudzu a successful invader. Many exotic plant species form symbiotic relationships with soil microorganisms to establish, and this symbiosis with local soil microbes can

Figure 7.6 Kudzu invading a native forest and forming a dense thicket.
(© Danny E. Hooks/ShutterStock, Inc.)

(continued)

Case Study (continued)

facilitate successful invasions. In general, if the exotic plant's new habitat contains indigenous species that are taxonomically related to the exotic one, then the soil may likely contain suitable symbiotic partners. Studies have shown that invasive legumes obtain nutritional benefits from interactions with *Rhizobia* bacteria and mycorrhizal fungi.

Rhizobia are nitrogen-fixing bacteria that form nodules on the roots of legumes such as kudzu. In exchange for fixing atmospheric nitrogen for the plant, *Rhizobia* receive carbon from the plant. The diverse strains of *Rhizobia* species actually have varying effects on plant growth. Although some legumes can partner with only a few rhizobial genotypes, other legumes can form a beneficial symbiosis with a wide range of rhizobial strains. It is likely that the ability of kudzu to be a successful invader may depend on its ability to form symbiotic associations with a wide variety of *Rhizobia* species.

Invasive plants also form symbiotic relationships with arbuscular mycorrhizal fungi (AMF), which facilitate phosphorus (P) uptake and plant growth. Although AMF can associate with a wide variety of plants from various geographical regions, the magnitude of the response in plant nutrition depends on the specific combination of plant and fungus genotypes. The success of kudzu may depend on its ability to form effective symbiotic relationships with resident AMF populations of native ecosystems. In actuality, kudzu's distinct advantage may be attributed to the tripartite *Rhizobia*–AMF–plant symbiosis.

Aside from being a scientifically daunting task, controlling kudzu also presents an economic challenge. The forest industry, utility companies, farmers, and state and federal parks are all impacted by the cost of controlling kudzu. Dr. Coleman Dangerfield of the University of Georgia estimates a cost of $500 per ha per year over 5 years to control kudzu on forest land, which is apparently higher in value than the lumber. According to Dr. James Miller of the USDA Forest Service, power companies alone spend over $1.5 million per year repairing damage caused by kudzu. Because kudzu poses such a significant threat to native ecosystems and industry, drastic measures must be taken to control and eradicate it. Current control methods involve mechanical, chemical, and biological options.

Impractical on a large scale, mechanical methods involve mowing or cutting vines and destroying the cuttings. Mowing must be repeated multiple times over several years to exhaust the roots' carbohydrate stores. Prescribed burning is another mechanical method that can be used in combination with chemical methods. Burning can allow easier access to heavily infested areas, making chemical application simpler and more effective.

Chemical control methods usually involve the use of herbicides, which kills both foliage and roots. Picloram is highly effective at controlling kudzu when applied from July to October. Other options include metsulfuron methyl, clopyralid, and triclopyr. Although successful, herbicides must be used with extreme caution, because they can damage nontarget species and persist in the soil. Choosing the most appropriate herbicide to use depends on factors such as the age and density of the infestation, the proximity of nontarget species, and physical features of the area such as slopes and streams. Repeated applications over several years are necessary for the complete eradication of kudzu.

Perhaps the oldest natural biological control of kudzu is overgrazing. Livestock will happily graze on kudzu, which is high in protein. Three to 4 years of continuous grazing is usually enough to exhaust root stores and eliminate a kudzu stand. Several introduced biological controls are also being investigated. *Pseudomonas syringae* pv. *phaseolicola*, a bacterial pathogen native to the United States, has been shown to kill kudzu seedlings by causing halo blight. Unfortunately, established kudzu can quickly recover from any damage. Another possibility is *Myrothecium verrucaria* (Albertini and Schwein) Ditmar: Fr., a fungal pathogen isolated from sicklepod (*Senna obtusifolia* [L.] Irwin and Barneby). This fungus attacks kudzu's leaves and stems, acting as a bioherbicide. Field tests showed that when *M. verrucaria* was applied with a surfactant, near total control of a kudzu stand could be achieved in 2 weeks. In fact, *M. verrucaria* proved so successful at kudzu control that USDA scientists Clyde Boyette, Harrell Walker, and Hamed Abbas were issued a patent in 2001. One possible obstacle to its use is the fungus' production of mycotoxins, which are toxic to mammals. However, less toxic strains are currently being developed. Controlling kudzu is difficult at best, but the success of native ecosystems in the South depends on it.

7.2 Case Study

Pale Swallow-Wort: An Emerging Threat to Natural and Seminatural Habitats in the Lower Great Lakes Basin of North America

Dr. Antonio DiTommaso, Department of Crop and Soil Sciences, Cornell University, Ithaca, NY

The invasive perennial vine *Vincetoxicum rossicum* (Kleopow) Barbar. [syn. *Cynanchum rossicum* (Kleopow) Borhidi] (pale swallow-wort or dog-strangling vine) has become a major concern in natural and seminatural areas of central New York State and the Lower Great Lakes Basin of North America within the last 15 to 20 years. This herbaceous vine was introduced into North America from the Ukraine region of Eastern Europe approximately 120 years ago and is currently expanding its range at an astounding rate. In its native range, pale swallow-wort can be found in forest-steppe and steppe zones where it is relatively uncommon. This species was first collected in North America near Toronto, Ontario, Canada, in 1889 and in the northeastern United States in New York State in 1897. It is now distributed from the Atlantic coast west to southern Michigan and northern Indiana and from southern Ontario, Canada, south through southern Pennsylvania (**Figure 7.7**).

(continued)

Case Study (*continued*)

Figure 7.7 Pale swallow-wort distribution in North America. (Reproduced from USDA/NRCS. *The PLANTS Database*. Baton Rouge: National Plant Data Center, LA, March 30, 2010 [http://plants.usda.gov].)

This member of the periwinkle family (Apocynaceae) has demonstrated the ability to form dominant, monospecific populations in many upland habitats and readily adapts to a wide range of light and moisture conditions, from full sun in open sites to full shade in mature forest understories. This related species to the milkweeds typically invades disturbed sites and exhibits aggressive growth on lime-derived soils, which are particularly vulnerable to invasion. Anecdotal evidence from private landowners in affected areas suggests that this destructive species can infest new areas over a relatively short period of time once it becomes established (i.e., 3–5 years).

Impact in Its Introduced Environment

Pale swallow-wort is of concern to managers of natural and seminatural lands throughout its range. The species threatens several of New York State's unique or rare ecosystems. For instance, its displacement of the nearly 5,000 ha of globally rare alvar habitats (i.e., shallow limestone barrens) of Jefferson County, New York, threatens 54 rare species of plants, insects, birds, and land snails (**Table 7.1**). Pale swallow-wort is also invading habitats in Onondaga County, New York, where the U.S. federally listed hart's tongue fern, *Phyllitis scolopendrium* var. *americana* occurs. In Connectictut, pale swallow-wort is overgrowing the only New England population of *Asclepias viridiflora*,

Table 7.1 Some of the Rare Species Expected to Benefit from the Management of Pale Swallow-Wort in the Lower Great Lakes Basin

Species/Community	Common Name	Federal or NYS Protected Status
Rare animals		
Lanius ludovicianus migrans	Migratory loggerhead shrike	E
Asio flammeus	Short-eared owl	E
Spizella pallida	Clay-colored sparrow	
Circus cyaneus	Northern harrier	T
Bartramia longicauda	Upland sandpiper	T
Ammodramus henslowii	Henslow's sparrow	T
Rare vascular plants		
Aster ciliolatus	Aster	E
Bouteloua curtipendula	Side-oats grama	E
Ceanothus herbaceous	Prairie redroot	E
Carex garberi	Elk sedge	E
Carex nigromarginata	Black-edge sedge	E
Castilleja coccinea	Indian paintbrush	E
Dracocephalum parviflorum	American dragonhead	E
Epilobium hornemannii	Alpine willow-herb	E
Lilium michiganense	Michigan lily	E
Sphenopholis obtusata var. *obtusata*	Prairie wedgegrass	E
Carex molesta	Troublesome sedge	T
Cypripedium arietinum	Ram's head lady's slipper	T
Carex backii	Rocky mountain sedge	T
Carex crawei	Crawe sedge	T
Corydalis aurea	Golden corydalis	T
Draba reptans	Carolina whitlow-grass	T
Geranium carolinianum var. *sphaerospermum*	Carolina cranebill	T
Geum triflorum	Prairie smoke	T
Hedeoma hispidum	Mock-pennyroyal	T
Panicum flexile	Panic grass	T
Sporobolus heterolepis	Prairie dropseed	T
Stellaria longipes	Starwort	T
Zigadenus elegans sp. *glaucus*	White camas	T

E, endangered; NYS, New York State; T, threatened. Used with permission of Sandra Bonanno and the Nature Conservancy (http://www.nature.org/).

(continued)

Case Study (*continued*)

a listed endangered species in Connecticut. The resultant loss of native plant species may reduce biodiversity and delay or redirect succession as well as reduce the value of croplands or wildlife habitats. Many Christmas tree growers and nurseries in affected areas report increased pressure by pale swallow-wort, especially during the past decade. This climbing vine not only competes effectively with favorable species for essential resources such as light and nutrients, but it also makes the plants more susceptible to uprooting and damage during strong winds. Some landowners in New York State have gone as far as abandoning horse pastures after 5 to 10 years of unsuccessful control efforts against pale swallow-wort. Old field locations in Ontario, Canada, colonized by pale swallow-wort have been shown to have much lower arthropod diversity compared with old field locations that have intact native vegetation. Research has also demonstrated that this invasive species may negatively impact reproduction in Monarch butterflies (*Danaus plexippus*) by attracting some ovipositing female butterflies to lay their eggs on this introduced species, but the immature stages of the butterfly are unable to survive on this plant. An added threat of pale swallow-wort to Monarchs is that this invasive may be outcompeting and displacing the native common milkweed (*Ascelpias syriaca*), the natural food host of this butterfly in many old field habitats where the two species co-occur. Pale swallow-wort also has the ability to form symbiotic associations with resident AMF in invaded soils and can alter the composition of AMF populations after colonization, therefore, likely limiting the growth and persistence of other species, including native plants.

Biology

Pale swallow-wort is a twining herbaceous perennial vine that grows 1 to 2 m in height, often in one season. In New York State, flowering begins in late May, peaks in mid-June, and ends in mid-July (**Figure 7.8**). The fruit pods (follicles) release seeds from mid-August to early October. Pale swallow-wort reproduces primarily via seeds, each bearing a coma of silky hairs, which facilitates long-distance wind and animal dispersal. This species may also expand clonally from tillers produced in the root crown, but range expansion appears to occur chiefly by seed production. Seeds are often polyembryonic, a condition in which some seeds give rise to multiple seedlings (up to eight). Polyembryonic seeds are more likely to develop into successfully established seedlings than non-polyembryonic seeds and may present an advantage to pale swallow-wort invasion by facilitating the establishment of new populations from a single seed. In some heavily infested sites of central New York State, seedling densities can be as high as 64,000 seedlings m^{-2}. Pale swallow-wort seeds typically germinate in the spring after production and generally persist in soil for less than 3 to 4 years. Seeds germinate and have high emergence rates (>50%) when buried at soil depths of less than 2 cm or when located on the soil surface. Depending on the level of embryony, pale swallow-wort seedlings have extremely high survivorship (71–100%) relative to most other plant species. This feature may partly be responsible for the high establishment rates and rapid expansion of this species in its introduced range in North America.

Figure 7.8 Pale swallow-wort flowers. (Courtesy of Antonio DiTommaso, Cornell University.)

Another characteristic of this invasive vine that may facilitate its establishment and range expansion is the production of a large and extensive below-ground root system.

Management Tactics

Control of this highly aggressive invasive vine using currently available methods has been difficult, and no single strategy has emerged as most promising. Presently, the control of pale swallow-wort in natural areas is best accomplished through the use of herbicides such as triclopyr. Despite the initial success of herbicide use, follow-up treatments are required to control the plant because of newly emerging seedlings from the seed bank. Thus, the duration and cost of subsequent control efforts largely depends on the size of the seed bank and on the longevity of the seed in soil or near the soil surface. Moreover, the success of restoration efforts within both managed and unmanaged areas will also be influenced by the quantity and type of plants found in the seed bank. Cultural controls have had limited impact upon further establishment or control of pale swallow-wort infestations. Repeated mowing can provide some reduction in plant height but has little impact on overall cover of pale swallow-wort. Cultivation will likely not kill established plants because root crown pieces remaining after cultivation can reroot, even under dry soil conditions.

(continued)

Case Study (*continued*)

Grazing and trampling can stimulate sprouting in pale swallow-wort, from leaf axes of stems or from root crown buds. Manual removal of follicles from established plants may also assist in prevention of further seed dissemination, but multiple harvests may be necessary. Pale swallow-wort has few natural enemies in its introduced North American range and has limited impact on its growth and reproductive potential. However, the fact that this species occurs infrequently or sparse in its native range suggests that natural enemies in its native range may be limiting its growth and distribution. Biological control of this invasive vine using effective and host-specific natural enemies collected in its native range may, therefore, provide the best prospect for sustainable long-term management of pale swallow-wort in North America and aid in restoration of affected habitats (**Figure 7.9**).

Figure 7.9 Dr. DiTommaso in pale swallow-wort–infested central New York State site. (Courtesy of Antonio DiTommaso, Cornell University.)

Key Terms

Biological control   161
Biotic resistant   164
Containment   161
Endemic   154
Eradication   158
Invasive species   150

Invasiveness   153
Native species   150
Naturalized   153
Niche preemption   162
Non-native species   150
Preventive measures   157

Key Questions

1. Describe the steps in the invasion of non-native species.
2. What are the main factors behind the rapid growth of invasive non-native species?
3. What are the main effects of invasive non-native species on ecosystem function?
4. What preventive measures can be taken against non-native species?
5. List methods of eradication.
6. What is integrated restoration management.
7. How can restoration of degraded ecosystems inhibit invasive non-native species?

Further Reading

1. Dahlsten, D. L. and Garcia, R. (eds.). 1989. *Eradication of exotic pests*. New Haven, CT: Yale University Press.
2. Mooney, H. A., Mack, R. N., McNeely, J. A., Neville, L. E., Schei P. J., and Waage, J. K. 2005. *Invasive alien species*. London: Island Press.
3. Pimentel, D. 2002. *Economic and environmental costs of alien plant, animal, and microbe species*. Boca Raton, FL: CRC Press.

8

SOIL

Chapter Outline

- 8.1 Soil Erosion
  - Global and Local History of Soil Erosion
  - Wind Erosion
  - Water Erosion
  - Factors Affecting Soil Erosion
- 8.2 Desertification
  - Regional Effects of Desertification
  - How Desertification Affects Us
  - How We Affect Desertification
  - Efforts to Slow Desertification
- 8.3 Soil Conservation
  - Soil Conservation Methods
  - Irrigation and Saline Soils
- 8.4 Restoration of Soil Nutrients
  - Nitrogen
  - Phosphorus
- 8.5 Soil Microorganisms
- *Case Study 8.1*: Importance of Soil Microbial Communities
- *Case Study 8.2*: Role of Arbuscular Mycorrhizal Fungi in Restoration of Mine Tailings

Soil is critical for the functioning of terrestrial ecosystems. It can be defined as the topmost layer of weathered rock material mixed with decomposed organic matter. From an ecological point of view, soil is a medium for microbes, animals, and plants. Soil stores water and nutrients that is available to plants. In addition, **decomposition** of organic matter and nutrient cycling takes place in soil. Decomposition of organic matter

> **Box 8.1 Various Sources of Pollution Affecting Soil**
>
> - Atmospheric deposition of pollutants (acid rain)
> - Infiltration of contaminated surface water
> - Land disposal of solid and liquid waste materials
> - Stockpiles, tailings, and spoil
> - Dumps
> - Salt spreading on roads
> - Animal feedlots
> - Agrochemicals (fertilizers and pesticides)
> - Accidental spills

is controlled by activities of soil fungi and bacteria. The rate of decomposition is also influenced by the level of humidity and temperature.

Land degradation is usually the first step in **soil erosion**. It is caused by a variety of factors, including deforestation, improper agriculture, mining, widespread pollution, and even invasive species (**Box 8.1**). The exponential growth of the human population, however, is regarded as the underlying reason for widespread land degradation. With the increase in the human population, more land is needed for agriculture and rangelands, which leads to intensified land degradation.

Problems of land degradation are found worldwide. About 24% of terrestrial land is undergoing degradation. This affects about 1.5 billion people worldwide who depend on ecosystem services derived from land that is facing degradation. For example, in Africa about one-third of the continent is facing serious land degradation that is beyond restoration. Also, in Mexico and Central America soils that are seriously degraded make up about one-fourth of agricultural land.

Serious land degradation can lead to soil degradation, which includes substantial loss of macronutrients (N, P, K) and **soil organic matter (SOM)** as well as damage to the biological compartment of soils. Soil degradation, which is on the rise worldwide, is caused by various factors and can result in declined soil functioning, acidification, and soil compaction. The U.N. Global Assessment of Soil Degradation found that nearly 11% of agricultural soil worldwide has been seriously degraded to the point where crops cannot be grown.

Soil degradation can lead to soil erosion, which is the quantitative loss of soil by the force of wind or water. Soil erosion often results from the removal of vegetation cover, leaving the soil vulnerable to torrential rainfall and high wind. Soil erosion is a serious matter, and it has an impact the livelihood of millions of people worldwide.

Disturbed soils typically lack soil horizons. Soil horizons are observed as layers of soils of different colors when holes or trenches are dug in undisturbed soils. Fertile soils close to the surface with high content of organic matter where most soil nutrients and soil microbes reside are referred to as A-horizon. The B-horizon is where clay accumulates. The C-horizon contains oxidized parent (bedrock) material, and the D-horizon contains unoxidized parent material.

Soil conservation uses techniques that slow down or halt the advance of soil erosion. The advance of soil conservation techniques has resulted in a significant reduction in soil erosion in the United States. Various soil conservation techniques and their application in ecological restoration are discussed below.

Soil degradation reduces SOM and nutrient levels. The decomposition and mineralization of SOM is regulated by various soil microorganisms. Restoration of soil nutrients builds up nitrogen and phosphorus pools in soils. Soil microorganisms play a pivotal role in enhancing nitrogen pools through the activities of rhizobium bacteria and in phosphorus cycling through the activities of mycorrhiza. Soil restoration involving the introduction of key soil microorganisms is also discussed below.

8.1 Soil Erosion

Global and Local History of Soil Erosion

Soil erosion directly impacts the standard of living of people worldwide and is probably the primary threat to food security for many developing countries. In some parts of China soil erosion has been catastrophic, threatening the lives of 400 million people. Drastic soil erosion due to massive logging on the island of Haiti has forced about 20% of the population to emigrate. Soil erosion has also had an impact on agriculture in developed countries. In the United States, for example, soil erosion and associated water runoff cost about $43.5 billion each year. During the last two centuries about one-third of arable topsoil in the United States has been lost. Specifically, Iowa has lost half of its topsoil, and Kansas has lost about 30% of its soil-N and more than one-third of its SOM. After the "Dust Bowl" era in the 1930s in the southern Midwest of the United States at least 80 million hectares were damaged and 20 million hectares of farmland were consequently abandoned. In an effort to combat the increasing problem of soil erosion, the U.S. government founded the U.S. Soil Conservation Service in 1935. This organization has relentlessly advocated for the use of proper technologies in soil conservation (**Box 8.2**) and has been successful in reversing this trend. In further efforts to mitigate soil erosion and with specific emphasis on agricultural lands, in 1985 U.S. Congress established the Conservation

Box 8.2 Focus of Soil Conservation

- Reducing soil erosion
- Soil stabilization
- Soil structure
- Soil fertility
- Soil organic matter
- Soil biodiversity

Box 8.3 Factors Contributing to Soil Erosion

- Soil erosivity
- Soil erodibility
- Land form (slope)
- Land management

Reserve Program. Under this program farmers were paid to restore agricultural fields that are highly prone to erosion. About 14 million hectares of formerly cultivated lands are currently being resorted under this program.

Soil erosion most often takes place where vegetation cover is reduced and soils are exposed to the action of wind and water erosion. The causes of soil erosion over the years include improper agricultural practices where removal of plant cover exposes soil to the elements and also exhausts the soil of nutrients. Excessive cultivation and trampling by livestock have also dislodged soil particles and accelerated erosion (**Box 8.3**).

Wind Erosion

Wind erosion occurs when vegetation or litter no longer covers soil and is particularly severe in arid and semiarid regions (**Figure 8.1**). Wind erosion involves the removal and transport of loose soil particles. Small soil particles can be transported long distances by wind. Such removal of small soil particles requires minimum levels of wind velocity. During wind erosion the larger soil particles move on the ground, bouncing on each other and initiating a chain reaction of sand grain movement that can eventually form dunes. The larger soil particles can form sand dunes along prevailing wind direction and, in turn, may inundate

Figure 8.1 Shelterbelts reduce the impact of wind erosion on agricultural fields in Michigan. (Courtesy of Erwin Cole/NRCS USDA.)

vegetation and accelerate the process of soil erosion. This can be regarded as a "domino effect" that accelerates soil erosion. Suspended soil particles may abrade and destroy plants, thereby further accelerating the process of erosion.

Water Erosion

Water erosion usually takes place when soil is exposed to torrential rain. Soils on agricultural fields that are left without vegetation cover are especially vulnerable to water erosion. It is estimated that 75 billion tons of soil are eroded by water from agricultural fields worldwide each year. The soil typically ends up as sediment in rivers, ponds, lakes, wetlands, and, eventually, the oceans.

The mode of water erosion is classified into several types. Sheet erosion, the most insidious, occurs when the whole surface of a field is gradually eroded by water in an almost uniform way. Once topsoil is removed, the subsoil is exposed to continuous erosion. Erosion of the subsoil can result not only in drastic reduction of nutrient levels but also in deterioration of the physical properties of the soil. However, even if soils are not eroded to a great depth, a high amount of soil can still be lost. For example, if a field loses 1.5 cm of topsoil, about 190 tons of soil are lost per hectare. Also, topsoil erosion is a serious issue because that is where most of the valuable SOM and macronutrients are stored.

A second type of water erosion, rill erosion, occurs on steep lands where water carves small channels into the soil. The rills can further develop into deep gullies that cut channels through the landscape. Just as is sheet erosion, so too is rill erosion regarded as one of the most serious forms of soil erosion.

Box 8.4	Methods to Reduce Soil Erosion

- Minimum tillage of soil
- Maintaining dense cover of vegetation
- Adding organic compost or mulches
- Contour farming
- Terracing

Factors Affecting Soil Erosion

Several factors determine the potential risks of soil erosion. These include soil's erosivity and erodibility. Soil's erosivity is the risk of erosion that a particular soil type faces. It is a function of the intensity, duration, timing, and amount of rainfall or wind velocity. Soil's erodibility is the vulnerability of soil to erosion, which is mainly determined by the proportion of soil particles (sand, silt, clay, and SOM). Soils with less than 2% SOM and less than 5% clay content are most vulnerable to erosion. Another factor that affects rates of erosion is the moisture content of soil. Dry soil is at a higher risk for soil erosion.

Land management also affects the rate of soil erosion. The condition of vegetation cover determines exposure of soil to wind and rain. Various land management practices, such as the cropping method, tillage (minimum till), the use of mulches (chopped straw), and especially maintenance of dense vegetation cover, protect the underlying soil and, therefore, reduce the rate of erosion (**Box 8.4**).

8.2 Desertification

Desertification refers to the spread of desert-like conditions, especially in arid or semiarid regions (**Box 8.5**). Desertification follows civilization. Its main causes are overcultivation, overgrazing (**Figure 8.2**), deforestation, and salination of irrigated fields. After the loss of vegetation cover, wind erosion can also induce desertification.

Climate fluctuations, including short spells of drought, play a role in the desertification process. These fluctuations can accelerate loss of vegetation cover, especially in drylands bordering natural deserts. Drought conditions result in decreased evapotransportation (movement of water through plants) and less cloud formation. Increased sun reflection (albedo) from the land soon follows. Once the land is desertified, drought conditions may persist. For example, the Rajapuna desert in India is most likely the result of a desertification process that may have started in the seventh century AD. In the Rajapuna region overgrazing resulted in land degradation that led to changes in regional climate and persistent drought conditions.

| Box 8.5 | Causes of Desertification |

- Overpopulation (human)
- Overgrazing by livestock
- Overcultivation
- Fuel gathering (dung and wood)
- Deforestation

Figure 8.2 Overgrazed pasture in the Mediterranean basin. This land was once a forest. (Courtesy of Sigurdur Greipsson, Kennesaw State University.)

Desertification followed early civilizations. Early inhabitants in Egypt cleared forests along the river Nile almost 5,000 years ago. An intensive cattle grazing was followed by desertification.

A desertified landscape may reduce regional rainfall and change seasonal rainfall patterns. These changes can advance global warming and reduce rainfall even further. Land degradation worldwide has reduced the capacity of ecosystems to fix atmospheric carbon by almost 1 billon tons between 1980 and 2003. Restoration efforts, however, can potentially counteract the process of global climate change by enhancing an ecosystem's function and its ability to act as a sink for excessive atmospheric carbon, as was discussed in Chapter 1. For example, through restoration programs vast degraded rangelands in semiarid regions could become a significant sink for large amounts of carbon. This depends, however, on restoration of these lands to their full ecological productivities.

Regional Effects of Desertification

Arid and semiarid drylands are especially vulnerable to desertification. Today, about 47% of terrestrial land can be classified as dryland. This includes arid, semiarid, and extremely arid lands. The largest proportion of dryland is found in Africa (37%), followed by Asia (34%), Australia (13%), North America (8%), and Europe (5%). Most of these drylands are still used as rangelands. However, vegetation of drylands is fragile, and soil is generally deficient in SOM and macronutrients. It is, therefore, not surprising that about 70% of all dryland is affected by land degradation.

Desertification is a serious environmental problem that affects about 65 million hectares of land that previously served in agriculture. It is alarming that desertification threatens the lives of more than 1 billion people worldwide. Countries that border natural deserts are especially vulnerable to desertification. For example, 97% of Yemen shows some form of desertification. Two of the most populous countries in the world, China and India, are also dealing with serious problems of desertification. In fact, desertification has affected one-third of both countries. In both countries natural deserts have been extending at alarming rates. In China desertified land is currently increasing by 360,000 hectares each year. The situation is even more serious in the much smaller country of Nigeria, where each year 351,000 hectares are lost to desertification. In Africa about 85% of drylands (26% of the total land area) is facing or has undergone some form of desertification.

In Europe, Spain is the only country facing serious problems of desertification. To date, about 30% of the country is desertified. Almeria, Spain was covered with oak forests 150 years ago, but now the area has been desertified. Climate change has accelerated this desertification process. Increased forest fires due to the rise in the human population after 1950 has also played a role in accelerated desertification in this region of Spain. With this accelerated desertification has come an adverse impact on the hydrological cycle in the Almeria.

In the United States desertification has been associated with rangelands in arid and semiarid regions. About 12% of rangelands in the United States are in poor condition. Overgrazing of these rangelands has been linked to the inadequate control of those that are federally owned. This exemplifies the "tragedy of the commons," as was discussed in Chapter 1.

How Desertification Affects Us

It is clear that desertification can directly affect the human population. In his novel, *The Grapes of Wrath*, John Steinbeck described a dramatized account of the Dust Bowl era migration away from desertified areas of Oklahoma and Texas to the once "promised land" of California. The most terrible consequence of desertification is famine, or a serious shortage of food and/or water, which affects public health and leads to the abandonment of farms and, in the worst cases, civil conflict, which often displaces people further into marginal lands. In fact, civil strife in many instances has been the direct result of famine on desertified lands. In the 1970s consistent drought in the south Sahara region (Sahel) resulted in more than 250,000 deaths. This episode of drought caused a massive migration of rural people. In Ethiopia between 1982 and 1986, drought coupled with civil war resulted in the death of more than 1 million people. Tragic cases of human desettlements due to desertification are ongoing.

The serious problems of worldwide desertification were addressed in 1977 in Nairobi at the U.N. Conference on Desertification (see Chapter 14). In 1986, to further study and alleviate this worldwide problem, the Desertification Control Programme Activity Center of U.N. Environmental Programme was founded. The main objective of this organization is to establish a database on desertification. More than 190 nations are a part of this convention. The U.N. Earth Summit in 1992 also made some improvements on these activities by involving local communities in solving these problems. The shift of emphasis from scientifically based solutions to social issues for solving desertification, however, is of concern.

How We Affect Desertification

The underlying cause of desertification is the recent exponential growth in the human population (see Chapter 1). Overpopulation in arid regions has led to overcultivation of marginal land that exposes soil to wind and water erosion. Also, the increase in the human population results in an increase in livestock population and, consequently, an overgrazing of marginal lands, which is widely regarded as a prime cause of desertification. Overgrazing results in critically low vegetation cover, which is a precursor to wind and water erosion. Livestock has increased dramatically in the developing world. For example, within a 20-year period (1955–1975) it is estimated that livestock (cattle, sheep, and goats) increased by 33% in developing countries. This increase in free-range livestock has far exceeded the carrying capacity of the marginal land. Communal lands have been particularly badly treated, exemplifying the "tragedy of the commons," as discussed in Chapter 1. Also, herders today use more efficient methods such as trucks to transport livestock rapidly to new pastures. This practice has resulted in intensified grazing pressure on marginal lands. In addition, deforestation and the clearing of shrubs and tress intensify the process of desertification. In grasslands disturbing shrubs quickens the desertification process. When shrub cover is disturbed, soil becomes susceptible to wind erosion, which increases the rate of desertification.

Halting desertification, therefore, mainly depends on the slowing of the rate of the growth of the human population and the associated increase in livestock population. Other strategies that can be implemented to mitigate desertification may include appropriate management of rangeland, forestry, and dryland cropping agricultural systems.

Efforts to Slow Desertification

Although the problem of desertification may look overwhelming, it should be possible to halt this process and restore degraded lands by using a variety of restoration techniques. This can especially be accomplished if dense vegetation cover can be established rapidly. Even though such work is expensive, the benefits of preventing lost productivity of the land are much higher than the potential costs of allowing it to stay unproductive or degenerate even further. The cost of such operations should be weighed against the cost and social consequences of continuous land degradation.

The first step toward halting desertification is to stabilize drifting sands. Sand stabilization in desertified land includes methods of depositing drifting sand. Among these are mechanical approaches that reduce the sediment load of the wind by decreasing its sand-transporting capacity. This can be accomplished by digging trenches, which allows a temporary protection until they get filled by sand. Even this temporary protection, however, might allow enough time for dune-building plants to establish. Another mechanical approach involves the erection of barriers and fences that reduce the wind's sand-carrying capacity on the lee side of the fence. With this method the potential of the wind to transport sand is sharply diminished. The accumulated sand may, however, give rise to additional problems. Establishing tree belts for shelter can provide a more permanent protection. This approach can be successful if tree species are specially selected for arid ecosystems.

The control of mobile dunes is challenging because of the amount of sand involved. One method of dune removal is mechanical excavation and transportation to a new location. This method, however, is expensive and often creates additional problems with the sand in the new location. An alternative solution is to facilitate the removal of sand and destruction of dunes "naturally." This can be achieved by disrupting the aerodynamic profile of the dune. Dunes can be reshaped to alter their aerodynamic form and retard their movement. Surface chemical treatment can be applied to stabilize the dunes temporarily. Some of the sand-stabilizing methods discussed in Chapter 9 can also be applied to desert conditions.

8.3 Soil Conservation

Soil conservation uses applicable techniques to reduce or halt soil erosion (**Figure 8.3**). Soil conservation plays an important role in securing food supplies from agricultural lands. The importance of efficient soil conservation is derived from the fact that about 97% of all human food and livestock feed depends on soil productivity. Though roughly 40% of terrestrial land is occupied by agriculture or rangelands, only about 3% of terrestrial land is covered with highly

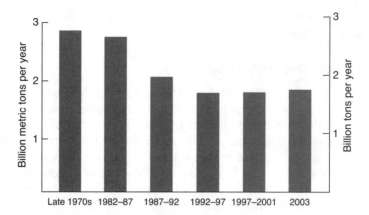

Figure 8.3 Decline in soil erosion in the United States.

productive soils. Soil conservation is a cost-effective strategy to maintain sustainable productivity on agricultural lands. Other benefits of soil conservation include a reduced need for land conversion (e.g., turning forests into agricultural fields). Effective soil conservation, therefore, aids in the conservation of native ecosystems.

Conservation Methods

It is critical in soil conservation to maintain dense cover of plants or litter on the ground. Initial ground cover can be made by using crop residues (mulching). Dense cover of plants and litter can intercept and dissipate raindrops that could otherwise cause erosion of exposed soil. Also, dense vegetation cover reduces the velocity of wind at the ground level. Enhancing vegetation cover is an effective strategy in controlling erosion in various landscapes, including rangelands, pastures, forest plantations, and wildlife areas. Establishing dense vegetation that forms permanent cover (including shrubs and trees) is the most efficient method to halt soil erosion permanently.

Other effective conservation technologies to prevent or slow soil erosion include proper soil tillage, building appropriate soil ridges (contouring), terracings and erecting wind barriers (**Figures 8.4** and **8.5**). These control measures can be used strategically alone or in combination to effectively reduce erosion rates. To determine the optimal strategy of soil conservation, many environmental factors must be taken into consideration, such as soil type and intensity of wind and rain.

Conservation tillage has been practiced successfully in agriculture to reduce soil erosion. Conservation tillage was part of the Food Security Act proposed in the United States in 1985. This includes no-till practices where soil is left undisturbed and only slightly disturbed during precision seed drilling (**Figure 8.6**). In the United States soil erosion on croplands has been reduced 40% by the non-till method over the last 20 years. In 2004 no-till croplands covered about

Soil Conservation 187

Figure 8.4 Soil conservation: contour ridges and strip cropping of corn and alfalfa in Iowa. (Courtesy of Tim McCabe/USDA ARS.)

Figure 8.5 Soil conservation: terraces on steep hills. **(c)** A cornfield in Iowa; **(b)** rice paddies in Asia. (Part a courtesy of Tim McCabe/NRCS USDA; part b © Kim Pin Tan/ShutterStock, Inc.)

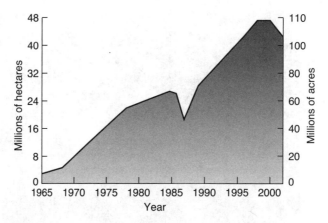

Figure 8.6 The advance of the minimum tillage practice in the United States.

25 million hectares in the United States, 24 million hectares in Brazil, 18 million hectares in Argentina, and 13 million hectares in Canada. Also, mulch-till is used where crop residues are tilled into the soil, still leaving a high proportion of mulches on the ground surface.

Building soil ridges (contouring) is effective in halting water erosion on agricultural fields. Soil ridges can be built with ground-based machinery to reduce water erosion and prevent the formation of gullies on lands with steep slopes.

Erecting wind barriers can effectively control wind erosion. Permanent structures such as wood fences can serve as wind breakers, especially on sites where it is difficult to establish trees. Where trees are more easily established, planting them in rows at right angles to prevailing winds is another effective method in controlling wind erosion. In addition to the benefits they provide in soil conservation, wind barriers also effectively increase soil moisture. Drifting snow in temperate regions is collected during the winter on the lee side of wind barriers. This adds to the soil moisture in the spring when it is often in critical supply. As a general rule, wind barriers conserve soil outward 10 times the distance from them as they are tall. The density of the barrier, however, can affect this equation. Wind barriers (hedges or trees) can be cut down or incorporated into the landscape after permanent vegetation cover has been established. Much lower grass barriers can be established by using tall perennial grasses. For this purpose hardy sand-stabilizing grasses can be used to stabilize drifting sand and halt erosion, as outlined in Chapter 9.

Long-term soil conservation involves appropriate land use practices. This includes cost-effective strategies such as a reduction in grazing pressure by livestock, the adoption of a suitable crop rotation, the implementation of judicious tillage methods, and the maintenance of soil nutrients.

The U.S. Conservation Reserve Program has aimed at conserving highly erodible agricultural lands. This program has outlined the economical benefits of practicing soil conservation. Also, education and awareness play a critical role in advocating for soil conservation. One movement, organic farming, focuses its practices on optimal soil conservation. With the increased market share of organic farms, the extra benefits of soil conservation used in organic farming practices will follow.

Irrigation and Saline Soils

The irrigation of agricultural lands with water that has minimal amounts of salt (NaCl) causes **salinization** of soil in the long term. Salt affects plants by inducing osmotic disturbances and also by specific ion (i.e., Na^{2+} or Cl^-) toxicity. Irrigation of agricultural fields is an ancient technique. Early civilizations in Mesopotamia bordering the Tigris and Euphrates rivers practiced irrigation. Today, most of this ancient agricultural land is just a salty wasteland or desert. Modern irrigation practices are widespread, especially in arid and semiarid regions, and more than 90 million hectares are irrigated worldwide. About 30% of this irrigated agricultural land is affected by saline soils, which results in yield reductions. Soil salinity is a serious problem in many countries. Today, problems of soil salinity are particularly serious in the arid regions of Argentina, Egypt, India, Iraq, Iran, Pakistan, and Syria. Agriculture in these countries relies to a large extent on irrigation. For instance, about 65% of agricultural land in Pakistan is irrigated, and about one-third of this land suffers from salinization. About 50% of irrigated land in Syria's Euphrates River Valley is extremely saline. An increase in the groundwater table is another problem of excessive irrigation, which, in turn, can solubilize salty sediment and bring it to the surface. Such problems have been experienced in Australia.

An alternative way of using saline soils in agriculture is to grow salt-tolerant crops. Salt tolerance is especially important during seed germination and seedling development. Poor yields are frequently the result of low germination and low seedling establishment. Although crops show salt tolerance during later stages of growth, they can be quite sensitive to salinity during early growth stages.

Saline soils can be reclaimed by improving drainage and by soil drainage with pure water to remove excess salt. Soil drainage can also be improved by planting deep-rooted trees that transpire the soil water. Leaching of saline soils is only successful when purified desalinated water is used. However, excessive leaching can deprive soil of essential nutrients, which need to be replenished. In addition to salt leaching, the restoration of sodic soils, therefore, involves the replacement of excessive adsorbed sodium by calcium or magnesium.

Another method of restoring saline soils is by implementing salt phytoremediation where halophytes (salt-tolerant plants) are grown in these soils. Halophytes can accumulate salt ions in high concentration in their foliage. The harvesting of halophytes gradually reduces the salt concentration of the soil. Halophytes with widespread distribution, such as *Sueda* sp. and *Atriplex* sp., are used for this purpose.

8.4 Restoration of Soil Nutrients

Loss of nutrients from agricultural soils in the United States is estimated at $20 billion per year. Degraded soils are typically deficient in SOM and macronutrients and have low cation exchange capacity (CEC). The CEC is a measurement of the positive charged ion holding capacity of a soil. The CEC is low in sandy soil but increases as the clay content increases. The CEC also increases as the SOM increases in soils. Restoration of severely degraded soils involves the restocking of nutrient capital into the soil. In this line it is important to hasten the development of SOM and the accumulation of macronutrients along ecosystem succession. The SOM is one of the most important factors in determining the functioning of soils. Increased levels of SOM are directly related to the ability of soil to store nutrients. Also, as the SOM accumulates, the water-holding capacity of soil improves. The SOM can be managed simply by increasing litter input from the ecosystem. An alternative method to increase the soil's SOM is to import organic matter or topsoil from a surrogate site.

As SOM increases, **mineralization** processes are developed. These processes ensure efficient nutrient cycling that supplies nutrients steadily to plants. The SOM of soil contains large amounts of nutrients, which is available to plants via mineralization—an important component of nutrient cycling. Soil nutrient cycling depends on mineralization of SOM and is performed by various soil microorganisms. Also, quantity and quality of litter inputs influence the rate of mineralization, which is usually in proportion to the nutrient capital in soil. However, more research is needed on the interactions of these soil processes and their implications in restoration.

Nitrogen

Nitrogen (N) plays a pivotal role in the early development phases of degraded soils. Nitrogen is essential for plant growth and is generally required in larger amounts by plants than any other soil macronutrient. Degraded soils, unfortunately, usually have serious shortages of this important macronutrient. It is possible to monitor the N buildup in soil by sampling soil and analyzing its chemical contents at regular intervals. The target N capital in soil restoration is ideally between 1,600 and 5,000 kg N per hectare. By comparison, intact soils can contain 2,000 to 10,000 kg N per hectare. Nitrogen accumulation on degraded sites is a very long process without organic or inorganic fertilizer amendments or input from biological N_2 fixation. Nitrogen accumulation in most soils is concomitant with the buildup of SOM as nitrogen is recycled in the ecosystem mainly by decomposition and mineralization of SOM. Decomposition of SOM is generally slow in temperate climates, resulting in slow N cycling. In the restoration of primary soils in temperate climates, steady supplies of inorganic N are, therefore, needed for the first years to build up the SOM and provide plants with a continuous supply of N.

Nitrogen accumulation and efficient recycling of N are important factors in soil development. There are, however, only two ways by which N accumulation can be accelerated:

1. Through amendments of fertilizers (inorganic or organic)
2. Through biological N_2-fixation such as free-living bacteria and symbiotic bacteria, including *Rhizobia* (on legumes) and *Frankia* (on alder trees)

Symbiotic **Rhizobium** (pl. *Rhizobia*) bacteria are the most efficient in N_2-fixation and, therefore, are the greatest provider of fixed N_2 into soil. *Rhizobium* form N_2-fixing nodules on the roots of most legumes. Using nitrogen fixation capacity of legumes in improving N budget of soil is a cost-effective alternative to using chemical fertilizers alone. It is a common practice in ecological restoration to use a variety of N_2-fixing legumes (mostly commercial ones) to build up the N capital in degraded soil on restoration sites. By properly managing legumes during restoration, N accumulation rates can be expected to be as high as 300 kg N per hectare per year.

Phosphorus

Phosphorus (P) is another important macronutrient that must be considered in soil development on degraded sites. Plant growth is usually severely limited unless P is available in combination with N. Phosphorus typically builds up slowly through the gradual weathering of rocks and atmospheric deposition (as dust). The accumulation of P is usually less than 5 kg P per hectare per year. The symbiotic association of plants with mycorrhizal fungi enhances efficient P uptake and recycling of P within the ecosystem. Degraded soils gradually become colonized by mycorrhiza and a variety of mycorrhiza increases as plant succession advances.

8.5 Soil Microorganisms

Microbes play an important role in the functioning of soils (see **Case Study 8.1** on page 197). They improve soil aggregation (which reduces erodibility of soil) and are especially critical for soil fertility. Microbes ensure nutrient cycling (C, N, P, S) through decomposition and mineralization and by facilitating nutrient uptake. Soil microorganisms contribute to the biological functioning of soils and also to the maintenance of soil quality, as outlined in Chapter 2 (**Box 8.6**). In addition, microbes play an important role in bioremediation, as discussed in Chapter 10. It is challenging to establish the biological component of soil after severe degradation because of the high complexity of this system. Establishing populations of beneficial soil microorganisms that can perform key ecological functions of soil such as N_2-fixation and P-cycling is essential for successful restoration. It is possible to restore these functions by the introduction of appropriate N_2-fixing bacteria and mycorrhizal fungi.

| Box 8.6 | Beneficial Activities of Soil Microorganisms |

- Decomposition of organic material
  - Enhance mineralization (N, S, and P)
- Increase plant nutrients for plants
  - Mycorrhizal species
  - Phosphate-solubilizing bacteria
- Nitrogen fixation
  - Free-living bacteria
  - Cyanobacteria
  - Frankia
  - Symbiotic *Rhizobium*/legume
- Plant growth promotion
  - Production of plant hormones
  - Protection against pathogens
  - Enhance nutrient uptake
- Biocontrol
  - Against diseases
  - Soil nematodes
  - Pathogenic fungi
- Biodegradation
- Soil aggregation

Although it is difficult to isolate and directly identify most soil microorganisms, recent progress in molecular biology has improved this process. For example, the use of real-time polymerase chain reaction (PCR) allows quantification and identification of DNA profiling of soil microorganisms. Soil microorganisms show tremendous diversity, and less than 1% of these species may potentially be cultured on agar plates. It is, therefore, almost impossible to restore populations of soil microorganisms to a perfect replica of previous conditions. Because of this high complexity of soil microorganisms, restoration must rely on passive recolonization of soil microorganisms during succession. For this, restoration efforts should focus on methods that facilitate vectors of microorganism colonization.

Decomposition of plant litter is performed by a group of microorganisms that are collectively called decomposers. Decomposers are frequently found on disturbed soils on restoration sites. They are essential in recycling the nutrients found in the plant litter and are, therefore, important in ecological restoration. Soil microorganisms involved in decomposing litter include soil fungi as one of the largest groups. Soil invertebrates (microfauna), including amoeba, nematodes, mites, and collembolan, also play an important role in decomposing litter.

Mineralization takes place when immobilized nutrients (found in the litter) are released back into soils and are subsequently made available for plants. Mineralization is achieved through organic processes such as ammonification, nitrification, denitrification, phosphorylation, and decarboxylation. In this respect the nitrification process that supplies plants with N is especially important. Mineralization processes are typically controlled by soil bacteria.

In restoration, it is valuable to follow up the development of microbial processes. For this purpose, microbial analyses can be used to assess critical soil processes. These include various methods of biomass estimation, enzyme activities (e.g., ATPase and dehydrogenase), soil respiration (CO_2 released), estimates of nutrient cycling, and changes in microbial populations with time (assessed by using real-time PCR). Microbial analysis can be used to compare degraded and intact soils and can also be used to follow up success of soil restoration by assessing changes in populations of soil microorganisms with time.

Changes in microbial populations are especially useful in indicating changes in ecosystem functioning, including nutrient cycling. Populations of soil organisms (e.g., earthworms, nematodes, mycorrhizae, *Rhizobium* bacteria, and various other groups) can be used as indicators of soil quality on restoration sites. Groups of soil organisms can serve as indicator species for restoration success. Indicator species should ideally respond to adverse changes over a short time and should accurately reflect the functioning of the ecosystem. Such indicator species should be ubiquitous on the restoration site.

Colonization of degraded soil by free-living microorganisms such as soil mites, nematodes, fungi, and bacteria takes place gradually. Colonization of symbiotic microorganisms such as mycorrhizal fungi is usually more erratic and takes longer time. Mycorrhiza colonizes degrade soil via several vectors such as migrating animals or by wind and surface runoff. Also, **arbuscular mycorrhizal fungi (AMF)** spread on site via root-to-root contact. Several ectomycorrhizal fungi develop mushrooms above ground. The spores of these fungi can disperse by wind over long distances. It is possible to manage colonization of key soil microorganisms such as *Rhizobium* and mycorrhiza by introducing these microorganisms strategically into soil on restoration sites. Colonization of key soil microorganisms can be managed, for instance, by inoculating plants with effective symbionts and introducing them to restoration sites. Methods of inoculating N_2-fixing *Rhizobium* with legumes are well established. Inoculating plants with mycorrhiza, however, is a more recent practice that is still being improved. Strategic outplanting of such material is important to maximize dispersal of soil microbes within restoration sites.

Mycorrhiza (root fungi) is a mutualistic symbiosis between plants and fungi. Several types of mycorrhizae are recognized. The most common one is the AMF, forming symbiosis with about 85% of flowering plant species. The second most common mycorrhizal type is the ectomycorrhiza formed between many trees, shrubs, and ascomycota or basidomycota fungi. Other types of mycorrhiza include orchidaceous mycorrhizae and ericaceous plants that form ericoid or

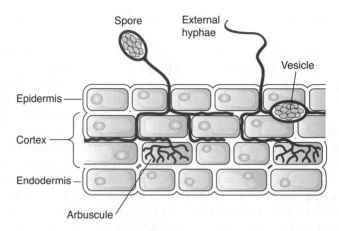

Figure 8.7 Arbuscular mycorrhizal fungi in root.

arbutoid mycorrhizae. The AMF colonize internal cells (cortical) of roots and at the same time form an extensive network of external hyphae (**Figure 8.7**). In turn, the absorptive area of the root system is increased by fungal external hyphae for nutrient and water uptake. The benefits of this symbiosis are established by facilitating nutrient uptake, mainly P. Additional benefits are alleviation of drought and protection against plant pathogens. The AMF symbiosis not only facilitates nutrient uptake but also regulates plant communities and ecosystem functioning. Mycorrhizal fungi are important in ecosystem restoration because they play a critical role in establishing nutrient cycling, especially the P cycle. In addition, mycorrhizae play an important role as carbon sink in the soil. It is estimated that as much as 40% of C that is cycled in terrestrial ecosystems is processed by AMF.

Knowledge of mycorrhizae functioning is essential for restoration of severely disturbed lands (see **Case Study 8.2** on page 201). Severe soil disturbances, including topsoil erosion or cultivation (plowing and harrowing), reduce or eliminate mycorrhizal fungi. These severe disturbances break up the common mycorrhizal network that links AMF and plants together. Drastic reduction in AMF activity is, for instance, experienced after long-term storage of stockpiled soil. Reducing plant cover on stockpiled soil reduces AMF activity because AMF are obligate symbionts on plants. Also, an increase in ruderal plant (that do not harbor AMF) communities reduces AMF activity in soil. Activity of AMF on degraded soils, however, can be improved by maintaining high cover of symbiotic plants on these soils. Restoration of mine wastes, therefore, emphasizes careful management of stockpiled soil and associated soil microbes, especially mycorrhiza, as discussed further in Chapter 10.

Recovery of AMF activities after drastic soil disturbances may take several years or decades. In disturbed ecosystems, AMF can be found dispersed in patches. These patches act as "bootstraps" in reassembling the ecosystem. The "bootstrapping hypothesis" is important for restoration ecology because introduction of AMF inoculated plants into disturbed sites could play a critical role in the ecosystem reassembly. Strategic inoculation of mycorrhizal plants on restoration sites could initiate the establishment of desirable mycorrhizal fungi.

The AMF play a role in regulating plant communities along successional trajectories. During primary succession the dependency of plants for AMF symbiosis changes from nondependent AMF plants to facultative mycorrhizal plant species to obligatory mycorrhizal plant species. Therefore, succession can potentially be managed by introducing AMF-dependent plants along with their AMF symbionts (see Chapter 4). Facultative AMF plants are commonly found as dominant pioneering species in primary succession. For example, in primary succession on coastal sand dunes, the pioneer grasses are usually facultative AMF plants. The establishment of AMF is essential to the stabilization of sand dunes and to the assurance of desirable succession trajectories. During succession, communities of different mycorrhizal types change where AMF plants are common in earlier stages and ectomycorrhizal plants are prevalent in later stages.

Free-living bacteria and symbiotic bacteria play important roles in dinitrogen (N_2) fixation. Diazotrophic bacteria can live in free association with plants and have access to C source (root exudates, secretions, lysates, and sloughed off cells) in exchange for fixed N_2 in the rhizosphere. The N that is fixed by diazotrophic bacteria can be taken up by plants. Most free-living diazotrophic bacteria are found in the rhizosphere, but some may reside inside root tissue. Aerobic conditions around plant roots inhibit N_2-fixation of free-living diazotrophic bacteria. Also, high levels of N in soil inhibit N_2 fixation. Diazotrophic bacteria are less effective in N_2-fixation than symbiotic N_2-fixing bacteria. However, diazotrophic bacteria could potentially be used as inoculum for nonleguminous cover crops. To evaluate inocula, it is possible to compare N_2-fixation of different diazotrophic bacteria by using mass spectrometry and the isotope ^{15}N technique. For this purpose soil is exposed to air labeled with $^{15}N_2$, and after incubation in an enclosed system under controlled condition the content of ^{15}N in tissue of diazotrophic bacteria is measured by mass spectrometry. High concentration of ^{15}N indicates effective nitrogen fixation.

Symbiotic N_2-fixing systems include *Rhizobia* nodules on legumes, actinorhizal plant and *Frankia*, *Azolla* (water fern and microsymbiont *Anabena*), and lichen symbiosis involving cyanobacteria. Plants that possess *Frankia* and actinorhizal symbiosis are used in reclamation of eroded areas and stabilization of sand dunes. *Frankia* is a gram-positive actinomycete that forms N_2-fixing vesicles in vitro. It forms nodules on host plants such as *Alnus* (alder trees), *Casuarinas* (Australian pine), myrtle, and mountain mahogany.

Legumes are commonly used in restoration. Legumes and *Rhizobia* bacteria form a specific symbiotic association. Specific *Rhizobia* species recognize host roots and establish and grow within these roots in root nodules. In exchange for direct C source from plants, it provides the plant with fixed N_2 as NH_3. Legumes are important keystone species in the restoration of nutrient-deficient soil. To optimize N_2-fixation of legumes, it is important to select and test under field conditions different strains of N_2-fixing bacteria. For this purpose strains can be isolated from nodules of native legumes, and their capacity to fix N_2 is compared with known strains. An effective strain should form highly effective N_2-fixing root nodules on legumes. Effective strains should tolerate adverse environmental conditions and survive in the soil before formation of root nodules. It is important to test potential strains in the field, and their performance should be followed up in the subsequent growing season.

The infection process of *Rhizobia* and legume roots involves chemotaxis (flavonoids) and attachment of bacteria on the root, which is followed by penetration through root hair. This infection process is genetically controlled. The nodule initiation involves root hairs that curl around *Rhizobia* and act as infection threads that reach the plant's cortical cells. *Rhizobia* become enclosed within the peribacteroid membrane and become bacteroid. The multiplication of bacteroids takes place within the root nodule. Effective N_2-fixing nodules are red in the center due to leghemoglobin (iron complex). Exclusion of O_2 is essential for the enzyme nitrogenase to be active. Ineffective nodules are easily recognized by their green or brown color in the center of the nodules, indicating no nitrogenase activity.

Nitrogen fixation depends on interaction between plant genotype and rhizobial strain. Environmental factors also play a role in the N_2-fixation, such as temperature (optimal at 10–37°C) and soil acidity. Nitrogen fixation is demanding on P supply for large amounts of ATP. Most legumes form tripartite symbiosis (*Rhizobia*, AMF, and legume association) to enhance P uptake.

Inocula of efficient *Rhizobia* strain can be mass produced under controlled conditions. For this purpose a carrier material is needed for inoculum production. Carrying material typically includes soil, peat, pumice, and vermiculite. The inoculum must include adhesive material, lime, sugar-based nutrients, and water (to keep inocula saturation adequate). The carrier material is autoclaved (sterilized) and placed in breathable plastic bags. Then, each bag is injected with rhizobial concentration (of selected strain). Inocula must have a shelf-life of at least 6 months. It must also contain a minimum number of bacteria at time of use. Inocula bags are opened in the field and the content is mixed with legume seed at the time of sowing. An alternative inoculation method is to use seed pelleting where *Rhizobia* inoculum is coated on each legume seed. This method is, however, more expensive. Inoculants can also be directly included in seed pelleting before sowing. Pellets must contain strong adhesives rolled in lime or rock phosphate.

Summary

Soil plays a pivotal role in the functioning of terrestrial ecosystems. Soil is made up of both inorganic and organic matter and harbors a myriad of microorganisms. Agricultural practices including fallow farmland overgrazing by livestock can result in soil degradation. Soil degradation can result in loss of SOM and nutrients. This is often followed by soil erosion, which is the removal of soil particles by wind or water. Land degradation in arid regions can result in desertification, which is the expansion of desert-like conditions. Desertification directly affects the living standard of millions of people in the developing world. Restoration of degraded and eroded land that has a potential in agricultural production should be a high priority worldwide. To control soil erosion, the establishment and maintenance of dense vegetation cover is essential. In restoring degraded soil, buildup of the soil nutrient pool is essential. Furthermore, the role of soil microorganisms should not be overlooked in establishing nutrient cycling, especially the role of N_2-fixing bacteria and mycorrhizal fungi to enhance P cycling.

8.1 Case Study

Importance of Soil Microbial Communities
Dan Buckely, Department of Crop and Soil Sciences, Cornell University, Ithaca, NY

Although apparently dominated by plants, terrestrial ecosystems contain more than 26×10^{28} microbes, and these invisible constituents of our soils contain approximately as much nitrogen and phosphorus as is found in all land plants. Soil microbes are more than just a reservoir of essential elements, however, because the processes they regulate are of central importance to the productivity and health of terrestrial ecosystems. Decomposition, nitrogen cycling, and bioremediation of contaminants are only a few of the processes mediated by soil microbes, and their power to influence terrestrial ecosystems is derived from both their sheer numbers and the diverse array of biochemical reactions they can catalyze. Whether through their interactions with plants, their influence on the export of soluble soil constituents to our water resources, or their consumption and production of greenhouse gases such as CO_2, CH_4, H_2, N_2O, and NO, the activities regulated by soil microbial communities have global consequences.

Assessing Microbial Diversity

Although most microbiologists would agree that the diversity of microbial communities in soil is extraordinary, there would likely be less agreement as to how best to measure that diversity. Diversity is composed of two factors, richness and evenness, so that the highest diversity occurs in communities with many different species present (high richness) in relatively equal abundance (high evenness). There are, however, fundamental difficulties associated with determining the richness and

(continued)

Case Study (*continued*)

evenness of communities composed of microbes whose morphological traits are generally useless for identification purposes. The traditional way to study microbes is to grow and study them in the laboratory; however, this approach has severe limitations for studying soil microbial communities because less than 1% of the bacteria that are visible under the microscope grow on standard microbiological media. As a result, soil microbial communities are usually studied by examining the presence of microbial biomarkers in the soil. These biomarkers are frequently molecules such as lipids, proteins, or nucleic acids that convey information about the microorganisms from which they originate.

The DNA sequences of ribosomal RNA (rRNA) genes have proved to be a valuable biomarker for studying the diversity of microbial communities in natural environments. A technique known as the PCR can be used to make many copies of DNA from individual microorganisms without the need to grow them in the laboratory. PCR can be used in number of different ways to examine the composition of microbial communities. Other techniques use oligonucleotide probes, which are short fragments of DNA that can be labeled with either fluorescent or radioactive molecules, to target rRNA molecules from specific types of soil microorganisms and make it possible to identify and enumerate particular microbial groups in soils. These oligonucleotide probes can be used to quantify the abundance of specific microorganisms by targeting nucleic acids extracted from the soil, or they can be used to visualize individual microbial cells as they occur in the environment. In this latter technique, called fluorescent in situ hybridization, the probes are labeled with a fluorescent dye which makes it possible to selectively stain and then visualize particular microorganisms using an epifluorescent microscope.

Scale of Diversity Measurements

The phylogenetic scale (e.g., strains, species, or genera) used to measure microbial diversity is no less important a consideration than the physical scale at which diversity is measured. For example, because the sequence of the rRNA genes change very slowly over time relative to the rate at which other genes evolve, organisms with very similar rRNA gene sequences can have different ecological characteristics. As a result, studies of microbial diversity that focus solely on the rRNA genes can underestimate community richness. This does not mean that rRNA genes are a poor biomarker to use in evaluating diversity in microbial communities. On the contrary, their slow rate of change is what makes rRNA genes a useful measure of the evolutionary history of an organism. Rather, the traits that can be predicted from relationships among rRNA genes are those that take a long time to evolve, such as whether or not an organism is capable of photosynthesis or the production of methane. In contrast, the DNA sequences of protein-encoding genes generally evolve more rapidly than rRNA genes and can potentially provide a higher level of resolution for diversity studies that focus on particular microorganisms or microbial groups. Any measure of diversity based on a single gene, however, will always underestimate the actual diversity present in a soil community because organisms that have identical DNA sequences at one locus can have multiple differences in other loci. As a result, the apparent microbial diversity that is observed in a particular system is a function of the phylogenetic resolution of the method that was used to assess diversity. Different ecological questions pertaining to microbial diversity may need to be addressed at different levels of phylogenetic resolution, and so it is important to choose a particular method for assessing microbial community composition that provides ecologically meaningful results.

Effects of Disturbance on Soil Microbial Communities

Alteration of the physical or chemical characteristics of soil can lead to changes in the structure and function of microbial communities. For example, chemical pollutants when present at various concentrations can have tremendous impacts on the diversity and function of soil communities. Management changes that alter surface topography, water distribution, soil organic matter content, soil structure, and pH all have the potential to affect the soil environment and the behavior of the microbial community. For example, changes in soil topography can influence water distribution within fields, which in turn can alter the frequency of soil saturation. Changes in saturation patterns can influence soil oxygen levels and result in an overall increase in denitrification rates and the subsequent loss of mineral nitrogen from the soil.

Management approaches that alter plant community composition can also have impacts on the soil microbial community. Plants influence the structure of soil microbial communities by consuming resources and adding root exudates and dead plant matter to the soil. Microbial biomass in the soil displays a positive linear relationship with annual net primary productivity, demonstrating that microbial communities are influenced by plant-derived carbon inputs to the soil. Looking more closely, plant community composition and spatial distribution within fields can influence the composition of microbial communities, but the dynamics of this relationship and the potential for feedbacks between plant and microbial communities remain poorly defined.

Changes in the composition of soil microbial communities can have consequences for their ability to perform ecosystem services. Litter bag experiments, in which plant litter is buried in mesh bags and decomposition is monitored over time, when performed in a range of ecosystems suggest that the composition of the community in the soil where the bag is buried can influence the decomposition of plant materials both quantitatively and qualitatively. In addition, the composition of denitrifying communities in soil can influence both denitrification rates and the ratio of nitrous oxide to nitrogen gas produced by soils. Thus, if soils harbor different denitrifying communities, perhaps as a result of past management practices, then those soils can respond differently to changes in the soil environment, resulting in different amounts of nitrogen loss and different production rates for nitrous oxide, a greenhouse gas 320 times more potent than carbon dioxide, the main culprit for global warming.

Microbial succession can also impact the function of microbial communities. An examination of microbial community changes that occur in soils over a

(continued)

Case Study (*continued*)

successional gradient encompassing sites abandoned for as many as 100 years indicates that late successional microbial communities are more metabolically efficient (in regards to the amount respiration needed to support a given amount of microbial biomass) than more recently disturbed sites. As a result, late successional microbial communities should tend to store more carbon in microbial biomass than communities in disturbed or recently (<20 years) abandoned sites. Thus, when integrated across landscapes and regions, changes in soil microbial community composition can have consequences for global climate in terms of both greenhouse gas production and carbon storage.

Recovery of Soil Microbial Communities from Disturbance

Microbial communities associated with historically disturbed sites can take years to return to their predisturbance states. The Kellogg Biological Station Long-Term Ecological Research site in southwestern Michigan contains fields under a range of agricultural practices, fields that have been abandoned from cultivation, and fields that have never been cultivated. We found that the composition of the microbial communities in fields under cultivation for decades was not appreciably different from that found in fields with the same management history that had been abandoned from cultivation for 9 years. In contrast, the microbial communities in both of these treatments differed significantly from those in nearby fields that had never been cultivated. Only after abandonment for more than 45 years did fields regain microbial communities that were similar in structure to those found in fields that had never been cultivated. The differences we observed in microbial community composition with respect to management history also had an impact on the composition and activity of denitrifying communities and the composition of the nitrifying community. Furthermore, microbial methane consumption is much greater in fields that have never been cultivated than in abandoned agricultural fields even after nearly a decade of abandonment. Clearly, soil microorganisms are influenced by historical soil characteristics that are retained long after changes in land management.

The fact that microbial communities require a great deal of time to recover from the effects of cultivation indicates that soil microbial communities are sensitive to soil characteristics. Long-term cultivation of the soil can significantly deplete soil carbon and nitrogen levels and can cause major changes in the distribution of soil resources and soil structure. The depletion of nitrogen and organic matter in agricultural fields can influence microbial activity in soil for many years after abandonment, and soil carbon and nitrogen levels can impact microbial community composition as determined by a range of techniques. Recovery of microbial communities from disturbance likely depends on the recovery of soil characteristics to predisturbance levels. Because models of nutrient cycling suggest that the unmitigated recovery of soil carbon and nitrogen pools to preagricultural levels can require hundreds of years, it is likely that microbial succession in fields abandoned from agriculture is a process that can span decades.

8.2 Case Study

Role of Arbuscular Mycorrhizal Fungi in Restoration of Mine Tailings
Katarzyna Turnau, Institute of Environmental Sciences, Jagiellonian University, Krakow, Poland

Under natural conditions all living organisms interact with numerous groups of microbiota that allow them to thrive under harsh conditions. Most plant species on Earth are able to form a symbiosis with AMF. This kind of symbiosis dominates ecosystems inhabited by herbaceous plants, although it can also support the growth of trees. The AMF belong to the Glomeromycota group, which was recently separated from Zygomycetes. These fungi colonize the cortical cells of the roots, forming tree-like structures called arbuscules and form an extraradical hyphal net that strongly increases the volume of the soil available to the plant roots for water and nutrient absorption. While water and nutrients are transported to the plant from the fungus, the carbohydrates produced during photosynthesis are being transported in the reverse direction. The bidirectional transfer is facilitated by the arbuscular branches lined with the plasma membrane of the plant cell, forming the interface between the symbionts. The ability to form an extensive hyphal network penetrating the substratum is of utmost importance for soil aggregation, structuring and stability, the development of other microbial consortia, and for the establishment of diverse plant communities. The importance of this association in the restoration of damaged ecosystems is presently well recognized.

Revegetation of the postflotation wastes, containing potentially toxic metals, is one of the best examples where the use of the AMF plays the key role in successful management. Spontaneous establishment of mycorrhizal fungi in such places is a long process spanning over two or more decades, because mining affects the composition, structure, and function of the whole ecological system. Soil has to be rebuilt from the beginning, because the original substratum is devoid of microbial life and poorly suited for plant growth. Additionally, the high density and low porosity result in unfavorable air–water conditions, restricted water infiltration during rainfall, and decreased water recharge by capillary rise from deeper layers during dry periods. Such factors favor wind erosion in dry periods and water erosion during rainfalls. The lack of the vegetation cover enables the spreading of toxic metals into the surroundings, creating a serious health hazard for all components of the ecosystem, including humans.

Conventional restoration practices include grading and leveling of spoil piles, topsoiling, adding soil amendments, and introduction of plants such as grasses, which can slow down erosion and stabilize the soil. The rate of restoration and land rehabilitation may be increased by introducing mycorrhizal inoculum or, if the fungi are already present, by stimulating the growth of the mycorrhizal fungal population. The number of AMF propagules depends on many factors, such as soil nutritional status, host plant, AMF propagule density, effectiveness of AMF species, and competition between them and

(continued)

Case Study (*continued*)

Figure 8.8 Inspection of experimental plots established on zinc tailings and evaluation of seedling formation. (Courtesy of Katarzyna Turnau, Jagiellonian University.)

other soil microorganisms. There are wastes that are over 40 years old where only few arbuscular mycorrhizal plants have so far been observed (**Figure 8.8**).

Phytoremediation gives better results if the introduced plants are associated with efficient AMF isolates that tolerate pollution. Fungi isolated from polluted areas were shown to be more effective than those originating from unpolluted sites, indicating that fungi can adapt to persistent soil toxicants. In the case of many wastes, a practice called topsoiling is used. It consists of the application of soil a few centimeters deep layered on top of the waste material. This technique is, however, very costly. Depending on the origin of such soil, the AMF propagules present may not be suitable for effective growth and plant support when exposed to the toxic layer underneath, or it might take a long time for the fungi to develop tolerance to the toxic conditions and to the dryness that is one of the most important limiting factors in the revegetation of mine tailings. It is, therefore, crucial to introduce inoculum containing fungal strains that originate from old mine/industry wastes, where due to longer times of exposure to the pollution the fungal strains were subjected to natural selection. Such isolates, originating from plants growing in the wild on soil naturally enriched in heavy metals, were shown to confer heavy metal tolerance to many plant species, including maize, alfalfa, barley, and others.

To limit the transfer of toxic elements into the food chain, it is important to revegetate the area with plants that take up as little heavy metals into their shoots as possible, because such areas are often inhabited by diverse groups of animals. Fungi that originated from polluted areas were shown to decrease further heavy metal uptake into shoots of heavy metal nonaccumulating plant species. Both plants and fungi display a considerable diversity in their metal exclusion ability, even between strains or cultivars belonging to the same species. Conventional phytoremediation practices often use grass cultivars that have been grown for a long time in strongly fertilized areas and, therefore, might not be responsive to mycorrhizae. Their growth might be supported for some

Figure 8.9 *Brachypodium pinnatum*, a plant species that originated from xerothermic grasslands and was introduced into zinc tailings as seedlings preadapted to harsh substratum by a few weeks of cultivation in the mixture of soil and tailing material; mycorrhizal inoculum was supplied at the beginning of plant cultivation. (Courtesy of Katarzyna Turnau, Jagiellonian University.)

time by water supply and fertilizer treatments, but if these practices are stopped such plants disappear fast and the further existence of the plant cover depends on spontaneous establishment of other plant species. In many cases such spontaneously occurring plants could offer even faster phytoremediation than the commercially available grass cultivars. However, most of them need mycorrhiza for establishment and proper functioning. If such plants substitute the originally introduced commercial grasses, their success depends on the practices that had been used during the previous period of remediation (**Figures 8.9** and **8.10**). Certain amendments or agrotechnologies can limit the growth of AMF. The growth of nonmycorrhizal plants, or for example, sulfur amendments, may strongly decrease the number of AMF propagules in the substratum. The fungi not only have to produce spores but should be effective plant colonizers and should stimulate plant growth in the sites where inoculum is to be introduced.

AMF can also enhance phytoextraction, where fast-growing plants with high accumulation properties toward heavy metals are used to clean the soil. *Ricinus communis* and *Sonchus oleraceus* have been proposed as mycorrhizal plants suitable for phytoextraction of lead (Pb) and cadmium (Cd). Also, a number of transgenic plants have been obtained, either by transferring appropriate genes from bacteria or yeasts or by generated somatic hybridization between plants. Most transgenic plants have so far only been tested under artificial conditions, and they still need further research before the application phase begins. Also, the transformation of AMF has been approached. The identification of genes

(continued)

Case Study (*continued*)

Figure 8.10 *Primula veris* introduced into zinc wastes from xerothermic grasslands after the period of preadaptation; the photosynthetic activity of this plant is similar on both sites. (Courtesy of Katarzyna Turnau, Jagiellonian University.)

with similar functions can be very important for the understanding of the mechanisms of resistance and tolerance to heavy metals and for the selection of the fungal strains most suitable for phytostabilization and phytoremediation.

Mycorrhiza does not necessarily directly stimulate phytoextraction, but it can increase the biomass of the plants, improve soil conditions, and protect the plants from pathogens. Phytoextraction is limited by the lack of economically feasible ways to extract metals from plant material. In most cases it seems to be more practical to keep the metals in the places where they are but to decrease the health hazard that can originate from their release into the surroundings. Phytoextraction supported by AMF may have, however, the practical role in the case of precious metals such as nickel (Ni). Only recently well-developed arbuscular mycorrhiza were reported in a few hyperaccumulating species belonging to the Asteraceae family growing on nickel-enriched ultramafic soils in South Africa (**Figure 8.11**). Among them the most important for phytomining is *Berkheya coddii*, which is capable of accumulating up to 3.8% of Ni in the dry biomass of leaves under natural conditions and produces a high yield exceeding that for most hyperaccumulators. *B. coddii* inoculated with native fungi (*Gigaspora* sp. and *Glomus tenue*) had not only higher shoot biomass but also significantly increased Ni content (over two times) compared with noninoculated plants. This finding greatly contrasts with the conventional opinion that the presence of AMF reduces the uptake of trace elements if they occur in excessive amounts. This finding should be taken into consideration when designing phytoextraction practices. After several seasons the Ni content in the top layer of the soil is usually reduced and, therefore, removed to expose the layers with higher metal contents. At the same time, however, the microbial communities that had developed in the top layers are destroyed and the process is considerably slowed down and its affectivity reduced.

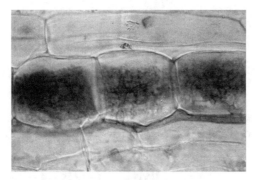

Figure 8.11 Arbuscules of mycorrhizal fungi developed within roots of nickel-hyperaccumulating plants. (Courtesy of Katarzyna Turnau, Jagiellonian University.)

Key Terms

Arbuscular mycorrhizal fungi (AMF) 193
Decomposition 176
Desertification 181
Land degradation 177
Mineralization 190
Mycorrhiza 193
Rhizobium 191
Salinization 189
Soil erosion 177
Soil organic matter (SOM) 177
Water erosion 180
Wind erosion 179

Key Questions

1. How does land degradation induce soil erosion?
2. How does wind erosion proceed?
3. What are the modes of water erosion?
4. How does desertification manifest itself?
5. How does desertification affect the living standard of people?
6. What are the main groups of soil microorganisms that are needed in establishing nutrient cycling on severely degraded lands?

Further Reading

1. Barrow, C. J. 1991. *Land degradation*. Cambridge, UK: Cambridge University Press.
2. Goudie, A. S. 1991. *Techniques for desert reclamation*. New York: John Wiley & Sons.
3. Jeffrey, D. W. 1987. *Soil-plant relationship*. Portland, OR: Timber Press.
4. Sylvia, D. M., Fuhrmann, J. J., Hartel, P. G. and Zuberer, D. A. 1999. *Principles and applications of soil microbiology*. Upper Saddle River, NJ: Prentice Hall.
5. Morgan, R. P. C. 2005. *Soil erosion and conservation*. Malden, MA: Blackwell.

9

Sand Dunes

Chapter Outline

9.1 Dune Formation
9.2 Ecological Processes
9.3 Disturbances
   Loss of Vigor
   Human Disturbances
   Invasion of Non-Native Plant Species
   Rising Sea Levels
9.4 Restoration Strategies
   Beach Nourishment
   Dune-Building Fences
   Native Plants
9.5 Long-Term Management
Case Study 9.1: Coastal Erosion at Dauphin Island, Alabama
Case Study 9.2: Ecological Effects of Sandy Beach Restoration in Northeast Norfolk, United Kingdom

Coastal sand dunes form fragile ecosystems that are frequently disturbed. Coastal dunes have a worldwide distribution. Consequently, coastal dune ecosystems show great biogeographical variation. Dunes are typically found on a relatively narrow hinterland behind the shore, but they may form continuous ecosystems along coastlines.

Coastal dunes have received considerable attention, mainly because of successional chronosequences that are easy to demonstrate. Also, their ecosystem has been described as being continuously in the phase of primary succession due to intrinsic disturbances. Coastal dunes have, therefore, served as a model system for studying primary succession. Furthermore, information derived from restoration of this ecosystem can be applicable in other severely degraded ecosystems such as desertified landscapes (including inland dunes), as outlined in Chapter 8.

Varied landscapes within sand dune ecosystems support a high biodiversity. Frequent disturbances associated with salt spray and sand instability are important factors in maintaining high biodiversity, although major disturbances such as tsunami or hurricanes are devastating.

Coastal dunes are valuable because they serve generally as a buffer against the impacts of strong wind and sea waves. For example, hurricanes typically reduce sharply in strength right after they come ashore.

Dunes are considered a valuable natural habitat, with high amenity and recreational value. The coastal dune ecosystem is especially valuable for nature conservation because this ecosystem often harbors a significant proportion of a region's particular flora and serves as a sanctuary for migratory birds.

Considerable funds are needed for coastal dune restoration. For example, the state of Florida spends about $160 million annually on such projects. Restoration of coastal sand dunes involves reconstructing the entire coastal landscape. This includes **beach nourishment** where sand is dredged offshore and deposited on the beach. Restoration programs center on the use of native coastal plants. Restoration strategies are discussed further in this chapter.

9.1 Dune Formation

Sand dune formation is affected by the proximity to rivers or sea currents with high sediment load. The net sediment transportation by sea currents away from river mouths along the shore affects sand deposition on the shores. The sand is typically made up of weathered rock minerals (quartz, silica, basalt, and feldspar) and carbonate sands of marine origin (animal skeletons). Sand dunes are generally formed where sand deposition exceeds the forces of erosion. When the sand is washed ashore its movement is affected by factors such as dune elevation, beach slope, beach width, coastline orientation, and local topography. Sand dune formation is also influenced by factors such as the embayment size and the prevailing winds.

Sand dunes form complex ecosystems, which show a variety of landscapes, including the shore, fore dunes, main dunes with windward and leeward slopes, wet dune slacks, and back dunes, with plateaus and hollows that may support grasslands, shrub lands, and even forests. Although a great variety of dune systems exist, a simplified system with only three ecozones is often used instead: the pioneer zone, the intermediate or scrub zone, and the back dune or forest zone.

Vegetation on the dunes plays an important role in dune formation and stabilization. Dune grasses are keystone plants in stabilizing drifting dunes. These grasses accumulate sand around their foliage, and their ability to grow upward through sand deposition also affects dune formation. Foliage of dune grasses slows wind erosion on the dunes that in turn increases sand accretion on the lee side of the dune. In general, the initial buildup of a dune is rapid. After it reaches an elevation of 5 to 10 m, its size usually stabilizes. The height of dunes varies according to sand supply, climate, and exposure related to local topography.

9.2 Ecological Processes

The environment on coastal dunes is characterized by strong winds, sand movements (accretion and erosion), high evaporation, salinity, and limited availability of macronutrients. A gradient is usually found such that environmental extremes diminish away from the shore. Sand dunes are, therefore, by any standard stressful habitats, and plants have evolved special strategies for survival in this harsh environment. Sand accretion on fore dunes may fluctuate to a great extent, and sand movement is considered among the most important factors that affect the distribution of plant communities on sand dunes. Exposure to salt (NaCl) and salt spray from the sea can limit the distribution of plants on dunes. Because airborne salt spray is greatest along the shore, declining further inland, only the most salt-tolerant plants grow on the shore and fore dunes. Dune plants that are able to tolerate airborne salt spray have evolved a protective wax layer on their leaves.

Soils on the dunes contain generally low amounts of macronutrients, especially nitrogen (N) and phosphorus (P), which limits vegetation growth. Nutrient input into coastal dune ecosystems mainly depends on the rate of atmospheric deposition, the abundance of free-living N_2-fixing soil microorganisms, the abundance of symbiotic N_2-fixing plant species, input by seawater (flooding and airborne spray), and organic debris (algae) that drifts ashore (**Box 9.1**). Nitrogen losses from sand dune ecosystems are attributed to denitrification, leaching, and litter removal (grazing). Phosphorus may also be a limiting nutrient on coastal sand dunes, especially at low soil pH where P can be complexed with iron (Fe)

Box 9.1 — Sources of Nitrogen Inputs into Various Coastal Dune Ecosystems

- Seawater can add little N to the ecosystem or just 0.26 g N m^{-2} $year^{-1}$.
- Atmospheric N input from dry and wet deposition has been estimated at about 1.5 g N m^{-2} $year^{-1}$.
- Nitrogen fixation by free-living soil bacteria has been estimated to be about 6 g N m^{-2} $year^{-1}$ in dune slacks.
- Nitrogen-fixing plants can add as much as such as 17.9 g N m^{-2} $year^{-1}$.

and/or aluminum (Al). In addition, the nutrient pool of sand dunes and their nutrient retention are very poor.

Succession has been demonstrated along chronosequences of sand dunes. Plant succession can take place on prograding sand dunes such as on Lake Michigan, where replacement of species was demonstrated from fore dune grasses to evergreen shrubs and bunchgrass after 100 years to a mixed pine tree forest after 345 years. The facilitation model of succession, where pioneer species improve environmental conditions for later arriving species, applies particularly to coastal sands where the substrate has not been influenced by organisms beforehand (see Chapter 4). Models of deterministic succession are particularly useful in long-term management of sand dune ecosystems to predict the patterns of vegetational changes after initial restoration efforts. Various factors influence the succession. For instance, changes in populations of soil microorganisms may influence plant succession on sand dunes. A model was proposed whereby differences in plants' susceptibility to soil pathogens were the driving factor of succession on the dunes. Changes in ecosystem functioning are more commonly demonstrated along succession on dunes. These changes are manifested as increased net ecosystem production, increased nutrient budget (N and P), and enhanced nutrient cycling.

Arbuscular mycorrhizal fungi (AMF) play an important role in the functioning of sand dune ecosystems, as outlined in Chapter 8. The AMF form a symbiotic relationship with roots of most vascular plants inhabiting coastal sand dunes. For instance, in tropical dunes of the Gulf of Mexico, 97% of the vascular plants formed AMF symbiosis. Exceptions to this high occurrence of AMF symbiosis can be found, particularly in the Arctic coastal region. In addition, species that belong to the plant families Brassicaceae and Chenopodiaceae (including their coastal representatives) generally do not form AMF symbiosis. The benefits of AMF to sand dune plants include enhancement of growth, mainly by improving nutrients (especially P), and water uptake. In field plantings of dune grasses the benefits of AMF were related to improved establishment, more vigorous growth, high number of inflorescences, and generally increased survival.

In addition to direct nutrient improvements, AMF may protect sand dune grasses against harmful soil pathogens. The AMF protection of plants against pathogens depends on the synergistic effects of soil pathogens. As a result of accumulation of species of pathogens in the rhizosphere, the balance between AMF and soil pathogens may be compromised, causing decline of vigor in dune plants.

Total AMF root colonization is usually high on sand dune grasses. Several factors appear to influence AMF colonization on roots. First, it is seasonal, typically with the highest values in the summer. Second, AMF root colonization generally increases with dune stability. The density of AMF propagules in the soil influences root colonization. The density of AMF spores may be seasonal, typically with highest values in the fall. Furthermore, density of AMF spores generally increases with dune stability.

9.3 Disturbances

Frequent disturbances on coastal dunes are common and cause ecosystem degradation or destruction (see **Case Study 9.1** on page 222). Natural disturbances include high tides, storm surges, hurricanes, and tsunamis. Hurricanes can have a great impact and even deflate coastal dunes. The natural recovery process after such drastic disturbance in the Gulf of Mexico takes about 5 to 10 years (**Figure 9.1**). Furthermore, the distribution of plants on the dunes has been linked to natural disturbances.

Loss of Vigor

Loss of vigor in sand dune grasses is a common problem and is manifested as a decrease in height and density of foliage, a reduced number of tillers, and a reduction in seed yield. Loss of vigor is often experienced in transplanted dune grasses within a few years and can result in the regression and erosion of the dune ecosystem. Loss of vigor and the consecutive die back of dune grasses are usually followed by colonization and eventual replacement by other plant species on the dunes. Various factors have been associated with the loss of vigor, such as competition between plants, plant age, soil pH, soil microorganisms, soil aeration, and reduced root growth. Competition for soil nutrients is likely to be responsible for the decline of even the most vigorous dune plants. Factors responsible for loss of vigor may vary in importance between sites. For example, loss of vigor of dune grasses represents a management problem where dune grasses are used in restoration. Loss of vigor of dune grasses, therefore, must be taken into consideration in the long-term management of sand dune ecosystems.

Figure 9.1 Impact of a hurricane on the fore dune at the Gulf of Mexico. (Courtesy of Andrew Roberts, Troy University [2004].)

Human Disturbances

Human disturbances include overgrazing by livestock, uncontrolled development, recreational activities, intentional burning, deforestation, and cultivation. The development of roads, tracks, car parks, buildings, residential buildings, and other structures near coastal sand dunes generally disrupts dune processes, particularly sand deposition on dunes and buildup of pioneer dunes. Developments on fore dunes restrict the amount of sand available to the back dunes. As a result dunes erode, and there may be a general loss of habitat. Sand extraction from the dunes for building material adds to this problem. Service pipelines, including storm water and sewage pipes that cross dunes, also create disturbances.

The major impact that recreational activities have on coastal dunes is dune erosion, caused most often by direct trampling on vegetation. Other recreational impacts include disturbance of fauna, especially birds, during their breeding season, and pollution (litter, oil, etc.) that can contaminate soils or surface water. Human disturbances on dunes can be assessed by the use of aerial photographs recording disturbance indicators such as path length, vegetation cover, and extent of bare sand. Consequently, voluntary zoning or seasonal restrictions can be implemented for certain parts of dune ecosystems. Otherwise, building formal public access ways or simply fencing off sensitive areas helps to protect the dunes. The location of access ways and protective fences needs to be chosen with consideration given to dune dynamics and human use of the dunes and beaches. Use of motor vehicles on dunes should always be prohibited. Education and awareness through signposts providing information and directions to the access ways may successfully influence human behavior on the dunes.

Sand dune vegetation is vulnerable to damage from grazing and trampling by livestock. Grazing reduces rooting depth, which may be critical in drought conditions. Trampling creates paths, which can initiate blowouts and, eventually, mobile dunes. Uncontrolled grazing of livestock on coastal sand dune vegetation has resulted in the destabilization and eventual erosion of dunes. Restoration sites need to be protected from livestock grazing, especially after initial restoration efforts, such as **transplanting,** when plants are still small and vulnerable. Protective fencing is often required to enclose restoration sites, preventing grazing by livestock and even public access.

Though uncontrolled overgrazing is almost always detrimental to coastal sand dunes, managed grazing can be used as an effective tool. For instance, where eutrophication on the dunes due to excess nitrogen deposition (derived from air pollution) has resulted in encroachment of grasses, grazing by livestock has been suggested to improve biodiversity and to maintain open landscapes on coastal sand dunes.

Invasion of Non-Native Plant Species

Sand dunes are usually characterized by open habitat that is frequently disturbed and are, therefore, prone to invasion by non-native plant species. Disturbed coastal sites are more prone to invasion of non-native species than naturally undisturbed sites. Non-native plants can alter ecosystem functioning by changing the groundwater table, by increasing nutrient status of the soil, or by out-competing native plants. In turn, this can change vegetation composition, the biodiversity and even the geomorphology of the dunes. Sand-stabilizing plants have been introduced into different parts of the world. For instance, marram grass (*Ammophila areanaria*) was introduced from Europe to the West Coast of the United States in the late 1800s to stabilize coastal sands. Today, this grass dominates most of the dunes north of San Francisco Bay, where it has altered the coastal geomorphology by creating new fore dunes. In addition, sea-dispersed rhizome fragments of marram grass colonize new coastal dunes, and it has been estimated that viable rhizomes can be transported up to 500 km along the coast in less than 2 weeks. Invasive N_2-fixing species have been introduced to stabilize drifting sand and may have a competitive advantage over native vegetation in N-limited dunes. For instance, N_2-fixing *Acacia cyclops* and *A. longifolia* (coastal wattle) were introduced from Australia to the Cape, South Africa. These species have now invaded the coast of South Africa widely and have increased the N content of the sand, which has resulted in a shift in species composition on the dunes.

Rising Sea Levels

The most serious threat to coastal ecosystems in the near future is the rise in sea level due to thermal expansion of seawater as a result of global warming and the addition of melt water from glaciers and the polar ice (**Figure 9.2**). It is currently not known how coastal vegetation will cope with this rapid rise in sea level. Some indications on how coastal ecosystems react to increasing sea levels can be found on coastlines that are retreating due to isostatic movements of land. For instance, a coastline retreat ranging from 1 to 50 m per year has been experienced on the Louisiana delta of the Mississippi River. In this case the sand

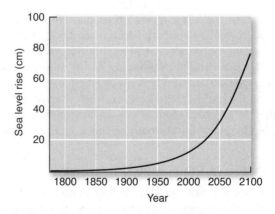

Figure 9.2 Estimation of global sea level rise in the near future.

dune grass sea oat (*Uniola paniculata*) has not been able to keep up with the pace of this rapid coastal retreat. Part of the unsuccessful colonization rate of sea oat was related to its low seed production. Global warming will negatively affect sensitive processes such as vernalization and seed production. If sea levels rise rapidly, the existence of coastal dunes is threatened, unless there is an available "escape route" for vegetation on back dunes. **Sea level rise** is also expected to result in sea-water intrusion, which, in turn, influences dune-slack vegetation. The rapid **coastal erosion** in Louisiana has also been related to decrease in the size and number of barrier islands that form the first defense against powerful sea waves and high winds.

Changes in precipitation patterns associated with global warming could also affect coastal vegetation. For example, vegetation on dune slacks on the tropical coast of Veracruz, Mexico, has shown dynamic relations with precipitation patterns. Years of exceptionally high precipitation resulted in high water tables in dune slacks, which killed trees and shrubs and increased plant cover on mobile dunes. Changes in precipitation patterns, therefore, affect coastal vegetation, and such information needs to be incorporated into restoration programs.

Elevated ambient temperatures and atmospheric CO_2 levels may also have direct or indirect effects on coastal vegetation. Elevated CO_2 levels could have potential impact on the dynamics of plant populations where competition and distribution pattern of species could be altered.

9.4 Restoration Strategies

Before restoration action a degraded site should be carefully investigated to determine appropriate restoration strategies and, at the same time, a long-term management plan (see **Case Study 9.2** on page 225) (**Box 9.2**). **Dune reconstruction** is often the first step in restoration, but it must consider the landscape of previous dunes, availability of sands, and goals of the work (**Boxes 9.3** and **9.4**). Site preparation involves reconstruction of sand dunes that have eroded. The shape of the reconstructed dunes should conform to existing nearby dunes. This can be achieved by using bulldozers, initiating a beach nourishment program, or erecting dune-building fences. Bulldozers and other heavy machinery can, however, leave behind destructive tracks on the dunes and should be used with caution.

Box 9.2 — **Planning the Restoration Work**

- When can the restoration work begin?
- What labor, machinery, and materials are required?
- How much funding is available?
- Is the restoration compatible with the desired land use?

> **Box 9.3 Factors to Consider During Dune Reconstruction**
>
> - Availability of sand (is transportation required?)
> - Type of sand required
> - Location and shape of the previous dune system
> - Location of the fore dune
> - Characteristics of any remnant dune
> - Available funds

Beach Nourishment

Where coastal sands are in short supply, beach nourishment is implemented. This involves offshore dredging and depositing the sand ashore. On average 1 km of eroded shoreline needs about 300,000 m^3 of dredged sand. The impact of dredging the sand must be monitored. A nutrient analysis on dune soils is beneficial before restoring the dunes because the sand/soil composition affects the type of vegetation that can establish. Beach nourishment of dredged sand is one method of long-term maintenance of shores experiencing erosion. This operation is expensive, however, and must be carefully evaluated. Sand that has been added via beach nourishment (i.e., adding dredged materials) needs to be treated before restoration. Treatments of the sand may include desalinization, addition of chemicals to alter the acidity levels, and addition of organic or inorganic fertilizers.

Dune-Building Fences

Using **dune-building fences** to reconstruct dunes is cost effective compared with using bulldozers, especially in remote areas. The rate of dune formation using dune-building fences, however, depends on the amount of sand blown from the beach. Moreover, dune-building fences must be supported by new fences when they become buried by sand and, eventually, by establishing dune vegetation. Dune-building fences can be erected to trap sand and reconstruct new dunes. Large dune ridges can be built up by a program of sequential fencing carried out over several years. Construction of dune-building fences depends on the dune system in question. Fences may have to be designed to incorporate adaptive management, as discussed in Chapter 14. The material for the fence should preferably be cost efficient, disposable, and biodegradable, because fences will become buried with sand. Sand accumulation is facilitated by constructing fences with a material of an optimum porosity of 50%,

> **Box 9.4**   Sequence of Logical Steps to Consider in Restoration of Damaged Coastal Sand Dunes
>
> Disturbances
>   Natural: storm surge, sand accretion or erosion, etc.
>   Anthropogenic: livestock, mining, trampling, etc.
> Reconstruction
>   Use of dune-building fences
>   Use of bulldozers
>   Beach nourishment
> Stabilization
>   Selection of keystone native plants
>   Collection of seed of native sand dune plants
>   Propagation in nurseries of clonal offsets of sand dune plants
> Transplanting
>   Seeding
>   Clonal offsets
>   Layout design of transplanted material
> Use of soil stabilizers
>   Mulches
>   Chemical stabilizers
> Fertilizers
>   Slow release
>   Fast release
>   Rate and timing of fertilizers
> Transplanting secondary species
> N_2-fixing plants
>   Shrubs and trees
> Protection
>   From grazing
>   From trampling
>   From vehicular traffic
> Long-term management
>   Control non-native invaders
>   Monitor die back of dune grasses
>   Monitor succession

and such fences can accumulate as much as 3 m of sand in 3 months. However, it is essential to follow up the dune process by transplanting dune grasses onto the newly reconstructed dune to stabilize the surface and to allow the dune to grow further. Dune grasses provide further shelter for other plant species to colonize the dunes.

Native Plants

An ecologically sound approach to dune restoration is enhanced by the use of native plants, especially where the goal is to establish a self-sustainable ecosystem. As with any restoration project, cost analysis is essential before a sand dune restoration that involves native plants. For example, the cost of different planting methods of marram grass was ranked in the order: culms with rhizomes > bundles of culms > rhizomes > seeds. Comparisons should also be made of available restoration techniques. Detailed restoration designs or management plans tend to be site specific. Determining the availability of seed of native plants and/or **clonal offsets** is also critical. Dune restoration should include the use of local genotypes of native species and should avoid any establishment of invasive non-native species, which have potential to alter the functioning of the ecosystem. Use of fast-growing cover crops to stabilize dunes can, however, be justified, though it may provide competition for native plants or enhance the establishment of invasive non-native plants. Use of native plants in restoration practices requires a comprehensive ecological understanding of the plants involved. Information on seed formation, seed germination, growth of seedlings, and biology of mature plants is needed.

The advantages and availability of plant propagules (**seed** or clonal offset) must be determined before implementation of a restoration program, because the lack of propagules of local genotypes of native plants is often a limiting factor. Furthermore, the scale of use dictates whether planting must be from seed rather than transplantation of clonal offsets. Careful planning ahead is essential to the successful use of clonal offsets, because it is difficult to hold planting stock in the nursery for several years.

Availability of seed is often a limiting factor as sand dune grasses produce generally sparse amounts of seeds and populations are maintained principally by clonal growth. To help mitigate the potential hindrance this can create, fertilizer application on the dunes can increase density of flowering spikes. For instance, a moderate application of a nitrogen fertilizer (54 kg per hectare) on sand dunes in Oregon resulted in a threefold increase in density of flowering spikes of marram grass. Similarly, seed yield of lymegrass (*Leymus arenarius*) can be expected to be 334 kg per hectare on semicultivated fields in Palmer, Alaska.

Seed collection can be accomplished by hand or by special harvesting machines (**Figure 9.3**). The latter can, however, have a detrimental impact on the sand dune environment and must be used with great care. Seed collection by hand has much less impact on sand dunes and is an ideal method for collecting seed of small populations. After the harvest seeds are dried and cleaned before storage in a dry place.

It is important during restoration to maintain genetic variability within the propagated species, and seed of locally adapted dune species should be used whenever possible. It has been suggested that seeds of local populations are preadapted to the local condition by obtaining maternal effects such as high Na^{2+} concentration in seed that improves their germination through a process called "ion enhancement."

Figure 9.3 Mechanized seed harvesting of *Leymus arenarius* on dunes. (Courtesy of Sigurdur Greipsson, Kennesaw State University.)

The use of seed of native dune grasses is effective in restoration of large areas, especially where seeding can be accomplished mechanically and where sand accretion is not rapid. Using seed could be disadvantageous, however, when germination is erratic and initial growth of seedling is slow. Seedling establishment on sand dunes is generally erratic, and this has been related to adverse water relations, lack of nutrients, and sand movements. Seed of dune grasses should be sown just below the sand surface where it can germinate and the seedling can emerge to the surface. The optimal position for germination is at a shallow depth where a buried seed can imbibe water readily and sense high diurnal temperature fluctuation (which may act as a trigger for germination). Emergence potential varies between species of dune grasses. For instance, optimal germination of seed of marram grass is at shallow depth of 2.5 to 3.75 cm. Conversely, drilling seed of lymegrass at 5 to 10 cm depth ensures good emergence of seedlings and avoids high temperature fluctuations and severe desiccation risk of the surface sand layers.

Burial with sand is a major hazard for plants on the dunes. Seeds can be buried beyond their capacity to emerge at the surface and seedlings may be unable to elongate as fast as the sand accretes. The potential of etiolation depends partly on seed mass and partly on the inherent elongation ability of the species. It is, therefore, beneficial to select relatively large-seeded populations for restoration of areas with potentially high sand accretion. Mechanized seeding can be accomplished by using an ordinary seed drill. Such operation requires clean seed of uniform size.

Seeding can be accomplished in the spring or fall. Seeding rates (i.e., kilogram seed per hectare) of dune grasses should be based on germination tests. Rapid germination is advantageous in almost all forms of restoration. Seed of sand dune grasses generally shows dormancy. However, a cold period (stratification)

> **Box 9.5** **Problems Associated with Collection of Clonal Offsets**
>
> - Lack of ready availability during the spring, when most needed
> - Unreliable quality of plant material
> - Disturbances to the surrogate dunes
> - Transplanting one or two species is usually not suitable for all sites on the dunes

pretreatment of seed can alleviate seed dormancy. Other methods of improving germination of dormant seed include treatments with plant hormones such as gibberellic acid. Surface seeding is not likely to be successful for dune grasses. After seeding is accomplished, the ground should preferably be firmed with machinery. All seeding techniques should be tested on a small scale before being implemented on a large scale.

Among the advantages of using clonal offsets are their availability and adaptability (they can be used on sites where sand is accreting rapidly or where drastic events such as flooding are expected, especially on fore dunes). The disadvantage of using clonal offsets is mainly associated with the high manpower requirement and subsequent high cost of the operation (**Box 9.5**). Clonal offsets of dune grasses can be dug up mechanically or by hand from nearby dunes that act as surrogate sites. Great care should be taken in minimizing the environmental impact of such an operation. The surrogate supply site should be as close as possible to the restoration site to decrease costs of transportation, and the whole site should be fertilized to ensure rapid regrowth of dune grasses. Clonal offsets that are dug up can be used directly on the restoration site or grown for 1 or 2 years in a nursery to increase their size before being transplanted. Establishing a mobile local nursery ensures a constant supply of plant material when needed. A clonal offset is derived from the smallest division of a rhizome that can grow into a new individual. A viable clonal offset should contain foliage with at least 15 to 30 cm of rhizome attached. Pieces of rhizomes can also be used as offsets. It is important to avoid damaging the foliage of clonal offsets during transplantation. For instance, care should be taken in storing clonal offsets in damp sand during transportation. The restoration site should be prepared by mechanically digging shallow planting trenches (20–30 cm deep) to receive the clonal offsets. After the clonal offsets are placed into the planting trenches, they are filled with firmly packed sand.

Transplanting can be carried out in most seasons except the peak of the summer. Planting can be mechanized, but transplanting by hand is preferred on steep hills or on small restoration sites. Offsets should be transplanted by spacing them up to 1 m apart in staggered rows or triangular arrays to increase efficiency of sand stabilization. The layout design of the transplanted offsets can influence dune formation: even planting on deflated sands promotes uniform sand deposition, whereas clustered planting can result in dune formation. Transplants should be fertilized to increase growth and the tillering rate.

It is important that the sand surface is stable to ensure establishment after seeding or transplantation. For this purpose mulches can be used to stabilize the dune surface temporarily. Mulch keeps the surface of the sand moist and increases the soil organic matter of the sand as the mulch breaks down. Mulches that can be used on dunes include chopped straw, peat, topsoil, leaf litter, and wood pulp. **Mulching** is particularly useful in large-scale restoration because it can be applied mechanically using a mulch spreader and is usually incorporated into the sand surface by disk harrowing.

Chemical soil stabilizers can be applied to restoration sites to stabilize the sand surface temporarily, decrease evaporation, and reduce extreme temperature fluctuation on the sand surface. Soil stabilizers are not an alternative to a restoration program but are usually applied on dune restoration sites after seeds have been sown or clonal offsets transplanted as they improve the chance of seedling establishment. Care should be taken in using soil stabilizers that could be polluting or otherwise harmful to the environment. The disadvantages of using soil stabilizers include high cost, difficulties of application, increased runoff during rainfall, high maintenance cost, and possible leaching of polluting chemicals.

Appropriate inoculation of effective soil microorganisms should be evaluated for transplanted materials. Plants raised in nurseries for restoration purposes should be inoculated with AMF. This can be achieved by preinoculating nursery-grown offsets with effective AMF isolates and transplanting them at regular intervals on the restoration site. Colonization of AMF in roots of dune plants depends on the presence of AMF propagules in the soil. Propagules are usually lacking in unvegetated coastal sands, but they can spread to unvegetated sites via passive means (i.e., wind, water, animals, and insects). This natural process of AMF dispersal, however, can be very slow and erratic. An important factor in the inoculation process is the number of species of AMF to be used. In general, several species of AMF are found in any particular sand dune ecosystem. Native AMF isolates containing a variety of AMF species are, therefore, recommended for introduction on restoration sites. However, the effectiveness of AMF isolates must be carefully tested. Different AMF isolates were found to vary in their ability to enhance growth of sand dune grasses and should, therefore, be screened before they are considered for inoculum development. Use of exotic AMF inocula in dune restoration should be treated with caution.

9.5 Long-Term Management

Restoration efforts on coastal sand dunes should always be followed by **long-term management** (**Box 9.6**). This involves maintenance of vigorous growth of the dune grasses by judicious use of chemical fertilizers. It is important for economic reasons, however, to evaluate in a small-scale trial the responses of

> **Box 9.6** **Long-term Management of Sand Dunes**
>
> - Replanting sites where previous establishment failed or die back is experienced
> - Applying fertilizer when required to enhance growth and seed production
> - Transplanting shrubs and trees on the dunes to facilitate succession
> - Controlling invasive non-native plants

particular dune plants to different combinations and rates of fertilizer before its large-scale application. In this case adaptive management should be used (see Chapter 14). The type of fertilizer used depends on the growth response of the species in question. For instance, sand dune grasses generally require high proportion of N, but lupines, shrubs, and trees require a higher proportion of P. Appropriate application rates vary according to species, location, and season.

Timing of fertilizer application is important because nutrient retention is generally low on sand dunes and rapid-release fertilizers can leach away quickly. Slow-release fertilizers have the advantage of releasing the chemicals over a relatively long period of time. Slow-release fertilizers are usually less cost effective than the ordinary rapid-release types. Fertilizers can also enhance establishment of vegetation on dunes. However, fertilizers should be used with caution, because extensive application generally results in greater biomass productivity of the grasses, reduction in biodiversity, and enhanced establishment of weeds and non-native species. Long-term management of dune ecosystems, therefore, needs to consider the effects of fertilizers on the interactions between species and, hence, on the succession trajectories. Fertilizers are usually applied at the same time or immediately after clonal offsets are transplanted or seeds are sown to ensure high survival and vigorous growth of plants. Addition of N fertilizers is essential for establishment of dune grasses. For this purpose, high rates of fertilizers are usually added. For example, application of 560 to 680 kg per hectare of complete fertilizer (20-20-10, NPK) enhanced growth of newly transplanted clonal offsets or seedlings in Alaska.

Sequential introduction of species to sustain succession is needed in dune restoration, because colonization by native plants can take a considerably long time (**Box 9.7**). The restoration of coastal dunes usually involves programs where species are sequentially introduced on the dunes to promote acceptable successional trajectories. The establishment of sand-stabilizing grasses is only the first step. For example, in Oregon an initial planting of marram grass was followed after 1 year by transplantation of broom (*Cytisus* sp.) and a year later by transplanting native coast pine (*Pinus contorta*) onto the dunes. Shrubs and trees are usually planted as container-grown plants raised in nurseries.

> **Box 9.7 Assembling Coastal Dune Ecosystem in South Australia**
>
> - Use stabilizing native dune grasses, *Spinifex hirsutus* and *Ehrharta villosa* (veldt grass)
> - Transplant N_2-fixing species, *Lupinus digitatus*, *Lupinus arboreus* (tree lupin), *Carpobrotus cyanophylla* (pigface), *Chrysanthemoides monilifera*, and *Acacia sophorae*
> - Transplant shrubs and trees, *Leptospermum laevigatum* (tea tree), *Banksia integrifolia* (coast banksia), *Casuarina stricta*, and *C. equisetifolia* (horsetail oak)

Long-term management of restoration sites involves controlling and eradicating non-native species. This involves cutting or uprooting the non-native plants, burning the cuttings, and spraying new seedlings with herbicide, as discussed in Chapter 7.

Long-term management of restored coastal dunes requires monitoring of vegetation cover to avoid deterioration of vegetation and erosion of the dunes. Vegetation cover on coastal sand dunes can be monitored using false-color imagery by remote sensing. To standardize these images, vegetation cover is measured manually in small fixed plots and compared with color density ratios of the corresponding plots on the false color images. The false-color images are then converted into values of vegetation cover according to color density ratios. False-color imagery is sensitive enough to differentiate between cover of different plant species. Changes with time in vegetation cover have been studied by integrating remote sensing and geographical information systems. Furthermore, geographical information systems can be used to assess vegetation succession on the coastal dunes.

Summary

The most serious threats to coastal sand dune ecosystems in the near future are sea level rise and increased impact from recreational activities. Coastal dunes are valuable as a defense against sea surge and strong winds. They harbor a high biodiversity and form an important habitat for migratory birds. Restoration of deflated coastal sand dunes should consider the complex landscape of the ecosystem. The first step in restoration is often reconstruction of the dunes involving beach nourishment. The goals of dune restoration are to stabilize dunes by establishing plant cover and to build a self-sustaining and diverse ecosystem. Native plants are preferred in the restoration work. Seed or clonal offsets are commonly used, depending on the scale of the problem. Beneficial soil microorganisms are critical for successful dune restoration. Various techniques are used to manipulate succession on coastal dunes. Sequential introduction of keystone species is used successfully in dune restoration.

9.1 Case Study

Coastal Erosion at Dauphin Island, Alabama
Andrew Roberts, U.S. Army Corps of Engineers, Regulatory Branch, North Section Louisville District, Louisville, KY

Coastal dunes are considered fragile ecosystems that are highly susceptible to degradation. Many factors can lead to instability and degradation within a dune ecosystem. These include strong winds, sandy texture of soil, abrasion by sand, sand movements (accretion and erosion), low moisture with high evaporation, salinity, low nutrient availability, human developments, and natural disturbances such as storm surges, hurricanes, and flooding. Combination of these factors can also have synergistic effect on coastal erosion. In addition, the inevitable rising of sea levels due to global warming is a factor that can cause worldwide disturbances of shoreline erosion. Vegetation plays an important role in the formation and stabilization of coastal sand dune systems. Even though coastal dunes are considered extremely stressful for vegetation, a variety of plant species has adapted to this environment.

Coastal erosion is a serious problem on the Gulf of Mexico where about 61% of the shoreline is eroding at an estimated cost of almost $1 billion a year. Coastal erosion in this region is usually caused by a combination of factors such as land subsiding, reduced onshore sediment transport, sea level rise, destruction of barrier islands, and hurricanes. Hurricanes have a profound effect on the coast of the Gulf of Mexico. For instance, the coast of Alabama was severely degraded by Hurricane Ivan on September 17, 2004 (**Figure 9.4**). In spite of compounded impacts by hurricanes and sea erosion, coastal ecosystems can show tremendous resilience. The distribution of plants and the status of AMF was examined across the coastal dune ecosystem of Dauphin Island.

Dauphin Island is a barrier island about 4.5 km from the Alabama mainland and 56 km south of Mobile, Alabama. The island is about 24 km long and surrounded by Mobile Bay, the Gulf of Mexico, and Mississippi Sound. The soil is composed of fine to medium sand with a median diameter ranging from 0.29 to 0.37 mm. Dauphin Island has experienced severe beach erosion over the past 100 years and has shown some of the most dramatic shoreline recession in the United States in the past 30 years (**Figure 9.5**). Loss of sand from the coastline appears to be mainly a result of dredging activities for large cargo ships entering Mobile Bay. Hurricanes have a profound effect on the coast of this region, and the shore at Dauphin Island was decimated by category 3 Hurricane Ivan and the fore dunes were severely damaged. In particular, the coastal ecosystem was studied at the Audubon Bird Sanctuary (approximately 30° 14.7' N, 88° 5.1' W) on the eastern end of Dauphin Island. This 66.4-hectare area is considered to be the only preserved coastal dune system on the island. Most of the island is developed for homes, and the ecosystem is degraded from its original status.

In coastal dune systems around the Gulf Coast of the United States, sea oats (*Uniola paniculata* L.) and species in the families of Aizoaceae, Cruciferae, Euphorbiaceae, and Poaceae are considered the dominant plant species of the shore and fore dune

Case Study 223

(a)

(b)

Figure 9.4 Effect of Hurricane Ivan on the coast of Dauphin Island, Alabama. **(a)** A portion of the island's beach on July 2001. **(b)** The same beach on September 17, 2004 immediately right after Hurricane Ivan came ashore. The white arrows, pointing to two properties that were there before and after the hurricane, provide reference points for the damage done by beach erosion. (Reproduced from A. B. Tihansky, *Sound Waves: September, 2005* [http://soundwaves.usgs.gov/2005/09/fieldwork2.html]. Accessed June 22, 2010. Photos courtesy of USGS.)

ecozones. In addition, *Heterotheca subaxillaris* (Lam.) Britt. & Rusby, *Rubus trivialis* Michx., and *Panicum amarum* Ell. are associated with the shore and fore dunes. The main dunes include sparse covering of slash pines (*Pinus elliottii* Engelm.), scrub and live oaks (*Quercus virginiana* P. Mill.; *Quercus geminata* [Small], Sarg.), woody goldenrod (*Chrysoma pauciflosculosa* [Michx.], Greene), rosemary (*Ceratiola ericoides* Michx.), conradina (*Conradina canescens* Gray), and sun rose (*Helianthemum arenicola* Chapman). In the back dunes in addition to the previous species of the main dunes, saw palmetto (*Serenoa repens* [Batr.], Small) and sand cactus (*Opuntia humifusa* [Raf.], Raf.) can be

(continued)

Case Study (*continued*)

Figure 9.5 Coastal erosion of the shore of Dauphin Island. (Courtesy of Andrew Roberts, Troy University [2004].)

found. Pines such as *Pinus elliottii* and denser vegetation such as *Opuntia humifusa, Serenoa repens,* and *Quercus geminata* succeed the shrub stage and precede the climax maritime forest dominated by *P. elliottii* and *Quercus virginiana.* An evergreen maritime/oak forest is the climax community on Dauphin Island and is common throughout the Gulf Coast and on large Gulf Coast barrier islands.

Distinct ecozones could be identified along the coastal ecosystem at Dauphin Island: shore, fore dunes, main dunes, and back dunes. Abundance of AMF was examined in these ecozones throughout the season. In addition, the mycorrhizal infectivity potential (MIP) of the dune soil was examined by growing corn in dune soil for one month. The AMF root colonization of the corn was then assessed and used as an index of MIP.

The majority of plant species (73%) examined belonged to families that typically host AMF species. Unusually high AMF root colonization in typically non–AMF-dependent plants was found. Significant differences were not found among ecozones using the MIP test. Numbers of spores were significantly higher in the main dunes and lower in the fore dunes. Unusually high numbers of spores were found in the shore samples. These results suggest that the unusually high values of AMF root colonization, MIP, and spore densities of shore plants could be due to sea erosion; the shore was recently a fore dune and a high number of AMF propagules probably still exist in the shore sand.

After the coastal devastation by Hurricane Ivan, a coastal restoration will aim at stabilizing the coastal dune system by increasing vegetative cover. Soil microbial composition, sand mobility, and beach nourishment must be taken into consideration. Restoration of fore dunes on Dauphin Island should include the use of native dune-stabilizing grasses such as planting nursery-grown plants preinoculated with local AMF at regular intervals across the system to provide desired effects in fore dune stabilization and revegetation. It is important that sand-stabilizing grasses are preinoculated with appropriate AMF before outplanting.

This study showed that AMF should not be a limiting factor in restoration of the dunes. Recolonization of the fore dunes will depend on propagules of plants from the main and back dunes. The resilience of the fore dune is consequently modified by the fact that it shares plant species with other ecozones such as the back dunes. Because recolonization of the shore depends on seed dispersal from other intact coastal ecosystems, conservation of nearby coastal areas is important as well. Restoration management of coastal sand dunes in view of shoreline erosion should include conservation of the whole system, including the back dunes and neighboring coastal ecosystems.

9.2 Case Study

Ecological Effects of Sandy Beach Restoration in Northeast Norfolk United Kingdom

Martin Perrow, Ecological Consultancy Limited, Norwich, UK, and Environmental Change Research Centre, Department of Geography, University College London, London, UK

The effects of global climate change manifested as an increase in sea levels and generally increased storminess has meant that the rate of coastal erosion has increased dramatically in many areas of the world. Along the seaboard of eastern and southeastern England, change is also exacerbated by the downward tilting land at about 2 mm per year. Norfolk has seen rapid erosion (at an average of over 1 m a year) of the soft cliffs (composed of glacial sands and clays with chalk) typical of this coastline and the sand and shingle beaches seaward of them (70% are in retreat) over the last 20 years or so. This has resulted in the loss of many houses and much agricultural land as well as beach habitat used by people and wildlife, particularly shorebirds. Moreover, storm surges and the incursion of seawater threaten the unique wetland system of the Broads National Park (see Case Study 13.3) as well as the numerous nature reserves of national and international interest including the coastal dune systems and lichen and moss dominated heath of the Winterton Dunes candidate Special Area of Conservation (cSAC)[1].

Part of the strategy for coastal defense in the region implemented by the Environment Agency (the responsible statutory authority) was the installation of nine offshore rock reefs over a 5-km stretch of coast between Happisburgh and Sea Palling. These were designed to dissipate the wave power hitting the beach and seawall and promote

---

[1] A European conservation designation for habitats and species.

(continued)

Case Study (continued)

Figure 9.6 Offshore reefs and the beach after recharge. (Courtesy of Mike Page Aerial Photography.)

retention of beach material. Four reefs were completed by July 1995 with a further five completed by July 1997. After reef construction the beach was recharged by sandy material pumped from a few hundred meters offshore along a pipe to "kick-start" its development (**Figure 9.6**). The recharged material was then distributed and sculpted by machinery. A total of 300,000 m^3 of sand was recharged after the installation of the first four reefs with a further 1.3 million m^3 2 years later after the next five reefs (**Figure 9.7**). After the subsequent erosion of material to the south of the reefs, rock revetment and replacement of two groynes was followed by recharge of a further 900,000 m^3 of sand in early 2000. Groynes help stabilize the beach and promote the capture of material transported during long-shore drift. A further 480,000 m^3 was added to this area in early 2003 followed by replacement of seven groynes in January through April 2004, in two groups of two and five adjacent groynes, respectively, the latter also accompanied by further recharge (176,000 m^3).

Focus of Ecological Monitoring

Monitoring of the ecological impact of the works and restoration of the beaches was funded by the Environment Agency, managed by Halcrow Group Ltd. and undertaken by staff at ECON Ecological Consultancy (**Figure 9.8**). This was focused on breeding birds, particularly little tern (*Sterna albifrons*) and ringed plover (*Charadrius hiaticula*). These are the European equivalents of the very morphologically similar least tern (*Sterna antillarum*) and semipalmated plover (*Charadrius semipalmatus*), respectively, in the United States. However, with its

Case Study   227

Figure 9.7 Redistribution of recharged sand by heavy machinery. (© Ivan Cholakov Gostock-dot-net/ShutterStock, Inc.)

Figure 9.8 Dr. Martin Perrow, founder and director of ECON Ecological Consultancy Limited. (Courtesy of Eleanor Skeate, ECON Ecological Consultancy Limited.)

preference for nesting on coastal beaches in England, ringed plover occupies a more similar ecological niche to western snowy plover (*C. alexandrinus*) or piping plover (*C. melodus*).

The focus on little tern and ringed plover was largely a result of their conservation importance. Little terns have an unfavorable declining European conservation status

(continued)

Case Study (*continued*)

and are a conservation priority in the United Kingdom where they are in long-term chronic decline, reducing by some 27% between 1985 and 1987 and 2000. Little terns had been known to nest in the Winterton area, some 11 km south of the reefs, since 1919. In 1967 the colony of 90 pairs was the second largest in Norfolk and one of the largest in the United Kingdom. Unfortunately, with increasing human pressure the colony declined, and by the early 1980s less than 10 pairs were present, reducing to only sporadic occupation in the 1990s. Despite this, Winterton was included within the Great Yarmouth North Denes Special Protection Area (SPA)[2], the only one of its kind in the United Kingdom for little terns, which until recently regularly held over 200 breeding pairs, which was more than 10% of the U.K. total and around 2% of the European population. The North Denes colony, some 12 km to the south of Winterton (and, thus, nearly 25 km from the northernmost reef), is intensively protected (by 24-hour wardening and electric fencing to deter ground predators) and managed by the Royal Society for the Protection of Birds.

Although classed as being of favorable conservation status in a European context, the ringed plover is of concern in the United Kingdom, with a 25% to 50% decline in the wintering population. This "amber" concern status is mirrored by the marked declined in the breeding population (totalling some 8,540 pairs) in some areas, notably Norfolk (where it may have declined up to 64%) in the last 25 years. Ringed plovers nesting on the coast are highly vulnerable to disturbance by people and their dogs, and it is in coastal habitats where the species is in decline, contrasting with the expansion of their breeding range to include the edges of gravel pits and reservoirs and further use of wet meadows.

To compensate for the lack of control data before beach restoration, monitoring was undertaken from north to south over the restored beach (5.1 km in length) associated with reefs, the later (2000) restored length of beach (5.6 km), and a natural control length of beach (3 km) covering the extent of Winterton Dunes. A further year's data were added in 2004, which also had the goal of monitoring the impact of replacement of seven groynes followed by further beach recharge in the middle section. Neither the northerly section associated with the reefs nor the southerly control section was affected.

Monitoring of nesting shorebirds was undertaken by eight walkover surveys of the entire beach front (12.3 km) in the breeding season from May to August over a 4-year period from 1997 to 2000 (i.e., after all reefs were completed). An attempt was made to follow the progress of all nests of ringed plovers throughout the study area. However, not all nests of little terns could be monitored individually because this species nests in loose colonies, which are generally enclosed by fencing erected by staff from English Nature (the statutory conservation organization). In this case monitoring of a subsample of nests was undertaken under license.

Analysis of the effects of beach restoration using heterogeneity chi-square tests between sections and between years could be undertaken on ringed plovers, which nested throughout the study area in reasonable numbers (16–28 pairs from 1997 to 2000). Tests were performed using a comparison between the expected and actual observed numbers of pairs present, nests, eggs laid, chicks hatched, and chicks fledged,

[2] A European conservation designation specifically for birds.

after these had been adjusted for the length of beach in the different treatments. For little terns, which nested sporadically and patchily in colonies, only a more subjective assessment of general population trends and the distribution of colonies could be made.

A record of all other birds, seals (both common [harbour] seal [*Phoca vitulina*] and grey seal [*Halichoerus grypus*]), and other wildlife of interest was maintained throughout. The number and locations of all people and dogs was also mapped throughout the study to assess the levels of use of the beach and the potential effects of disturbance.

Monitoring of vegetation communities was also conducted through mapping of general vegetation types over the entire study area and the detailed sampling (% cover of all species) in 2 × 2-m quadrats in each of the four possible habitats across the beach dune profile (strand-line, fore dune, mobile dune, and fixed dune) along six equidistant transects in each of the three study sections ($n = 18$ transects and $n = 65$ quadrats). Further mapping of communities at Winterton Dunes using National Vegetation Community classification was also undertaken as there was concern that the recharge of alkaline sand could potentially impact on the acidic dune communities after the redistribution of some material on the wind.

Short-Term Response of Vegetation to Beach Restoration

Just 1 year after beach restoration, large patches of sea rocket (*Cakile maritime*), a pioneering herbaceous flowering annual plant of the strand-line, became established behind the northernmost reef. This appeared to assist the colonization of other species, including the dune-building grass sand couch (*Elytrigia juncea*) and the herb *Atriplex* sp., which initiated the formation of embryonic dunes below the sea wall. Colonization of plants was reflected in the highly significant decrease in the proportion of bare sand in quadrat samples. Further erosion of the beach continued to limit any sort of strand-line vegetation in the area immediately to the south of the reefs in this period. Contrary to the explosive recolonization behind the reefs, even after recharge of the area in 2000, little vegetation developed, perhaps illustrating that suitable inocula may be patchy in time and space and recolonization patterns may be difficult to predict.

Substrate samples taken from each quadrat revealed that the recharge of sand from offshore to the section behind the reefs significantly raised the pH of the sand along the whole beach front, including the control section several km to the south. This indicated that the sand was rapidly redistributed by the wind. Not only was the change marked, with an increase from pH 8 to 8.5 and pH 7.5 to 8 in the different sections, it persisted for 1 to 2 years before declining back to normal levels. Interestingly, there was also a significant decrease in the pH of substrate in the dunes in section below the reefs. This was best explained by the continued erosion of the beach, which allowed the incursion of waves to the seawall, thereby exacerbating sea spray onto the dunes beyond. Absorption of CO_2 on frequently wetted substrates and plants could conceivably have led to weak solutions of carbonic acid (H_2CO_3), thereby reducing the pH. Recharge of this section of beach in 2000 saw the resumption of more normal pH values in the dune behind the seawall.

Fortunately, there was no evidence of systematic change in pH of the dune systems in the control section at Winterton, which remained at natural, quite acidic (pH 4–5)

(continued)

Case Study (*continued*)

levels throughout the study. This suggested that although sand may be readily distributed along the beach, the interchange with the wider dune system is either not as great or simply diluted by the large amount of dune present. The plant communities of this valuable system showed no evidence of any systematic change beyond natural fluctuation.

Response of Little Terns to Beach Restoration

Breeding little terns lay up to three eggs in sand scrapes. They are perceived to prefer nesting amongst patchy relatively large stones in loose, small (5–15 pairs) colonies. However, in real terms they are less fussy about the type of nesting substrate than the prey supply of small 30- to 90-mm fish (most importantly young-of-the-year herring [*Clupea harengus*] and sprat [*Sprattus sprattus*]) as well as invertebrates within immediate reach of the colony (<6 km). A combination of patchily distributed food supplies in areas of reduced human disturbance, typically through deliberate protection in nature reserves, has led to the concentration of little terns in fewer, larger colonies.

During the 1990s there was a rapid increase in the number of pairs attempting to nest in the Winterton area from none in 1993, 2 in 1994, 6 (raising three young) in 1996, 14 in 1997, 16 in 1999, to 45 in 2000. High tides caused the loss of nests in 1997 and 2000, but otherwise predation and/or disturbance by people and their dogs causing losses at the egg stage were implicated in the failure of nesting birds. However, in 2002, 127 pairs fledged at least 58 chicks, with this success paling into insignificance in relation to the 233 pairs and 447 young fledged in 2003, which was the largest production of chicks in a single colony in the United Kingdom since records began in 1969. This astonishing resurrection of the Winterton colony was only possible as a result of the exceptional recruitment of young herring, particularly in 2003. The switch away from North Denes was thought to be the result of vandalism, which destroyed 98 nests in early season in 2002, with disturbance by helicopters in 2003, leading to only 10 nests being put down.

The changing fortunes of little terns in the area, however, were also reflected by their use of the restored beach and specifically the area behind the northernmost reef (the Eccles colony). As early as 1998, just 1 year after completion of recharge, birds were observed prospecting. Two pairs were observed nest scraping in 1999 but were not proved to nest, and then in 2000 a pair raised two chicks. In conjunction with the success at Winterton, Neil Bowman, the voluntary warden who "protected" the colony with a simple pole and string fence, reported that 11 pairs raised 12 chicks in 2002, with 37 pairs raising 58 chicks in 2003. The number of pairs increased further to 47 pairs in 2004. Although the Winterton colony of 150 nests failed as a result of mass abandonment due to lack of suitable prey, this was not the case with the Eccles colony, with chicks successfully hatching. This suggests prey was available around the offshore reefs, which offers a new habitat to be exploited by a range of fish and invertebrates that may fall prey to little terns. Unfortunately, the hatched chicks appear to have been systematically predated by kestrels and none fledged. In 2005 the 36 nests failed as a result of a violent storm coupled with predation by corvids (see below), again mirrored by abandonment at Winterton (83 nests).

Although the Eccles colony has failed in the last few years, this is not unusual for little terns, which typically exhibit widely varying annual fortunes. Far more important is the fact that a new offshoot colony has become established within the vulnerable northeast Norfolk population. This helps spread this most important of the U.K.'s subpopulations between several sites, which is perhaps a more natural reflection of what is the typical nesting pattern of this species—in smaller colonies, including on newly created habitat on a morphologically dynamic coastline. That this has also been on a restored section of beach has also illustrated the potential for creation of new nesting habitat for little terns in other areas of the United Kingdom.

Response of Ringed Plovers to Beach Restoration

Ringed plovers are even more adaptable nesters than little terns nesting directly on sand, among different sized shingle, on fine shingle itself, and even among vegetation in the dunes. Ringed plovers are known to return to their breeding territories from mid-February onward and begin nesting in early May. Birds may re-nest after the loss of a clutch of young chicks or even be double-brooded, which means nests may still be present in mid-July. With beach recharge extending over virtually the entire season in 1997, with heavy plant (machinery) redistributing sand after recharge, a short-term response was anticipated. In fact, although the number of pairs (four compared with six) and number of nests (five compared with nine) was lower than expected, this was not sufficient to cause a significant difference between sections over the longer period. Territorial pairs appeared to attempt to compensate for the disturbance by a high proportion of birds (41%) nesting away from the beach itself above (and even on) the seawall and in the dunes, compared with the usual values of 3% to 4% adopting this strategy in other years. Other means of compensating for disturbance were seen in 2000, when although the expected three pairs of birds held territory in the recharged section, only two pairs attempted to breed, with one nesting on the seawall and the other delaying breeding until June, long after the works had been completed. In 2004 additional monitoring in March and April before the nesting season showed that after initial territory formation, 43% of pairs then disappeared for a while, perhaps put off by the presence of machinery, only to return (or be replaced by other pairs) when the works were complete. Nesting then commenced within a few days. The short-term effects of beach restoration are thus subtle and the response of individual pairs to disturbance by plant and humans may be variable, masking the overall impact.

The medium-term effects, however, are graphically illustrated by the outstanding success of the birds nesting on the restored beach behind the reefs within a few (1–3) years. Here, 12 chicks fledged in 1998, threefold more than had fledged from the entire study area in 1997. By 2000, 28 chicks had fledged, some 65% more than was expected according to the length of beach. Even more remarkable is that some 20% of the restored beach behind two of the reefs closest to the access point onto the beach from the village of Sea Palling was not available to the birds as a result of the intense use by people, particularly later in the season. This agreed with the findings of Durwyn Liley (from the University of East Anglia) in North Norfolk who also showed birds simply avoided nesting in areas heavily disturbed by people.

(continued)

Case Study (continued)

As a significantly greater number of chicks hatched and subsequently fledged from the restored beach (which also significantly reduced the number of eggs laid as there was no need for replacement clutches), the rate of predation of either eggs and/or chicks was clearly low, at least in the early years. Predation of nests, especially by mammals including stoats (*Mustela erminea*), hedgehogs (*Erinaceus europaeus*), and red foxes (*Vulpes vulpes*), may be high, with 33% of all nests lost in Liley's study. Applying the findings from this study to our own, it seems that the significantly wider (and higher) restored beach behind the reefs provided a greater range of nest locations, reducing the chances of egg predation especially by stoats and hedgehogs. The recolonization of vegetation after restoration of the beach is also likely to have had the beneficial effect of providing cover for chicks from all manner of predators, including avian ones such as gulls and especially kestrels (*Falco tinnunculus*), which are also effective predators of little tern chicks (see above). When very small, chicks typically freeze at the approach of a predator (or human), crouch, and rely on camouflage; when slightly larger they are particularly adept at bolting for cover where they shelter beneath vegetation.

After increased production of young plovers and/or incursion of birds from elsewhere, there was a rapid increase in the number of pairs using the study area, from less than 20 pairs in 1997 and 1998, 28 in 1999 and 2000, to 44 pairs by 2004. This was explained by an increase in available breeding habitat, particularly as a result of the increase in beach width, which Liley had shown allowed a greater density of birds per unit length of beach to become established. Until the recharge of the beach to the south of the reefs in 2000, the width (and height) of the restored beach behind the reefs had been significantly greater and, thus, able to support a greater density of birds. After restoration of this section too, it may be that the full potential of the study area had finally been reached.

Despite the success of beach restoration in the medium term, in the longer term there was a worrying decline in the success of ringed plover pairs. In 2004 just 27% of clutches laid actually hatched chicks, with the loss of 89% of all eggs laid or chicks within a few days of hatching. Although this could simply be described as a poor season, perhaps as a result of climatic factors or food shortage, there was some evidence that this was indirectly linked to the success of beach restoration. In simple terms, human use of the beach frontage, especially the restored sections, escalated enormously, with around double the numbers of people encountered over the season in any other year (**Figure 9.9**). On just one date in 2004 over 1,300 people and 84 dogs were recorded. It seems that although birds had become increasingly clustered in the less disturbed sections of beach, losses of nests were still high. Indeed, in Liley's study 8% of all nests were still trampled, although this was predicted to be much higher (23%) if birds had distributed randomly. In our study there was also evidence of an increase in the rate of predation of nests from a novel set of opportunistic predators/scavengers encouraged by the increased presence of humans and their rubbish. Corvids, including carrion crow (*Corvus corone*), magpie (*Pica pica*), rook (*Corvus frugilegus*), and jackdaw (*Corvus*

Figure 9.9 Beaches are a popular destination for holiday-makers. (Courtesy of Sigurdur Greipsson, Kennesaw State University.)

monedula), were frequently recorded from the beach in 2004, whereas they had never been seen during three previous seasons of surveys. Crows and magpies in particular are notorious egg thieves and were observed to predate nests of both ringed plovers and little terns (see above) by warden Neil Bowman.

As a result of intense human disturbance, the future of ringed plovers in the study section now appears to depend on the protection of suitable areas, as has become the "normal" situation for little terns (see above). Indeed, ringed plovers now appear to actively select protected areas. However, this is something of a double-edged sword, as predators including crows and kestrels may also become focused on these concentrations of birds. The long-term solution for ringed plovers at least may be to protect a relatively large number of sites along the beach to allow nesting over a wider area, which has the added benefit of reducing the concentrated impact of predators. A similar strategy may be used to mitigate "in-kind" for habitat temporally affected by further recharge or other works to protect and restore the beach, when this is undertaken in the breeding season.

Response of Seals

Although a small number of both common and grey seals appear to have used the beaches to haul out for some time, grey seals began to use the section of beach immediately to the south of the reefs as a breeding site as recently as 1993 when two

(continued)

Case Study (*continued*)

pups were born in the breeding season of November to December. The numbers increased steadily in subsequent years with 6 in 1994, 6 in 1995, 15 in 1997, and 17 in 1998 and rapidly thereafter with 52 in 2002, 63 in 2003, 78 in 2004, and 106 in 2005. John Heseltine, the warden, suggests the seals prefer the new rock groynes installed in 2004 because they afford greater protection from the sea and windblown sand. It thus appears that beach recharge is also likely to have a positive effect on the breeding success of seals.

Conclusions

The recharge of the beaches and other works, such as the installation of offshore rock reefs and the replacement of groynes in northeast Norfolk, was effectively driven by the needs of people and the defense of property and land. Despite some short-term disadvantage for nesting ringed plovers where work has been undertaken in the nesting season and an increase in pH of the beach, such effects were temporary and easily outweighed by the medium to longer term benefits for vegetation development and the breeding performance of ringed plovers, little terns, and grey seals. For little terns in particular restored beaches have provided a further nesting site for the most important population in the United Kingdom.

Little terns also appear to have benefited from the new marine habitats provided by the offshore reefs, which appear to support small fish and other prey species for terns. The reefs have also encouraged other bird species such as turnstones (*Arenaria interpres*) on spring passage, none of which was seen in 1997, with 9 in 1998 and culminating in 43 in 2000 and 34 in 2004. The reefs are also now one of the handful of sites that support wintering purple sandpipers (*Calidris maritime*) in Norfolk.

Such has been the success of the scheme for humans, however, shown by the huge increase in visitors to the beaches in the study period, that breeding birds are becoming more and more confined to protected areas, a situation that will require careful management in the future if the positive aspects of beach restoration for little terns and ringed plovers especially are to continue.

Key Terms

Beach nourishment   207
Coastal erosion   213
Clonal offset   216
Dune-building fences   214
Dune reconstruction   213
Long-term management   219

Loss of vigor   210
Mulching   219
Sea level rise   213
Seed   216
Transplanting   211

Key Questions

1. Describe the main landscape features across coastal dune system.
2. Describe environmental gradients from the shore to the back dunes.
3. What are the main disturbances that coastal dunes face?
4. What are the advantages and disadvantages of using seed or clonal offsets in restoration of coastal dunes?
5. Describe long-term management and aftercare program of restored coastal dunes.

Further Reading

1. Brown, A. C. and McLachlan, A. 1990. *Ecology of sandy shore.* New York: Elsevier.
2. Carter, R. W. G., Curtis, T. G. F. and Sheehy-Skeffington, M. J. 1992. *Coastal dunes.* Lisse, The Netherlands: Balkema.
3. Jefferies, R. L. and Davy, A. J. 1979. *Ecological processes in coastal environments.* Malden, MA: Blackwell.
4. Martines, L. M. and Psuty, N. P. 2004. *Coastal dunes: ecology and conservation.* New York: Springer.
5. Maun, M. A. 2009. *The biology of coastal sand dunes.* Oxford, UK: Oxford University Press.
6. Packham, J. R. and Willis, A. J. 1997. *Ecology of dunes, salt marsh and shingle.* New York: Springer.

10

MINES AND POLLUTED SITES

Chapter Outline

10.1 Mine Waste Restoration
  Surface Stabilizers
  Metal-Tolerant Genotypes
  Stockpiled Soil
  Commercial Grasses
  Native Plants
  Monitoring
10.2 Phytoremediation
  Hyperaccumulators
  Metal Chelates
10.3 Bioremediation
  Impeding Factors
  Requirements for Bioremediation
  Landfarming
Case Study 10.1: Restoration of Gold Mines in Ghana, West Africa
Case Study 10.2: Phytoremediation of Lead-Contaminated Soil
Case Study 10.3: Bioremediation of a Pesticide: Hydroxylation of Bensulide

Metals play a critical role in plant and animal nutrition. Metals that are required in low concentrations are classified as micronutrients. These metals are essential in the metabolism of organisms. For instance, copper (Cu) forms a component of many enzymes that acts generally as cytoplasmic oxidases. Also, zinc (Zn) forms an integral part of a number of metalloenzymes, such as carbonic anhydrase, that regulates CO_2 exchange and is closely involved in the N metabolism of plants. Severe imbalance of micronutrients can cause death, whereas marginal imbalance contributes to poor health and stunted growth.

Heavy metals, such as lead (Pb), cadmium (Cd), and mercury (Hg), are classified as nonessential for living organisms and can be toxic at relatively low concentrations.

Exposure to excess levels of metals can result in toxicity. Toxicity may include both lethal and sublethal effects. The criteria for metal toxicity in organisms are acute mortality, growth cessation, and impaired reproduction with increased mortality of offspring. Introduction of a low concentration of a toxic chemical (pollutant) into the environment can have a neurological impact and may affect animal behavior.

Plants require few metals as micronutrients (iron [Fe], Cu, Zn, manganese [Mg]). These metals can, however, become toxic in excess amounts. The margin between sufficiency and toxicity is very narrow for most micronutrients. Moreover, plant species vary in their tolerance to metal exposure. Metal toxicity in plants usually manifests itself in the following ways:

- Interveinal chlorosis (yellowing) of the leaves, which resemble symptoms of iron (Fe) deficiency
- Stunted growth (**Figure 10.1**)
- Metal-specific effects such as necrosis (dead spots)

Figure 10.1 Metal toxicity of plants: stunted growth of foliage and leaf chlorosis. The plants on the right were given a toxic level of nickel (Ni) solution. The plants on the left were grown without Ni exposure and served as control. (Courtesy of Sigurdur Greipsson, Kennesaw State University.)

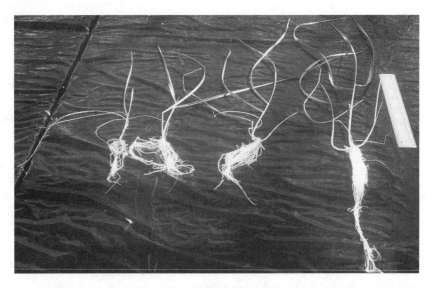

Figure 10.2 Metal toxicity of plants: stunted growth of roots. The three plants on the left were grown in a toxic copper (Cu) solution. The plant on the right was not exposed to the Cu solution and served as a control. (Courtesy of Sigurdur Greipsson, Kennesaw State University.)

Metal toxicity could also have an indirect effect on plants by inhibiting root growth and, consequently, reducing nutrient uptake (**Figure 10.2**). An excess of one or many metals may interfere with nutrient uptake, and, thus, the toxic effect may resemble those of deficiency of another metal. For example, an excess supply of metal can reduce iron absorption by plants and affect iron distribution in roots and shoots. Metal competes for and displaces Fe from active metabolic centers within cells. Heavy metals typically inhibit photosynthesis where Fe plays a crucial role. Also, heavy metals retard the flow of electrons in the electron transfer chain of plants' mitochondria and thus can be expected to have a detrimental effect on respiration.

Extraction of minerals, metals, and fossil fuels is an ancient practice that has escalated over the decades. Mining activities usually result in inhospitable mine wastes or a polluted local environment, although the scale of disturbances associated with mining varies tremendously. For example, surface mining can extend over a large area, whereas deep underground mining is usually more confined to the mine shaft and associated tailings. During either of these processes, however, heavy metals and other pollutants may become airborne and create widespread pollution in the environment. Widespread pollution of heavy metals such as lead, copper, nickel (Ni), and zinc is associated with metal extraction and smelting. The pollution that results from mining activities mainly impacts local environments by the disposal of mine waste (including mine tailings) and mine spoils (which may include topsoil, subsoil, and blasted rock). The spoils are usually coarse textured and low in macronutrients (N and P) and organic matter.

Mining activities also have environmental effects other than direct damage to land surfaces, such as damage to surface water and groundwater, impact on wildlife, and aesthetic damage. Also, acid drainage is a serious problem associated with mine wastes. Acid drainage of mine spoils can dissolve minerals and metals, resulting in saline conditions with high concentrations of metals. Plant establishment is especially inhibited in acid soils below pH 5.5. Acid drainage is typically collected in constructed wetlands, where pollutants are stabilized.

In the past mines were simply abandoned when resources were depleted. Today, this approach is no longer acceptable, and mining operations follow strict reclamation laws (see Chapter 14). Reclamation laws aim at stabilizing quickly the soil surface and often require the establishment of diverse native species assemblages. Legislation often requires that disturbed areas around mines (including mine tailings) be restored to an acceptable ecological state supporting wildlife or otherwise be reclaimed to an alternative use. For example, the revegetation of surface coal mines has a long tradition and is required by state and federal coal-mining laws and regulations. These operations require initial budgeting, thorough planning, and long-term monitoring.

Restoration ecology has a strong background in the restoration of mine wastes. The early work of Anthony Bradshaw and his students on mine wastes in the United Kingdom laid a foundation for restoration ecology as a science. His work was especially important in outlining appropriate uses of metal-tolerant genotypes of plants in the restoration of mine wastes. Restoration of mine waste may involve the use of metal-tolerant genotypes of plants or commercial species to establish an initial cover. Today, environmental laws require restoration of mine sites with emphasis on the use of native plants. These approaches are discussed further below. Remediation of polluted soil can be accomplished by using plants (i.e., phytoremediation) or soil microorganisms (i.e., **bioremediation**), techniques that are discussed in detail below.

10.1 Mine Waste Restoration

The initial steps in restoration of mines include treatments with heavy machinery and even the use of explosives (blasting) to grade and level the ground surface (see **Case Study 10.1** on page 252). Also, the level of soil contamination should be estimated in the early phases of restoration. Several options are available in remediation of contaminated soil. These include in situ and ex situ techniques. The goal of initial in situ remediation is lowering toxic levels in soil by using passive remediation technologies such as volatilization and leaching. Active in situ remediation techniques, such as bioremediation and phytoremediation, can also be used (discussed further below).

Engineering technologies using ex situ and in situ remediation are also used to treat contaminated soils. Ex situ remediation might involve the excavation of polluted soil with its eventual placement into a chemically secure landfill. Also, excavated soil can be transported to a secure location where landfarming is used to treat the soil (discussed below). In situ remediation includes vapor extraction, stabilization, and solidification; soil washing; soil flushing; critical fluid extraction; chemical precipitation; vitrification; thermal desorption; and incineration. These methods are usually expensive and are detrimental for soil structure and its biotic component.

Restoration of mine waste includes establishing plant cover on its surface. Without restoration efforts, abandoned coal mines may remain sparsely vegetated for several years. In more drastic cases mines with extremely acidic spoils (pH of 3.5 or less) may remain barren or unvegetated even after several decades. Mine waste provides a harsh microenvironment for plants. The harsh microenvironment includes high exposure to sunlight, drought, and lack of macronutrients. Mine wastes usually have very different chemical and physical properties from nearby intact soils and are generally inhospitable to native plants. The most frequent problems with mine wastes are the absence of soil organic matter and nutrient deficiencies. In addition, soils on mine tailings have low water retention capacity, which may induce drought conditions. Toxic concentrations of metals in the soil are also common.

Surface Stabilizers

Initial restoration of mine wastes involves the use of **surface stabilizers** (e.g., hydrogels). Surface stabilizers can be effective for the temporary prevention of wind and water erosion; however, their use does not provide a long-term solution. Mulches (chopped straw) are also widely used for this purpose. Mulches are particularly important because they increase the cohesion and aggregation of the surface. In addition, mulches provide surface roughness, improve water filtration and retention, and supply much-needed organic matter to the soil.

Metal-Tolerant Genotypes

A common approach in the restoration of mine wastes is to use **metal-tolerant genotypes** of plants as cover vegetation. This approach was pioneered by Anthony Bradshaw and his students. The use of metal-tolerant genotypes of plants is an alternative to the use of added or stockpiled soil and has been successfully used on mine wastes worldwide. This work was initiated by the finding that old mine sites in the United Kingdom had populations of grasses growing on soil that contained toxic amounts of metals. These populations showed specific metal tolerance, at least compared with commercial varieties of these grasses. For example, sparse populations of *Agrostis tenuis* growing on an old copper mine in the United Kingdom were found to be highly tolerant of toxic levels of Cu, but populations of the same species found on a nearby pasture were intolerant. In the United Kingdom another metal-tolerant Zn grass (*Anthoxanthum odoratum*) was found to have several characters different from those of an intolerant nearby population. These included smaller stature and the ability to grow in soils with low supplies of macronutrients. In fact, these local metal-tolerant populations were derived from nearby populations where metal-tolerant genes found in low frequency had been amplified through natural selection. For practical purposes the seed of these tolerant populations were collected and propagated on cultivated sites to increase seed production. Seed of these cultivars were successfully collected and used in revegetation of metal-contaminated sites. Then, seed of these new metal-tolerant populations were marketed to be used in revegetation of mine sites worldwide. They had a particular advantage because they persisted longer and produced an excellent stabilizing cover compared with intolerant commercial varieties of the same grass species. Three metal-tolerant cultivars were produced commercially: *Agrostis capillaris* cv. Goginan (tolerant to acidic Pb and Zn mine waste), *Festuca rubra* cv. Merlin (tolerant to calcareous Pb and Zn mine waste), and *Agrostis capillaris* cv. Parys (tolerant to Cu mine wastes).

Stockpiled Soil

As the initial step in a restoration effort, it is common practice to add soil on the surface of an abandoned mine. Usually, this soil has already been stripped from the mine site before the mine operation and stored nearby for some time. The soil typically includes a mixture of topsoil and subsoil material, referred to as **stockpiled soil**. Adding these materials on the top of toxic mine waste provides nontoxic growing conditions for plants. If the mine waste contains very toxic material, it is possible to prevent toxicity by adding a thick layer of overburden or even a more permanent layer of clay or rock before topsoil is added. By sealing toxic material in this manner, roots do not get in direct contact with the toxic waste. This method can be excessively expensive, however, especially if the overburden material has to be transported from a long distance.

State and federal coal-mining laws and regulations require the coal-mining industry to remove topsoil from the land before mining and to replace it back as the final cover. Careful management of stockpiled soil is a beneficial method for ensuring establishment of vegetation after mine operation finishes. In addition to improving the establishment of native plants and ensuring onsite succession, topsoil also limits water percolation into the underlying mine spoil and, therefore, reduces the generation of acid mine drainage.

Stockpiled soil may be stored for several years during the mining operation. This soil is then available later for restoration of abandoned mine waste. It is important to quickly establish plant cover on newly stockpiled soil. In turn, this practice enhances survival of soil microorganisms, especially the symbiotic ones. It is also important to keep stockpiles shallow so that roots can penetrate most of the soil volume and maintain cover of native vegetation.

The practice of stockpiling soil has direct effect on soil microorganisms. Mycorrhizal fungi are especially vulnerable because the soil hyphal network is easily damaged during the stockpiling process. As propagules of arbuscular mycorrhizal fungi (AMF) occur mainly in the topsoils, the numbers and infectivity of propagules of AMF in stockpiled soil is drastically decreased. Furthermore, as stockpiled soil is being stored, the soil microorganism may be exposed to unfavorable chemical or physical factors. Populations of symbiotic microorganisms can decline or vanish due to lack of appropriate host plants. However, if symbiotic soil microorganisms have declined drastically, it is possible to use topsoil from nearby intact sites in the reclamation effort.

Commercial Grasses

A common restoration effort of mine waste after the spread of stockpiled soil involves surface stabilization by the use of commercial grasses. The initial aim of establishing dense vegetation cover on mine waste is to stabilize the soil surface and prevent soil erosion that can spread pollutants further. Also, rapid surface stabilization prevents siltation in nearby streams. There are other benefits in establishing dense vegetation cover. For example, vegetation can curtail acid drainage because vegetation cover can decrease the rainwater available for deep percolation. Furthermore, vegetation cover facilitates accumulation of soil organic matter that can act as an oxygen sink. Decline of vegetation cover on mines is often associated with acidification and subsequent elevated levels of metals.

The use of commercially available species, especially grasses, for initial revegetation is a common practice to stabilize the ground quickly. The use of commercial grasses in revegetation is favored because this seed is readily available. Also, seeding techniques are well established. In addition, germination is rapid and fertilizer requirements are well known. Direct seeding of commercial grasses is commonly accomplished, for instance, by **hydroseeding** where seeds mixed with water-retaining chemicals (i.e., hydrogels) and nutrients are sprayed on the soil surface. Hydroseeding is especially appropriate for steep or inaccessible sites. These grasses provide rapid initial vegetation cover. Such cover can be maintained for several years by steady application of chemical fertilizers. Dense cover of commercial grasses declines rapidly, however, as application of chemical fertilizers ceases.

Native Plants

In 1977 the Surface Mining Control & Reclamation Act emphasized the establishment of a diverse and permanent cover of native plants to enhance succession. The long-term goal of restoration of mine waste is usually to establish native vegetation that is self-sustaining. Decline in the cover of commercial species is often associated with colonization of native ones. Commercial grasses may act as nurse crops and facilitate colonization of native species; however, dense cover of commercial grasses can also inhibit colonization of native species. Similarly, dense ground cover of lichens and mosses that may establish on the site can reduce tree seed germination due to chemical inhibition.

Commercial tree species are also used in restoration of mine waste. For instance, the use of commercial pine species as a nurse crop for later-successional forest species has been successful. Restoration practices would benefit, however, from transplanting a greater variety of native tree species.

Mine waste restoration has recently focused on the use of native plants. To facilitate this work a list of native plants that can potentially be used from surrounding regions should be evaluated. However, the use of native plants may depend on their ability to survive the harsh environmental conditions on the mine surface, even after stockpiled soil has been added.

To facilitate the establishment of native species, it is important to establish seed dispersal vectors by increasing retention of nearby seed sources. If dispersal barriers are seriously limiting colonization by native plants, direct seeding and/or transplanting of container-based seedling should be promoted.

Restoration of mine wastes should aim at ensuring succession, as discussed in Chapter 4. First, efforts should be made to establish pioneer species, which should ideally facilitate later successional stages. Second, planting strategies to accelerate succession can be implemented (e.g., wave approach and staggering succession), also discussed in Chapter 4.

As succession is enhanced, restoration programs for mine waste should focus on establishing nutrient cycling within the ecosystem. Long-term use of chemical fertilizers may be necessary to increase the N and P budget of the ecosystem. This, in turn, may facilitate nutrient cycling (see Chapter 2).

One problem with the application of fast-release chemical fertilizers is that the nutrients can simply leach away into the groundwater. It is, therefore, more economical to apply smaller amounts of fertilizers more frequently throughout the growing season. Alternative use of slow-release fertilizers may be appropriate, but such practice is usually cost prohibitive. Strategic use of symbiotic soil microorganisms is pivotal in enhancing nutrient cycling in the ecosystem, as discussed in Chapter 8.

Monitoring

Monitoring restoration success of mine wastes should involve frequent estimation of vegetation cover, species richness, and succession trajectories. This should also involve periodic measurements of concentrations of metals and acidity of soil. In addition, groundwater samples should be collected and analyzed for acidity and potential contaminants (heavy metals). Permanent wells to collect groundwater samples should be installed for this purpose. If dust particles derived from polluted mines are of concern, then air samples should be periodically collected. Monitoring methods should be carefully selected to fit the need of each site. Monitoring contamination derived from mine waste should be included in the overall design of mining operations. This may include the monitoring of smelter-stack plumes, dust, and surface water or groundwater. Also, monitoring programs should be installed to assess the long-term success of the work. This involves monitoring succession as was discussed in Chapter 4.

Biological monitoring is commonly used to assess environmental quality on mine waste and the surrounding environment. This includes bioassay (where target organisms are monitored) and installation of early detection systems where sensitive organisms and occurrence of indicator organisms (species that are tolerant to a particular stress) are monitored. Accumulation of target pollutants in different trophic levels (i.e., bioaccumulation) should be carefully monitored by collecting samples from organisms at different trophic levels. This is particularly important for pollutants that are known to accumulate in organisms that are found high in trophic levels.

Biological monitoring involves an acute toxicity test performed under controlled conditions (**Box 10.1**). This test determines the relative toxicity of a chemical to organisms (usually insects or animals). **Toxicity tests** can be performed in controlled laboratory experiments (ex situ) with a limited number of variables, or effects can be studied in a natural ecosystem (in situ). Toxicity tests are designed to estimate the median lethal concentration (LC_{50}) of a particular chemical exposed to an organism. The LC_{50} is the chemical concentration estimated to produce mortality in 50% of a test population over a specific time period (usually 24–96 hours). Effects other than mortality (e.g., behavioral or physiological) are measured by using median effective concentration (EC_{50}) (**Box 10.2**). The EC_{50} is the concentration of a chemical estimated to produce a specific effect in 50% of a population of test species after a specified length of exposure (usually 24 or 48 hours). Effect criteria typically include immobility, loss of equilibrium, failure to respond to an external stimulus, and abnormal behavior.

Box 10.1 — **Criteria for Selecting Species for Toxicity Test**

Those species should ideally
- Have a broad range of sensitivities
- Be widely available and abundant
- Be representative of the ecosystem in question
- Be amenable in the laboratory

Box 10.2 — **Criteria for Selecting a Behavioral Response for Toxicity Test**

Those behavioral responses should ideally be
- Amenable to laboratory conditions
- Sensitive to the toxic chemical in question
- Ecologically relevant
- Practical to conduct

10.2 Phytoremediation

Plants can be used to remediate contaminated soils. This practice, called **phytoremediation**, involves the ability of plants to absorb pollutants (usually metals) from soils and accumulate these pollutants in their tissues (see **Case Study 10.2** on page 257). Phytoremediation is a cost-effective technique compared with alternative methods. For instance, placing contaminated soil in a landfill can be up to 300 times more expensive than using phytoremediation.

Phytoremediation uses phytoextraction, phytodegradation, and phytostabilization. In **phytoextraction** plants are used to accumulate pollutants into their tissues (**Figure 10.3**). After plants have accumulated the target metal, they are harvested and removed for further processing. Processing contaminated plant material includes thermal treatments, microbial treatments, chemical treatments, and even the placement of the material into a secure landfill.

Plants and associated soil microbes are used to sequester pollutants in **phytodegradation**. Sequestration of pollutants can be enhanced either by chemical amendments to the soil or through the activities of plants and associated soil microbes. For example, plants and their associated microbes in the rhizosphere can facilitate biodegradation of notorious pollutants such as trichloroethylene (i.e., Cl_3C_2H).

Phytostabilization is the process by which pollutants are adsorbed or entrapped in either plant tissue (usually roots) or in the soil. Bioavailability of pollutants is further decreased by phytostabilization techniques. The role of plants in this process is to increase the sequestration of the pollutant in soil.

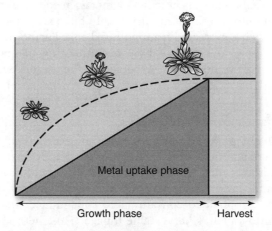

Figure 10.3 Continuous phytoextraction. Solid line represents metal concentration in the foliage. Dashed line represents biomass of foliage. (Reproduced from *Annual Review of Plant Physiology and Plant Molecular Biology* by D. E. Salt. Copyright 1998 by Annual Reviews, Inc. Reproduced with permission of Annual Reviews, Inc. in the format Textbook via Copyright Clearance Center.)

Plants can further sequester pollutants by precipitating them into an insoluble form and by incorporating organic pollutants into the plant lignin.

Hyperaccumulators

Heavy metals usually accumulate in plant tissue at low levels (0.1–100 ppm). Plants that accumulate unusually high concentrations of metals in their tissue are named **hyperaccumulators**. Hyperaccumulators are rare plants that grow in soils that have naturally high concentrations of metals. Such soils are usually found in remote regions that have exposed rocks containing metals. These plants tolerate much greater concentrations of metals in their own tissues than can be tolerated by most ordinary plant species. For example, the hyperaccumulator *Thlaspi caerulescens* can accumulate up to 4% of Zn in its tissue without showing any toxic effects. This plant can accumulate Zn at the rate of 125 kg per hectare per year.

The use of hyperaccumulating plants in phytoremediation involves growing the appropriate species on the mine waste and using the foliage of the plant to collect metals. The foliage of the hyperaccumulating plant is harvested. The dry biomass can be burned in special incinerators to reduce its bulk, and the residual ash can be either stored in a landfill or sold for further processing of valuable metals. The use of hyperaccumulators in phytoextraction is usually a long-term commitment. For instance, a polluted site containing 2,000 kg per hectare of Zn takes about 16 years to remediate to an acceptable level.

The use of hyperaccumulators to remediate mine waste is, however, currently of limited use. The limitations are mainly because these plants usually only accumulate one specific heavy metal. In addition, hyperaccumulators grow slowly and produce limited harvestable biomass, which leads to low amounts of metals being extracted from polluted soil. Furthermore, limited information exists on how to cultivate hyperaccumulating plants in various ecoregions. Finally, there are no known hyperaccumulators for metals such as Pb, Cd, and uranium (U).

Metal Chelates

Alternatively, phytoextraction can be induced by using ordinary high-yielding crops in conjunction with application of specific metal chelates (**Figure 10.4**). This method is termed **chelate-assisted phytoremediation**. Metals are usually recalcitrant or resistant in soil and, therefore, difficult to remediate. Metal chelates, when applied to contaminated soil, induce the uptake of the metal of interest into a plant's harvestable foliage. This method does not require metal-tolerant plants and commonly uses high-yielding crops such as corn, ryegrass, and Indian mustard. The high biomass of crops is an advantage compared with the low biomass of hyperaccumulators. Although concentrations of metals in crops are usually much lower compared with tissues of hyperaccumulators, the

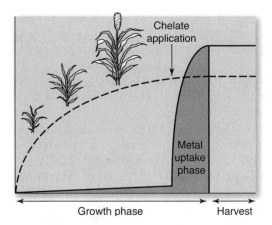

Figure 10.4 Chelate-assisted phytoremediation. Solid line represents metal concentration in the foliage. Dashed line represents biomass of foliage. Arrow indicates application of chelate. (Reproduced from *Annual Review of Plant Physiology and Plant Molecular Biology* by D. E. Salt. Copyright 1998 by Annual Reviews, Inc. Reproduced with permission of Annual Reviews, Inc. in the format Textbook via Copyright Clearance Center.)

high biomass of crop plants compensates for the relatively low metal accumulation. The advantage of using crop plants includes readily available seeds and cultivation technologies, including nutrient and weed management, that are well established under a wide variety of environmental conditions. In addition, most crops used for this purpose grow rapidly, produce large biomass, and can be grown in a wide variety of ecoregions.

In phytoremediation the ability of plants to accumulate metals is a function of their biomass and metal concentration in harvestable, aboveground tissue. It is, therefore, important to optimize the biomass before the induced influx of heavy metals. For this purpose the contaminated site is prepared by tilling the soil. The crop is cultivated until its growth is optimized by adding fertilizers, maintaining irrigation, and proper weed management. When the optimal growth is reached, the appropriate metal chelator is applied. After the application of metal chelators plants show signs of metal toxicity, usually after several days or weeks. At that time the aboveground foliage of the crop is harvested. Application of metal chelates such as EDTA (ethylenediaminetetraacetic acid) to contaminated soils can increase Pb accumulation in plants by 15 times. Using chelate-assisted phytoremediation, ordinary crops can remove more than 500 kg per hectare of Pb per year. Application of metal chelators such as EDTA results in high inflow of metals into plants. Because of the high influx of metals, plants quickly show toxic signs (chlorosis and necrosis) and at that stage can be harvested. For example, simultaneous accumulation of metals occurs in Indian mustard plants after application of EDTA to soil contaminated with various heavy metals. This is beneficial and gives an advantage compared with the use of hyperaccumulators that typically accumulate one metal at a time.

The application of metal chelates to the soil not only enhances the bioavailability of metals in the soil but also increases the translocation of the metal from the roots to the shoots. This translocation is essential in phytoremediation, where accumulation of metals in roots is often a limiting factor, especially when it comes to harvesting the aboveground foliage.

The use of chelators in phytoremediation is metal specific. A chelate-assisted phytoextraction of a contaminated soil needs to determine the appropriate chelate–crop combination. For example, EDTA or EDDS (ethylene-diamine-dissuccinate) are used for remediation of Pb-polluted soil, EGTA (ethyleneglycotetraacetic acid) for Cd, citrate for uranium (U), and ammonium thiocyanate for gold (Au).

10.3 Bioremediation

Soils can become contaminated by a variety of organic chemicals, either by accidental spill, storage leak, or widespread use of agrochemicals (e.g., pesticides, herbicides, and fungicides). Unfortunately, many agrochemicals have harmful effects on the ecosystem and even on human health. Some of these chemicals are recalcitrant and accumulate with time in soil.

Soils contaminated with organic and "xenobiotic" pollutants can be treated by various engineering methods such as vapor stripping, thermal desorption, soil washing, incineration, and even excavation. Alternatively, bioremediation can be used to treat these pollutants in situ (**Box 10.3**). Bioremediation has proved to be an efficient and cost-effective method for the reduction of pollutants in soil (**Box 10.4**). It uses the ability of soil microbes (fungi and bacteria) to catalyze (breakdown) pollutants (see **Case Study 10.3** on page 261). **Biodegradation** of complex organic pollutants relies on the metabolic versatility of soil microbes. Many pollutants (organic or inorganic) as they are broken down serve as nutrient and energy sources for soil microbes. These chemicals include single-carbon compounds, acyclic and aromatic rings, halogens, nitrogen-containing pollutants, and sulfur-containing compounds.

Box 10.3   Criteria for Bioremediation

- Microorganisms must have necessary enzymatic activity to catalyze the contaminant in question.
- Targeted contaminant must be bioavailable.
- Contaminated site must have soil conditions optimal for microbial activities.

> **Box 10.4   Pros and Cons of Bioremediation**
>
> *Pro*
> - Reduced hazard to cleanup personnel
> - Low environmental impact because no or little waste is generated
>
> *Con*
> - Limited use when bioavailability of contaminant is low
> - Some compounds can be toxic to microorganisms
> - Aerobic and anaerobic processes could be required at the same time
> - Acidic and alkaline conditions might be required at the same time

Impeding Factors

Numerous factors can impede bioremediation, including low levels of soil microbes or lack of specific enzymes promoting catalytic activity in the resident microbial population. To overcome these limitations, inocula of supplemental microbial population containing specific enzyme activity can be added to the contaminated soil. The inocula can also include genetically engineered microbes with specific catalytic capacity to break down a particular pollutant. Populations of these genetically engineered bacteria increase in the soil when the organic pollutant is in high concentration but decline rapidly and vanish as soon as the target pollutant is depleted. A variety of genetically engineered bacteria that target different pollutants is available commercially.

Biodegradation usually results in nonharmful products. For example, hydrocarbons found in gasoline can be broken down to carbon dioxide and water. Biodegradation of gasoline is usually rapid, and significant degradation of such hydrocarbons usually occurs under favorable conditions in 6 to 18 months.

Bioremediation is most effective for large volumes of high concentrations of pollutants in soils. This process is, however, limited where pollutants are heterogeneously dispersed in soils. Another factor that may limit bioremediation is if the concentration of a pollutant in soil is either very low or very high (toxic). Also, many pollutants are relatively recalcitrant to biodegradation. They may also require the combined metabolic efforts of several different microorganisms. Biodegradation of some pollutants may need additional nutrient and energy sources to build up and maintain populations of desirable microbes.

Requirements for Bioremediation

Bioremediation depends on optimal soil conditions where the target microbial activity takes place. A number of bioremediation strategies have been developed, but they must consider basic environmental conditions to optimize efficiency.

Several environmental factors can affect the success of bioremediation, and one of the most important ones is temperature. Biological activities of most microorganisms have an optimal temperature range. For example, the optimal temperature for bioremediation of gasoline is about 40°C.

Another critical factor is the oxygen level in soil. Aerobic microbial activities require oxygen in the soil. To facilitate oxygen supply in soil, leaky pipes can be installed in the soil for air to be pumped in. This technique is called "air purging." Another technique is to use hydrogen peroxide (H_2O_2) that is added into the soil; as it breaks down it provides oxygen (and water) for microbial activity.

Microbial activities also require moisture, which can be supplied to dry soils by irrigation. Acidic and alkaline conditions are usually detrimental for microbial activities. Bioremediation is usually optimal in a soil with a pH around 7.8. Soil pH can be manipulated by addition of alkaline or acidic material and buffers to stabilize the conditions.

Biostimulation with the addition of nutrients is often needed in bioremediation. Macronutrients, especially N and P, are added as required. Micronutrients and even simple C compounds (sugars) can be added to the soil through irrigation to stimulate microbial activities. Nutrients are usually delivered through irrigation water to soils in support of microbial activities.

Landfarming

Landfarming involves removal and transportation of contaminated soils to another site that has been specially prepared for bioremediation. The land receiving the contaminated soil is graded and contained by a soil berm to eliminate surface runoff within the landfarming area. If necessary, the ground can be sealed off by clay or plastic material to avoid any contamination of the groundwater. Subsequently, the contaminated soil is spread over the designated area to enhance the spontaneous initial remedial processes involving volatilization, aeration, and photolysis. The contaminated soil is then mixed with soil on the site that may act as natural inocula. Otherwise, specific inocula can be added to the soil. This activity initiates biodegradation, especially of simple hydrocarbon compounds. The contaminated soil is mixed into a 15- to 20-cm layer of soil to increase contact with soil microbes. The pH of the mixed soil is adjusted to create a neutral pH. If the soil is deficient in nutrients, biostimulation is initiated by adding chemical fertilizers. For example, as much as 880 kg per hectare of complete fertilizers (N-P-K, 13-13-13) have been used at one time in bioremediation of the widespread herbicide atrazine (2-chloro-4-(ethylamine)-6-(isopropylamine)-s-triazine) in soil. Monitoring the levels of the pollutants along with nutrient levels and soil pH is essential to optimize the success of bioremediation. As soon as the levels of the pollutants are significantly reduced, a new load of contaminated soil can be added to the site and mixed with the resident soil. Landfarming has been shown to be as much as five times more cost effective than some alternative methods (e.g., excavation and disposal of contaminated soil in a landfill).

An example of successful use of landfarming can be found in the remediation of crude oil lakes in Kuwait. After the defeat of the Iraqi army in Kuwait in 1991, one of the greatest environmental terrorist attacks took place as Iraqi forces destroyed and set ablaze hundreds of oil wells in Kuwait. After the oil wells were contained, more than 300 oil lakes were formed with a total area of about 50 km^2. The first step in the remediation was to retain the remaining oil in the lakes (about 85% of the crude oil was recovered). The remainder of the oil that was mostly mixed with sand represented a hazardous source of pollution, which could potentially have negative impacts on public health and long-term effects on valuable groundwater supplies in the region. Oil remediation of the polluted sand was, therefore, a pressing issue. Preliminary studies were conducted on the polluted site and on the level of pollution in the oil lakes. This step was necessary to select appropriate bioremediation strategies.

The unusual environmental conditions in the desert required specific bioremediation strategies. The crude oil did not migrate rapidly into the sand, and, therefore, excavation and nearby landfarming was selected as the most appropriate method. This was followed by a more detailed study to determine the concentration of oil in the sand, chemical characteristics of the oil pollution, and estimation of the volume of contaminated sand. The distribution profile and concentration of oil pollution in the sand layer was examined. Oil was observed as deep as 270 cm in the sand. The oil migration into the sand did not follow a clear pattern, and oil concentration in the contaminated sand ranged from 20% to 60%. The next step was to select an appropriate nearby site to serve for landfarming. Also, appropriate techniques to excavate, prepare, and handle the polluted sand were selected.

Bioremediation of crude oil commonly uses the ability of resident bacteria populations to break down C compounds in the oil. Crude oil is high in C but low in N and P. Strains of native bacteria found in the desert soil were isolated, and mass production of inocula was performed under controlled conditions. These bacteria could degrade oil molecules, which are known for their toxic and carcinogenic effects. The oil-polluted sand was inoculated with these bacteria strains to enhance biodegradation. The oil-contaminated sand was excavated by heavy machinery and transported to a nearby landfarm site. The contaminated sand was screened to remove large stones. This was followed by mixing of the contaminated sand with chemical fertilizers and other material such as sawdust and compost to improve its physical properties. The oil-contaminated sand was then put into soil piles on the surface of the prepared landfarm.

In this case landfarming involved two techniques, composting soil piles and static bioventing of the soil piles. Composting soil piles involved the provision of air through regular turning of the piles and steady supply of water and microbial inoculum through installation of leaky pipes. The pipes were positioned inside each soil pile at predetermined spacing to optimize distribution of water, nutrients, and microbial inoculum. Static bioventing of soil piles did not involve any mechanical mixing of the sand, which reduced operating costs. This technique involved the installation into the soil piles of special ventilation pipes (air purging) with holes at predetermined intervals. The pipes were installed through the soil piles and connected to air compressor to provide a steady supply of oxygen. This operation provided a rapid biodegradation of oil hydrocarbons where 70% to 80% of the pollutants were removed within 10 months.

Summary

Mine wastes are widely dispersed and provide a harsh environment for organisms. Not only is the microenvironment on these sites inhospitable to establishing vegetation, but toxic levels of metals often exist in the soil. It is important to revegetate mine wastes rapidly to avoid the further spread of pollutants. Environmental law typically requires mine revegetation. Metal-tolerant varieties of commercial grass species can be used for this purpose. However, removal of toxic metals from soil involving the use of plants is a technique referred to as phytoremediation. Synthetic chelators such as EDTA and EDDS can be used to facilitate the uptake of metals and translocation to the plant's aboveground foliage. The foliage is then harvested and disposed. Bioremediation involves the use of resident soil microorganisms to break down organic pollutants. The resident soil microorganisms can be bioaugmented by adding nutrients, water, or extra supply of air to the contaminated soil. If the desirable microorganisms are not found in the soil, then they must be inoculated with an appropriate strain of bacteria. Genetically engineered strains of bacteria can be used for this purpose. Landfarming involves transporting polluted soil to a site where the activities of soil microorganisms are optimized through tilling, watering, air purging, and addition of nutrients.

10.1 Case Study

Restoration of Gold Mines in Ghana, West Africa
Charlotte Tay, Golden Star (Wassa) Ltd., Accra, Ghana

The mining industry in Ghana is the leading foreign exchange earner and provides employment for over 18,000 people. The industry conducts its operations in an environmentally responsible manner; it seeks to comply with appropriate environmental protection and sustainable mining operations. According to the Environmental Protection Agency (EPA) of Ghana, the industry recently achieved over 75% compliance with mining environmental regulations and standards in 2002. Irrespective of the gains, there has been increasing awareness of the possible harmful effects on the environment by mining activities, and like in most parts of the world, the impact of mining on the environment has become a concern to the general public.

Case Study

Figure 10.5 Gold mine tailing in Ghana. (Courtesy of Charlotte Fafa Tay.)

The large quantities of waste generated and resources consumed in the mining process cause the most significant impacts (**Figure 10.5**). Waste generation during the ore excavation process is mostly in the form of waste rocks and may contain low concentration of other pollutants (such as arsenic and mercury) that are relatively toxic. Mine tailings (mixture of fine sand, silt, clay, and cyanide solution), generated in mineral processing, are typically high in potentially toxic metals and sulfur compounds that can hamper vegetation establishment. Dust, oxides of nitrogen, sulfur dioxide, and carbon monoxide emissions are the main constituents of air pollutants. Water quality may be affected by heavy metal contamination and leaching, processing chemicals, erosion and sedimentation, and in some areas acid mine drainage. This problem is often compounded by the activities of small-scale and illegal gold-mining activities. The physical impact of mining activities is a permanent scar on the land that destroys the natural topography indefinitely.

The mining industry together with the government of Ghana is mitigating the adverse impacts from mining activities by undertaking environmental management measures aimed at the prevention, reduction, or elimination of any degradation of the environment. One aspect of the measures relevant to this text is the restoration of mined fields. Reclamation is the term that best describes the restoration efforts of the mining industry in Ghana. To demonstrate the governments concern in achieving best practice in environmental management, in 2001 the EPA introduced financial assurance

(continued)

Case Study (continued)

regulations in the form of reclamation bond for ongoing mining operations focused at ensuring that mining companies implement their commitments as stated in their Environmental Impact Statements as well as in their Reclamation Plans.

The reclamation of mined lands in Ghana is exemplified by Goldfields Ghana Limited, Damang Gold Mine's reclamation efforts as presented in the following text. The Damang Gold Mine is located near the village of Damang in the western region of Ghana. The mine is an open pit mine, and the ratio of waste to ore mined is approximately 4:1.

The Damang Mine is located in the rain forest zone of southeastern Ghana, an extensively disturbed area due to land clearing for subsistence agriculture, timber harvesting, and artisanal (illegal) mining. Much of the area around the mine site is composed of secondary growth forest with dominant species like *Musanga* and *Ficus*. It is common to find old cocoa plantations beneath higher canopies. Wetter areas are typically populated with raphia palm, sedges, and bamboo. Most of the rainfall takes place between June and July and then in October. The annual rainfall is over 2,000 mm. The soils are slightly acidic with generally poor levels of nutrients.

The objectives of the reclamation plan are as follows:
- Provide a final land use that considers the needs of the stakeholders
- Provide landforms that blend with the natural topography
- Provide a site both chemically and physically stable
- Leave disturbed areas in a safe condition
- Ensure that potential long-term environmental liabilities associated with the closure of the site are minimized
- Restore as much of the mining area to a sustainable land-use capability as is practicable
- Provide rehabilitated areas that contribute to the long-term sustainability of the local economy

The company has shown its commitment to the reclamation of the mine by being the first to comply with reclamation bond policy in Ghana in 2001. With progressive reclamation, the bond was reduced by almost $1 million for some selected areas that have achieved primary completion. Presently, the guarantee for the reclamation of the mine site is approximately $4 milllion. The company recognizes the vital role played by stakeholders in the implementation and achieving set goals and targets in the reclamation program. This is emphasized in the company's environmental policy and further demonstrated in an annual event, "Open House Forum," that brings all stakeholders together to deliberate on issues and progress of the reclamation program. The stakeholders include chiefs, elders, and other representatives from the community and representatives from the District Assembly, EPA, and other mining companies.

The initial reclamation at the mine site was construction to stabilize slopes that had the potential to erode. However, as the mine development continued, so did the reclamation work, and the total rehabilitated land in 2005 was 226 hectares. The company has so far disturbed 780 hectares of land within its mine lease. A key to meeting its reclamation objectives is the development of a practical reclamation suite of tools that

Case Study

Figure 10.6 Revegetation of gold mine in Ghana. (Courtesy of Charlotte Fafa Tay.)

provide guidance on the three major components of the company's reclamation plan, namely topsoil management, landform design, and revegetation (**Figure 10.6**).

The importance of topsoil as a source of seeds, propagation material, nutrients, and symbiotic microorganisms is recognized. Wherever practicable, topsoil made available by land disturbance is preserved for future reclamation activities. The topsoil is usually stored in piles not more than 2 m deep for the establishment of cover crop to minimize erosion and maximize the long-term viability of the soil medium.

Major earthworks for reclamation take into account appropriate drainage. A drainage density at least matching preexisting density is reestablished. Any constructed drainage structures for significant flow is designed and constructed for an extreme storm event. Steep slopes of any waste dumps or stockpile areas are resloped to an average slope of no more than 20 degrees. Drainage interception ditches, drop structures, sculpted areas, and energy dissipaters are included as required. In addition to recontouring slopes for erosion control, landform design incorporates appropriate erosion control system and structures to stabilize soil. This may be in the form of rapid establishment of vegetative cover or use of synthetic stabilizing materials such as fiber mats and Hyson cells.

Wherever possible, the use of native plant species for revegetation is maximized. Exceptions are the use of non-native species or crops tolerant to extremes in soil chemical or physical conditions. Fast-growing species, such as vetiver, *Brachiara*, and Guinea grasses, are used in erosion control. Cover on slopes is enhanced with the goal

(continued)

Case Study (*continued*)

of promoting the incorporation of organic matter into the soil profile and providing a pool of nutrients for the development of the required land-use system.

For actual implementation of the reclamation program, the initial step is to identify lands that are no longer required for the continued operation of the mining project. This is followed by designing appropriate landforms that blends with the natural topography of the area. Earthworks include resloping of deposited waste rock to obtain the required slopes indicated in the plan and reshaping of areas of pits, borrow areas, ramps, tailings areas, and other mine infrastructure features.

Oxide (subsoil) material that is not economical for milling is used for capping of waste dumps. The oxide material is deposited on the rock and spread to a suitable thickness based on the final land use planned for the area. Drainage channels are installed to control storm water flow down rehabilitated slopes. All flat surfaces and sloped areas, which have been compacted either naturally or due to traffic, are deeply ripped before revegetation. Ripping is performed only when soil conditions are dry enough to allow shattering of the compacted soil to capture seeds moved by overland flow. The area is then finally capped with topsoil.

Placement and spreading of stockpiled topsoil is done during dry weather conditions to prevent compaction of the material. Generally, topsoil is spread on all surfaces of waste rock dumps or fill areas. The thickness of the topsoil layer provided over the oxide material is determined by resource availability and the end land use of the area. The minimum thickness where it is spread is usually 100 mm, although the desirable thickness can be up to 250 mm. Vegetative cover is immediately established to control erosion.

The next stage involves the provision vegetation, which entails establishing an initial vegetative cover to control soil erosion. This is usually planned to take advantage of the early rains of the wet season. Revegetation involves the use of species that contribute most to the stability and utilization of the ecosystem or farming system that will be developed in the area.

Initial stabilization of slopes may include but is not limited to seeding of the soil with legumes immediately after capping with topsoil so as to use the available moisture in the soil for early germination and to provide nitrogen fixing to improve soil conditions. Vetiver, Guinea, *Brachiara*, and citronella grasses are planted across slopes to control erosion.

Tree or crop planting is carried out during the rainy season. The areas to be planted are identified and the appropriate species for the end land use selected. Trees planted include legumes and native species as well as common crops grown by the local people. Selected areas for planting are then pegged out with the size of holes and distance between plants determined by species type and topsoil requirements for the area. Where topsoil spreading cannot be done, excavated holes are filled with topsoil. Finally, potted plants 2 months old or more are planted out manually. Revegetation of tailing dams may be carried out year-round and involves initial propagation of metal-tolerant grasses and direct seeding of legumes to stabilize the surface. This is followed by planting of trees and crops.

Maintenance of the rehabilitated areas is a basic requirement for achieving productivity and the progression to the final ecosystem or farming system. Enhanced development of the ecosystems may be achieved by, but not limited to, regular weeding to rid farms of weeds and rodents, tending, application of fertilizer, promptly repairing areas showing signs of erosion, and periodic spraying of pesticides and pruning of trees and crops.

Evaluation is an essential component of the reclamation program at the Damang site. Evaluations of the programs are regularly reported to the EPA. Regular field meetings with the EPA and other regulatory agencies, regular reporting to Goldfields Corporate Office, photo documentation of work progress over time, documentation of stakeholders' views during "Open House Forum," achievement of objectives and targets set in the Reclamation Plan by Goldfield's environmental management committee, and reports on field trial programs also provide a means for assessing the progress of reclamation programs.

The EPA's evaluation of the reclamation program is based on a reclamation criteria embedded in the bonding agreement. This involves the use of three stages to describe the level of reclamation activity and also provides an opportunity for a reduction in the amount of reclamation bond in the following manner:

- Upon achieving primary completion (the area should be in its final landform with a controlled drainage pattern that minimizes surface erosion), the bond for an area of disturbed land is reduced to 30% of the original bond for that area.
- Upon achieving land-use completion (level of vegetation establishment that, in large part, achieves the final land-use objectives of the disturbed area), the amount is reduced to 20% of the original amount.
- Upon achieving final completion (the criteria for land-use completion have been achieved for three seasonal cycles and the land requires no additional monitoring), a bond will no longer be required for a particular area of disturbed land.

After the final completion, the company will be released from all environmental responsibilities, obligations, and liability in relation to the mining lease land or the relevant tract of the disturbed land with the exception that the company will leave the cash deposit in place for an additional 3 years after final completion is achieved.

10.2 Case Study

Phytoremediation of Lead-Contaminated Soil
Anna Hovsepyan, Department of Environmental Engineering Sciences, University of Florida, Gainesville, FL

Over the past century industrial activities including mining and manufacturing have been major contributors to extensive soil pollution with heavy metals. Although on a global scale leaded (Pb) gasoline is the major source of increased Pb levels in soils, mining activities also generate waste materials that often contain excessive concentrations

(continued)

Case Study (*continued*)

of Pb. Numerous factors and processes influence concentrations of Pb in soils; therefore, its distribution in soils is considered to be complex. Soil represents a major sink for Pb, and most Pb is retained in the upper 2 to 5 cm of soil. Lead is geochemically immobile in the top soil layers and is maintained in soil primarily by adsorption onto the surfaces of mineral particles, complexation by humic acids, and precipitation reactions. Even though the total Pb concentration in many contaminated soils may be high, the phytoavailable Pb fraction is often exceedingly low. Thus, Pb is very persistent in soil and presents a long-term pollution problem that poses a threat to human and animal health. Very low levels of Pb in blood of children can affect their IQs adversely.

Phytoremediation is an emerging in situ remediation technology that uses plants to decontaminate polluted soil. It is an environmentally friendly and cost-effective alternative to physiochemical methods that can destroy soil microorganisms, thus affecting the overall quality of the soil as a growth medium. Phytoremediation of polluted soils can be used to remove, contain, or stabilize Pb and other toxic metals in the soil. Consequently, various phytoremediation strategies exist for metal decontamination of soils such as phytostabilization, phytodegradation, and phytoextraction. The choice of phytoremediation strategy depends on many variables that include the nature, speciation, and concentrations of the target metal in soil, as well as the environmental risk it represents and the subsequent use of the land. For instance, phytostabilization is used to immobilize pollutants through plant and soil chemical processes such as absorption or precipitation by plant roots. In phytostabilization, plants usually change the speciation of metals in soil in such a way that it results in reduced mobility and bioavailability of the target element, thus rendering it harmless. Furthermore, phytostabilization prevents metals from leaching into the groundwater and entering into the food chain. Phytodegradation involves the breaking up of organic pollutants such as hydrocarbons and pesticides through various plant metabolic processes. In phytoextraction, plants are used to extract metals from the soil and to accumulate them into their aboveground portions, which then can be easily harvested and disposed in proper landfills. Phytoextraction also offers the possibility of recovering expensive trace metals.

The success of phytoremediation in mine reclamation mainly depends on the phytoremediation strategy and consequently the choice of plant species. For example, phytostabilization would use metal-tolerant grasses that do not translocate metal pollutants above the ground. For phytoextraction to be efficient, a plant must be able to accumulate metal concentrations equivalent to about 1% to 2% of its dry weight and must also be tolerant to pollutants, grow quickly, and produce a high biomass.

Phytoextraction has a great potential for the reclamation of Pb-contaminated soil, particularly in instances when other remediation technologies are not efficient or too expensive. The use of metal-accumulating and high biomass–producing crop plants, such as corn, Indian mustard, or sunflower, grown in soil supplemented with synthetic chelators such as EDTA or EDDS has yielded promising results in the uptake of Pb from soil and translocation of Pb from roots to shoots (**Figures 10.7** and **10.8**). Lead

Figure 10.7 Chemical-assisted phytoremediation: grasses grown in pots in a growth chamber and subjected to various treatments. (Courtesy of Sigurdur Greipsson, Kennesaw State University.)

Figure 10.8 Chemical-assisted phytoremediation: use of EDTA (left). (Courtesy of Sigurdur Greipsson, Kennesaw State University.)

is not very soluble in soil. The use of synthetic chelators increases phytoavailability of Pb in the soil and consequently facilitates translocation of Pb from roots to the harvestable portions of plants. The chelator EDTA is thus one of the most effective and widely used in phytoextraction. As a result of the high affinity of EDTA toward Pb, EDTA forms a highly water-soluble Pb–EDTA complex. This complex is the major form of Pb that can potentially be taken up by plants.

(continued)

Case Study (*continued*)

The dosage and the time of EDTA application are very important considerations. The optimal EDTA concentration may vary between soil types and plant species. It has been suggested that Pb solubility and bioavailability may be increased by the addition of EDTA in the range of 0.1 to 10 mmol EDTA kg^{-1} soil. Very high doses of EDTA may enhance the stress on plant and soil microorganisms such as soil fungi because of the potential toxicity of the lead–EDTA complex. Additionally, a higher Pb-to-EDTA complex ratio in soil would achieve a higher Pb uptake by the plant. As a result the accumulation of high levels of Pb can be fatal to seedlings. To facilitate rapid Pb accumulation and avoid the toxicity effects of Pb, it is, therefore, recommended that a single dose of EDTA be applied to the root zone only after a sufficient amount of plant biomass has been established. Plants should be usually harvested several weeks after EDTA addition. The chelator EDTA is effective where Pb contaminated soil is treated in landfarming, but EDDS might be more appropriate in situ because it breaks down much faster in the soil. Any risk of Pb entering the groundwater level would be much less if EDDS were used in situ.

Plant and soil microbe interactions should not be overlooked when designing a phytoremediation strategy for polluted soil. A very important consideration is given to plant–AMF interaction, which is the most common type of mutualistic associations with plants. Plants benefit from this symbiosis because AMF supply plants with nutrients (especially phosphorus) and protect them from drought and root pathogens. The efficiency of AMF in metal uptake vary between different AMF and plant species. Some AMF species prevent metal uptake in plants as a protection mechanism against metal toxicity, whereas other AMF species facilitate metal uptake in plants by solubilizing metal complexes and, thus, increasing their bioavailability in soils. In addition, the role of AMF in heavy metal uptake is highly metal specific. For example, in a greenhouse study Pb and manganese uptake by corn was lower in mycorrhizal plants compared with nonmycorrhizal plants, whereas the uptake of zinc and copper was higher in mycorrhizal than in nonmycorrhizal plants. The role of AMF in the phytoremediation of polluted soil should be given a careful consideration.

Greenhouse experiments are the first step in setting optimal phytoremediation practices; however, field trials are necessary for the successful outcome of phytoremediation. Each site has unique characteristics; therefore, a preliminary assessment of contaminated sites is required before the implementation of phytoremediation practices. Site-specific factors that should be evaluated include climatic conditions, soil texture, moisture, pH, and soil nutrient supply. Selection of plant species, therefore, should not be only based on the nature of the target metal but also on the physiochemical characteristics of the contaminated soil.

Lead can be very persistent in polluted soils, thus causing long-term environmental and health problems. Phytoremediation of mine tailings promotes vegetation coverage of the area and minimizes environmental risks associated with Pb and other heavy metal pollution. Furthermore, phytoremediation not only offers economic and technological advantages over conventional remediation technologies but also fulfills aesthetic requirements.

10.3 Case Study

Bioremediation of a Pesticide: Hydroxylation of Bensulide
Richard Belcher, Department of Environmental Sciences,
University of California, Riverside, CA

Bioremediation centers on the use of microorganisms to reduce levels of environmental pollutants. Bacteria and fungi contain enzymes that catalyze reactions that can be used to manipulate contaminants in a beneficial way. The ideal reaction is through total mineralization, which converts the substance into harmless byproducts such as carbon dioxide and water. Another possibility is that the pollutant can be broken down into a source of energy, such as carbon from hydrocarbon contaminants, and then metabolized. Then, cometabolism creates compounds that cannot be used for energy but are less harmful to the environment. Environmental pollutants derived from pesticides include organic contaminants such as pentachlorophenol and benzene or heavy metals like mercury and cadmium. Most organic pollutants are accidentally released into the environment except for pesticides, which are used intentionally to exterminate pests. Pesticides are primarily (74%) used in the agricultural sector. In the global market the United States is responsible for use of 20% (500 million kg) of all pesticides.

The extensive use of pesticides creates many factors of concern due to the risk of toxicity. Residues can often be found on many crops that are sprayed with pesticides, posing a risk to fruit and vegetable consumers. There is also environmental concern because pesticides often affect nontargeted organisms that play an important role in the ecosystem. Pesticides are not limited to crops; they can be used for turf grasses, roadsides, industrial applications, the timber industry, and other non–crop-related uses.

Some of the most toxic pesticides widely available on the market are organophosphates (contain a phosphorus group), which have the ability to inhibit cholinesterase. Cholinesterases are a major class of enzymes including acetylcholinesterase, which is vital to the regulation of nerve impulses in vertebrates, such as humans. Impulses are received through cholinergic neuron stimulation by the neurotransmitter acetylcholine, which is then broken down by acetylcholinesterase into the inactive metabolites choline and acetate. When acetylcholinesterase is inhibited, the nerve impulses create an overstimulation, which lead to convulsions, paralyzed breathing, and, in extreme cases, death. Although most organophosphate pesticides are not persistent in the environment, one herbicide, bensulide ($C_{14}H_{24}NO_4PS_3$), is of concern. Bensulide is a preemergence (suppresses seed germination) herbicide applied to agricultural crops, turf grasses, ornamentals, and ground covers. It prevents mainly the germination of annual grasses and broadleaf weeds. Over 60% of its use is devoted to crops, and annual usage totals 250 tons. Geographically, bensulide applications are scattered along the Pacific coastal states and the Midwest region with major users being California, Texas, and Pennsylvania.

(continued)

Case Study (*continued*)

Bensulide is one of the few recalcitrant organophosphates. Persistence of pesticides is measured by its half-life, which stands for the time it takes for one-half (50%) of the substance to break down. Persistent pesticides have a half-life of more than 100 days. The half-life for bensulide can extend as far as 363 days. Also, bensulide has a tendency to adsorb, or bind, strongly to soil particles. Adsorption is estimated by mixing together water, pesticide, and soil. First, the distribution coefficient is calculated by dividing the pesticide concentration adsorbed to soil by the concentration remaining in the water. Finally, the distribution coefficient is divided by the amount of organic carbon (which is directly proportional to soil organic matter) found in the soil. This yields the adsorption coefficient, and the higher this value, the stronger the pesticide is adsorbed. Studies on bensulide's adsorption coefficient (K_d) range from 1,245 to 7,721 in soils like sand, sandy loam, and clay loam. Bensulide's high half-life and adsorption coefficient combined with its incorporated application into the soil affords grazing animals and birds the opportunity to ingest bensulide over long periods of time.

Bensulide also has a tendency to accumulate in organisms, so it can remain in animals as well. This process, referred to as bioaccumulation, is quite harmful because it results in increasing concentrations of a chemical in organisms that are high in the tropic levels compared with the amount found in the surrounding environment. Because of interrelationships within the food web, bioaccumulation can also allow these toxins to climb the food chain and affect different trophic-level organisms. An example of this can be found in the organochlorine pesticide dichlorodiphenyltrichloroethane, or DDT. DDT was introduced during World War II to protect the troops from malaria by effectively reducing mosquito populations. Because of its success, DDT was widely used in the United States. However, DDT can bioaccumulate and was carried up the food chain to birds, causing eggshell thinning. American bald eagles were one of the many birds affected by accumulated DDT. Comparative studies on eggshell thinning have shown similarities between DDT and bensulide. It also revealed that bensulide needs to be 10 times greater in concentration to have such an effect, but that can be reached because the application rates of bensulide are much greater than those of DDT. It is, therefore, reasonable to assume that bensulide can have a similar hindrance on eggshell thickness as DDT once had.

Considering the potential harm bensulide can cause, it is imperative that efforts be taken to remove it from lands that no longer benefit from its herbicidal properties. This is the case where bensulide was once used on farmlands that have now been converted into forest preserves or zoned for commercial use. Limited studies have been performed on bioremediation of bensulide, so I decided to use the well-researched area of bacterial remediation of pollutants. Like most pesticides, bensulide has a complicated structure that demands several reactions to degrade it, so the one reaction route of cometabolism is the most viable option. Considering this, hydroxylation was selected because it incorporates a hydroxyl group (–OH) to a compound. Hydroxyl groups, commonly found on alcohols, are known for making a substance more soluble. Hydroxylation could, therefore, assist in increasing bensulide's solubility, allowing it to exit the soil more freely. Cytochrome P450 is one of the most common enzymes that carry out hydroxylation and is not substrate specific. Although many organisms carry cytochrome P450, it is not common in soil bacteria. An exception is *Bacillus megaterium*, which is a gram-positive, aerobic, spore-forming soil bacterium. This rod-shaped bacterium is easily isolated from soil and considered to be one of the largest Eubacteria.

The working hypothesis was that the cytochrome P450 from *B. megaterium* can carry out hydroxylation on bensulide. To test the chemical reaction (ex vitro), it was important to simulate the basic environmental conditions. For instance, cytochrome P450 had to be suspended in an appropriate buffer to simulate the fact that the enzyme is intracellular (i.e., it is only found within the bacteria). Also, all cytochrome P450 reactions are fueled by NADPH (an oxygen donor), and, therefore, it was added to the buffer solution. Finally, the bensulide was added. An ultraviolet spectrometer, which uses light to measure liquid-suspended compounds, was calibrated to quantify NADPH concentration. This indicated when the reaction was complete. Because the NADPH gradually dissipated, this suggested a reaction (most likely hydroxylation) had taken place. The liquid was subsequently removed and analyzed using a high-performance liquid chromatograph, an instrument used to separate and identify compounds.

Based on this analysis, an unidentified peak appeared which implied that bensulide was probably altered to include a hydroxyl group. The results suggested that cytochrome P450 from *B. megaterium* can possibly carry out hydroxylation on bensulide. In turn, this could allow bensulide to become more soluble and to be effectively removed from the soil. However, more research must be conducted before any practical bioremediation techniques can be developed for bensulide-polluted soil. Because cytochrome P450 is an intracellular enzyme, future research is aimed toward extracellular enzymes to avoid bensulide being consumed first by the bacteria.

Key Terms

Biological monitoring 243
Biodegradation 248
Bioremediation 239
Biostimulation 250
Chelate-assisted-
 phytoremediation 246
Hydroseeding 242
Hyperaccumulators 246
Landfarming 250

Metal-tolerant genotypes 240
Phytodegradation 245
Phytoextraction 245
Phytoremediation 245
Phytostabilization 245
Stockpiled soil 241
Surface stabilizers 240
Toxicity tests 244

Review Questions

1. How does metal toxicity in plants generally manifest itself?
2. Outline the main problems with storing stockpiled soil.
3. What are the limitations in using hyperaccumulators?
4. Describe the process of chelate-assisted phytoextraction.
5. Describe how the optimal soil conditions can be enhanced to optimize bioremediation.
6. Describe the process of landfarming.

Further Reading

1. Alexander, M. 1994. *Biodegradation and bioremediation*. San Diego: Academic Press.
2. Bradshaw, A. D. and Chadwick, M. J. 1980. *The restoration of land. The ecology and reclamation of derelict and degraded land*. Berkeley, CA: University of California Press.
3. Chadwick, M. J. and Goodman, G. T. 1975. *The ecology of resource degradation and renewal*. Malden, MA: Blackwell.
4. Ripley, E. A., Redmann, R. E., Crowder, A. A., Ariano, T. C. and Farmer, R. J. 1996. *Environmental effects of mining*. Boca Raton, FL: Lucie Press.

11

FOREST

Chapter Outline

11.1 Forest Degradation
　　　Clear-Cutting and Selective Cutting
　　　Human Settlement and Ranching
　　　Land Mismanagement
　　　Pollution
11.2 Forest Restoration
　　　Passive Restoration
　　　Active Restoration
　　　Regeneration Niche
11.3 Tropical Rain Forest
　　　Threats to the Rain Forest
　　　Restoration
　　　Obstacles to Restoration

Case Study 11.1: Recovery of Forested Ecosystems After Management in Nova Scotia, Canada

Forests cover about 31% of terrestrial land and are still a dominant feature of the landscape worldwide. Large-scale conversion of forests into agricultural lands has taken place in different biomes. For example, temperate forests have been reduced by about 35% from their original size. Similarly, **tropical rain forests** (TRFs) have been reduced by about 50%. During the last decade (2000–2010) about 130,000 km^2 of forested lands were converted annually to other uses.

Most of this massive **deforestation** has taken place recently. The exponential growth in the human population is the main underlying reason for the recent escalation in deforestation. Human population pressure leads to increased demand for fuelwood, other forest products, and, most importantly, land for agriculture. About half of all wood that comes from forests is used as fuelwood. Large areas worldwide are still facing devastating deforestation, and today small forest fragments can indicate only what the larger coverage once was. For example, only about 10% of the original forests exist in Madagascar as fragments, and only about 2% remain as fragments of the once extensive Atlantic forests in Brazil. Deforestation is almost always followed by advances in human settlements. For example, about half of the forests in Australia were cleared by the European settlers.

Deforestation can alter ecosystem functioning, including nutrient cycling and hydrology. Also, deforestation can induce wide-scale soil erosion and sedimentation in streams and rivers.

Deforestation of vast areas can have impacts on regional climate patterns. For example, in West Africa large-scale deforestation has resulted in reduced evapotranspiration and, consequently, less moisture being brought inland by prevailing wind where drought conditions are not uncommon. Similarly, in the Amazon basin the gradual conversion of TRF to grasslands and savanna may lead to reduced rainfall further inland. About 50% of the rainfall in the TRF is derived from evapotranspiration within the forest. Destruction of the TRF will most likely lead to increased drought in the Amazon basin. Intensified droughts could also lead to more forest fires and may prevent reestablishment of the TRF on degraded sites.

Degraded forests and deforested lands can be restored using different approaches, depending on the scale of the damage. These restoration approaches are discussed below.

11.1 Forest Degradation

Clear-Cutting and Selective Cutting

About 30% of all forests are being degraded. Economic loss as a result of forest degradation is estimated to be about $2 trillion. Forest degradation can result from different methods of logging. For example, **clear-cutting** is the most widespread method of logging and, at the same time, is the most devastating because it clears the whole forest over a large area. Clear-cutting may alter the landscape and lead to permanent damage to the ecosystem. This includes increased soil erosion leading to excessive runoff and sedimentation in streams, loss of soil nutrients, and buildup of soil crust. Non-native species and weeds frequently invade clear-cut forests. Invasion of non-native species and weeds can prevent or seriously delay forest regeneration. Furthermore, symbiotic soil organisms such as *Rhizobia* and mycorrhizal fungi can be reduced by clear-cutting, especially if host plants are eliminated. Also, appropriate pollinators and animals responsible for seed dispersal are eliminated by large-scale clear-cutting, which can seriously halt forest regeneration.

Another more agreeable method of logging is selective cutting, where only the most valuable trees are logged. This method is often considered a better alternative to clear-cutting. There are also problems associated with this method of logging, however, such as potential genetic erosion of the targeted tree population. For example, if the tallest trees within a population are continuously being logged, the average height of trees in a population is lowered. Also, selective cutting of only 10% of trees within a forest probably results in damage to more than half of the forest. This damage takes place when large machinery is brought into the forest for cutting and hauling the logs out of the forest. Issues of soil compaction are also frequently reported, leading to poor forest regeneration.

Human Settlement and Ranching

Dirt roads are typically built during logging operations because they allow easier access into forests. When the logging is done, these roads serve to facilitate human settlements into the forests. Human settlements induce intensive cultivation, grazing by livestock, and regular burning (often in ecosystems that are not adapted to fire, which in turn prevents forest regeneration and leads to soil impoverishment). For example, human settlements in the TRF are usually plagued with poverty and extremely low standards of living and lack normal social infrastructure such as schools and a health care system.

Cleared forests are often converted to pastures (grasslands) and overgrazing can lead to soil erosion. After soil erosion, natural regeneration of forest can take a considerably long time. For example, it may take up to 30 years after artificial grasslands are abandoned for soil organic matter to reach predisturbance levels and even up to 60 years before native trees start to colonize. It may even take centuries for the full recovery of a native forest community.

Another aspect of logging and ranching in these forests is how adversely it affects indigenous people. In fact, both the culture and lives of indigenous people are endangered. For example, the population of indigenous people living in forested lands in Brazil were reduced by about 98% due to other settlements. Native knowledge on how to survive in these forests is rapidly being lost along with valuable ethnobotanical and ecological information. Ecological restoration should particularly pay attention to these native cultures. Empowering indigenous people and especially recognizing their ownership of the land is important in this respect. Native inhabitants benefit enormously from restoration activities on their land, and the restoration practice itself can benefit from their local or traditional ecological knowledge.

Land Mismanagement

Common use of forested land has, in many cases, meant a complete lack of responsibility on the management of this land. Such practices follow the trend outlined by the "tragedy of the commons," as discussed in Chapter 1. For example, inappropriate nationalization of forests in Nepal has resulted in near total destruction of forests in the country.

In a more detailed example, the government of Brazil has opened the Amazon basin for large-scale industrialized agriculture and settlements. A 3,200-km Trans-Amazonian highway was built in the 1970s to make it easier for logging companies and settlers to reach remote locations in the Amazon basin. Illegal logging and a network of dirt roads followed. Settlements along the highway were further encouraged by the government by providing each settler a 100-hectare lot, 6 months' salary, and easy access to agricultural loans. There are, however, many difficulties with this type of agriculture in the Amazon basin. For instance, shortly after the TRF is cleared, excessive erosion (up to 100 tons per hectare) usually takes place. The soils are quickly exhausted of available nutrients, and more forest has to be cleared almost annually to maintain this form of agriculture.

Brazil's government-assisted settlements of the Amazon basin is a tragic example of land mismanagement. They have resulted in economic and social failures for the peasant settlers. Moreover, they have had adverse impacts on the indigenous people of the Amazon basin. Most of these settlements have been abandoned because of lack of social infrastructure and adverse economic situations. Perhaps most serious are the long-term environmental problems that have come with the TRF destruction. It has been estimated that the annual TRF destruction in the Amazon basin has averaged about 2.6 million hectares. Over only a 5-year period (2000–2005) Brazil lost more than 13 million hectares of TRF. The rate of deforestation has been at an astounding 3,560 hectares per day. The deforestation has, however, slowed down in the Amazon basin. Between 2003 and 2004 about 2.7 million hectares were deforested compared to 0.7 million hectares between 2008 and 2009.

Another example of land mismanagement leading to widespread destruction of forests can be found in the history of the settlement of subarctic Iceland. It is estimated that at the time of the earliest human settlement (9th century), up to 40% of the country was covered with birch (*Betula pubescens*) woodlands. Today, only about 1% of the country is covered with birch woodlands (**Figure 11.1**). This is mainly the result of land clearing for farms and a continuous assault (burning and overgrazing) on these woodlands throughout the centuries. Today, scattered birch fragments are found widespread over the country, indicating the previous extent of these woodlands. The locations of many former woodlands are known, but today those sites are seriously degraded or just barren lands. In the few locations where complete clear-cutting has not occurred, genetic erosion followed the selective cutting of the largest trees. Also, massive soil erosion has resulted in the total loss of about 60% of the vegetated land in Iceland. This catastrophic soil erosion that has taken place in Iceland since human settlement has been linked to the destruction of the woodlands.

Pollen analysis indicates a shift in vegetation cover after the settlement from a birch-dominated canopy to open sward dominated by grasses. The woodlands included species resilient to small climate changes and even disturbances associated with volcanic activities such as widespread ash fallout. The low biodiversity implies that ecosystem functions may depend on only a few species. Certain species of this ecosystem may be resilient to changing climatic conditions, but the ecosystem may be vulnerable if key functions are disrupted to a great extent. Today, the degenerated state of the woodlands is no longer resilient to disturbances or small climate changes. A possible sequence of woodland degeneration is shown in **Table 11.1** and **Figure 11.2**.

Figure 11.1 A rocky hillside in Iceland that is surrounded by rocks and sloping grassland. A birch woodland to the left is separated from the grassy hill by a fence to prevent grazing. (Courtesy of Sigurdur Greipsson, Kennesaw State University.)

Table 11.1 Community States After Perturbation of Pristine Birch Woodland: Degradation Increases from Community State 1 to 6

Community State	Perturbation	Main Five Indicator Species (>25% Cover)
1. Birch: lush understory	None-pristine	Betula pubescens, Vaccinium uliginosum, Hierochloe odorata, Rubus saxatilis, Geranium silvaticum
2. Birch: understory of herbs and grasses	Grazing Logging Fire	B. pubescens, G. silvaticum, H. odoratum Deschampsia flexuosa, V. uliginosum
3. Willow: heath	Grazing Fire	Salix lanata, S. phylicifolia, D. flexuosa, Agrostis. tenuis V. uliginosum
4. Blueberry: heath	Grazing	V. uliginosum, Empetrum nigrum, Arctostaphylos uva-ursi, Calluna vulgaris, Betula nana
5. Crowberry: heath	Grazing	E. nigrum, V. uliginosum, A. uva-ursi, D. flexuosa, Carex bigelowii
6. Kobresia: moss-heath "tussocks"	Grazing Wind-erosion	Kobresia myosuroides, Racomitrium lanuginosum C. bigelowii, E. nigrum, Thymus arcticus

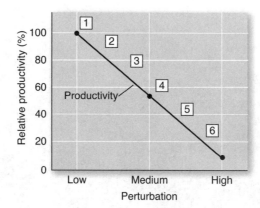

Figure 11.2 Predicted relative productivity during degradation of birch woodland/heathland ecosystem. Relative status of community states 1 to 6 is shown (see Table 11.1 for details).

Table 11.2 Sequence of Woodland Degradation Leading to Soil Erosion

1. Woodland destruction: birch is cut, burned, and overgrazed.
2. Degradation of the understory vegetation: overgrazing by livestock.
3. Loss of *Salix* shrubs: loss of plant litter.
4. Gradual change in plant communities toward degenerated state: tussocks covered with mosses.
5. Gully erosion initiated.
6. Wind erosion initiated.
7. Catastrophic soil erosion across the landscape.
8. Barren landscape: bedrock, gravel, or sand.

The conversion of birch woodlands to heathlands was usually the first step toward degradation of the ecosystem. Reduced cover of *Salix* shrubs was critical in this process, and open sward resulted in the formation of gullies. These gullies can cut right through the soil mantle and initiate wind erosion, leading to catastrophic soil erosion where the whole soil mantle can be lost to the bedrock (**Table 11.2**). The erosion of Andisol can be so intensive that entire woodlands are turned into barren landscape within few decades. This example demonstrates how compounded disturbances on the ecosystems of otherwise healthy, thriving species can result in ecosystem disintegration and collapse.

Pollution

Just as human settlements and land mismanagement have adversely affected forests worldwide, so too has industrial pollution. For example, acid rain associated with the burning of fossil fuels has adversely affected forests in northern

Europe, Canada, and the eastern region of the United States. Acid rain and its associated acidification of soil remain prevalent over large areas worldwide and will have unforeseen long-term effects.

11.2 Forest Restoration

Forests that have been logged usually have great potential for restoration if soils are not impoverished. This is especially true where forest fragments are found scattered throughout the landscape or where patches of intact forest still exist nearby. Such patches are also called biological legacies and contain portions of the original ecosystem. Restoration of forests aims at restoring the structure and biodiversity of these important landscapes. Pragmatic restoration approaches involve succession management, as discussed in Chapter 4. Forest succession usually takes a long time, measured in decades or centuries.

Passive Restoration

A passive restoration approach that facilitates natural colonization and secondary succession is best suited where forest degradation is not extensive, soils are not impoverished and especially where forest fragments still exist within the degraded landscape. Such strategy is cost effective but could involve long-term monitoring. Also, onsite management actions involving active control of weeds or non-native species and fire management is probably needed for a long time. Some ecological surprises such as invasion of non-native species are expensive to deal with, but careful monitoring can avoid the escalation of such problems, as discussed in Chapter 7. Invasive pathogenic species can have devastating effects on forests as was the case for the American chestnut (**Figure 11.3**).

An example of successful passive restoration can be found in the secondary succession currently taking place on abandoned farms in the northeastern United States. The forest was cleared for agriculture during early European settlement, but these first New England farms were abandoned in 1840s when the Midwest was being settled. Those abandoned farmlands are now mostly covered by deciduous forest. The structure and diversity of those "regenerated forests" are similar to the original forest. To initiate secondary succession, it was important to remove the main disturbing factors. Abandoning the farms and, therefore, reducing logging and livestock grazing was, in this case, enough to facilitate secondary succession. Also, secondary succession was rapid because the soil remained intact. Many forest fragments were present within the landscape, which also served to shorten the recovery time. Those forest fragments have also been a constant source of native seeds that provide seed rain into nearby degraded lands.

Land-use history was found to influence forest succession. For example, in Massachusetts white pines (*Pinus strobus*) now dominate fields that were previously plowed, whereas oaks dominate pastures that were not plowed. These secondary forests will, therefore, not always be identical to their predisturbance

272   Chapter 11 ■ Forest

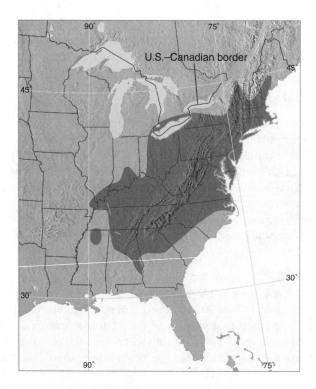

Figure 11.3 The original distribution of the American chestnut. The chestnut tree was almost decimated from its entire range by the invasive non-native fungus chestnut blight.

state, at least not within one century. It is clear, however, that secondary succession leading to natural recovery of degraded forests can occur over large areas, and passive restoration is a pragmatic management option in ecological restoration. Another example of a large-scale recovery of forests is given in **Case Study 11.1** on page 284.

Active Restoration

At the landscape level fragmented forests can be restored by using "expansion strategies." This could involve restoration of "buffer zones" where the size of forest fragments is gradually increased, as discussed in Chapter 6. Restoration of buffer zones around forest fragments is likely to be successful because of the close proximity to native seed source from the intact forest fragment.

Another approach is to link fragments together by using corridors or stepping stones (forest islands), as outlined in Chapter 6. Large swaths of open landscape act as dispersal barriers for seed dispersed by animals. Transplanting tree islands as stepping stones throughout the degraded landscape can, therefore, facilitate seed dispersal into large swaths of degraded lands. In an attempt to optimize this work, geographical information system models can be used to target concentrations of forest fragments to designate successful transplanting sites. Such techniques can guide restoration to effectively connect forest fragments.

Most tree seeds are dispersed by animals. Seed dispersal by birds into degraded sites can be facilitated by installing bird perches at regular intervals throughout the degraded landscape. Of course, seed availability depends on seeds produced in nearby forest fragments or intact forests. Continuous colonization of native species is, however, enhanced by the seed rain from forest fragments or nearby intact forests. Strategies to improve seed dispersal into degraded lands, therefore, should be the focus of restoration efforts.

Scattered tree plantings (as individual trees or small islands) throughout the degraded landscape attract seed-dispersing animals. Birds are especially active as seed dispersers when they have access to isolated or remnant trees. Transplanting single trees into degraded landscapes, therefore, can offer perches for birds, which might facilitate forest regeneration.

If establishment of native trees on a degraded site is slow or not taking place, transplanting native trees can overcome such dispersal barriers. If large intact forest fragments are found within the degraded site, then only few tree species should be targeted for transplanting. Using container-based trees raised in nurseries is, however, a more expensive method of forest restoration than direct seeding. Nursery-raised trees can be established from seed, wildings, or stumped saplings. This method is especially appropriate where limited amounts of seed are available, however. The initial goal in forest restoration by transplanting native trees is to augment the population of keystone species that facilitiate succession. Transplanting is often accomplished by skilled labor, although using tree-planting machines is cost effective. Spatial scaling in transplanting native trees, however, is yet another factor that needs to be considered. In forest restoration, planting design should imitate natural patterns of the species as they appear in the reference site, if possible.

Succession of degraded forest is often slow due to dispersal barriers of seed. This is especially a problem if forest fragments are absent or very scattered within the degraded landscape. **Direct seeding** of native trees, however, can be used to overcome dispersal barriers. Seed can be dispersed directly over a restoration site using various methods (i.e., sown by hand, ground-based machinery, or aircraft) depending on the size and isolation of the site. Direct seeding by aircraft in isolated sites that are inaccessible by ground-based machinery has shown mixed success. Large-scale restoration is probably impossible without direct seeding using machinery. Several problems, however, are associated with direct seeding. Usually only a small proportion of the surface-sown seeds germinate. In addition, seed predation is often a limiting factor. To enhance seed germination various substances (including plant hormones) can be coated on the seeds. Such an approach, however, adds substantial costs to the restoration work. Another potential problem is the erratic nature of seedling establishment of surface-sown seed. This approach also requires a large supply of seed. Dense vegetation cover on restoration sites can, for instance, inhibit germination and establishment of surface sown seeds. The sucess of direct seeding can be improved by implementing some ground work (disking, harrowing, or plowing) before the operation. However, such ground-based operations add substantial costs to restoration projects. Another method to improve surface seeding is to implement weed control before, during, and after seeding.

In addition to the above listed prohibiting factors for direct seeding, another major inhibitor may be the large amount of seed needed for such a program. To mitigate high costs, harvesting programs of native seeds should be considered in the early phases of forest restoration programs. Restoration of forests containing high biodiversity may require using a mixture of seeds from many native tree species. Seed collection from native stands can be challenging for many reasons, however. The fact that seed maturation is not synchronized for all tree species, for instance, might be one major challenge. The difficulties in mechanizing the harvest of many tree species might also put limits on these operations. To address some of these challenges, restoration projects might target seed collection of only few keystone shrub or tree species that facilitate succession.

Direct seeding is potentially the most effective method of large-scale restoration of birch woodlands in sub-arctic Iceland on restoration sites that have degraded soil. The birch can be one of the pioneer colonizers on newly exposed glacier moraine, emphasizing its strong potential in establishing on harsh microsites. The success of direct seeding is improved by disk plowing or harrowing the ground before the actual seeding. Seeds are harvested from local birch stands using handheld vacuum-harvesting devices. Seed quality varies between years, and it is important to monitor seed germination.

If direct seeding on degraded site fails, then another active restoration approach such as dense **enrichment planting** of container-raised trees may be implemented. Enrichment planting involves the transplanting of a variety of tree species. To enhance the success rate of this method, it is important to use tree species that mature early to produce seed. Degraded sites can be prepared before transplanting by implementing prescribed burning, weed control, and/or ground disking or plowing, but this adds cost to the restoration work. Also, aftercare programs involving weed control and nutrient amendments improve survival and ensure growth of the transplanted trees.

Instead of transplanting all tree species at once, planting in phases may be appropriate by using "assembly rules," if information on such an approach is available. For example, at sites that are severely degraded, a two-phase restoration strategy should be considered. The first phase involves the use of cover crops (trees or shrubs) that are fast growing and tolerant of the adverse environmental conditions on degraded sites. Using non-native trees or shrubs as nurse crops might be an option but should be considered carefully because of the trees' potential to invade or alter soil chemistry as well as populations of microorganisms. The second phase involves transplanting the dominant native tree species, but only after the harsh microenvironment is improved. Planting programs that aim at enhancing succession should also be encouraged.

On sites where weeds or non-native species could be a problem, a **dense transplanting** strategy should be considered. Such a strategy could lead to a closed canopy, which is beneficial in preventing (by shading) invasion of non-native species and weeds. Dense transplanting, however, adds enormously to the cost of restoration projects. A dense planting strategy accomplished by using limited numbers of tree species is also successful in attracting seed-dispersing animals. Such a strategy that uses many species facilitates restoration of biodiversity. This method is especially useful where natural recolonization is slow and the risk of invasion by weeds or non-native species is considered to be high.

Regeneration Niche

Development of restoration strategies in Mediterranean climates has focused on using facilitative effects of native shrubs for transplanted trees. Shrubs and native understory vegetation can facilitate reforestation in the dry and hot Mediterranean climate. In this region's reforestation efforts, clearing of existing vegetation, especially shrubs, is not necessary. In fact, such an operation can actually initiate soil erosion. Instead, existing vegetation can facilitate the establishment of transplanted trees. The existing vegetation buffers extreme microclimatic conditions, such as solar radiation and soil temperature, and helps in conserving moisture and soil nutrient content. Also, native shrubs provide transplanted trees with shelter from high wind and abrasion of sand or debris. Existing vegetation, especially shrubs, can, therefore, provide **regeneration niches** for transplanted trees by amelioration of the microclimate and by buffering adverse environmental conditions. As an example, native shrubs such as *Salvia lavandulifolia* have been used successfully as a nurse plant for reforestation of native target pines (*Pinus silvestris* and *P. nigra*) in the Sierra Nevada mountains of Spain. Other arid regions could follow this model of using native nurse plants in reforestation without the need for excessive groundwork.

A second example of a facilitative effect of native species to provide a regenerating niche for target native species is found in bald cypress (*Taxodium distichum*) stream swamp forests in southeastern United States (**Figure 11.4**). Bald cypress is a deciduous conifer that has its natural distribution extending from Delaware to southern Florida with northwestern limits in Tennessee and Illinois and southwestern limits in Texas.

The restoration of bald cypress forests is important because they form a refuge for a myriad of plants and animals and their streams are important spawning grounds for fishes. Bald cypress forests also play an important role in the hydrology of streams, especially in buffering floodwaters, trapping sediment from the water, and reducing sediment erosion and eutrophication. In addition, bald cypress forests have an exceptionally high primary productivity and can act as a sink for atmospheric CO_2. In fact, bald cypress forests make some of the most efficient carbon sequestering ecosystems in the world. Bald cypress grows

Figure 11.4 A logged bald cypress forest in Alabama. (Courtesy of Sigurdur Greipsson, Kennesaw State University.)

well at high stand densities and gives high yields. The annual aboveground production of biomass in bald cypress floodplain forests in Florida was 15,700 kg per hectare.

The stream swamp forest is dominated by the braided stream water and adjacent mud banks and small ponds. The initial restoration of degraded bald cypress forests should involve remediation of the hydrology on the site by stabilizing braided streams.

The bald cypress is a very flood-tolerant species that is able to withstand long periods of inundation. Seasonal but short term flooding is important to replenish the moisture and nutrients in the soil. Unfortunately, seasonal flooding along many rivers is curtailed due to a series of dams and other engineering structures that control the flow of rivers.

During the early 1930s there was a major logging operation in the southeast United States that resulted in clear-cutting of most of the virgin bald cypress stands. Bald cypress is valuable for its lumber, and its forests have declined at an alarming rate during the last century. The bald cypress is highly valued for its very decay-resistant wood. During drought periods bald cypress forests are more accessible by large logging machinery, and at that time they are especially vulnerable to clear-cutting. After logging these streams forests have generally low market value. Individuals and organizations interested in restoration should be encouraged to buy these lands at a low costs. Floodplains used for cattle grazing that are prone to flooding could also be targeted for bald cypress restoration.

Factors that could facilitate the regeneration of cleared bald cypress forests are site specific and should be carefully evaluated. It is critical to know if cleared forests can regenerate by natural seed rain from nearby stands of bald cypress, if available, or from individual trees that remain on otherwise cleared stand.

There are several methods available for propagating bald cypress for restoration purposes. These methods include direct seeding using container-raised plants, cuttings, and saplings. Transplanting nursery-raised seedlings on a large scale can be prohibitively expensive and is often impossible in remote areas. Direct seeding, however, might be a cost-effective method of restoration. Establishment from seed requires soil to be saturated with water but not inundated for a few months after seeding because seed germination is inhibited under water.

Cleared bald cypress forests are typically colonized by black willow (*Salix nigra*) that can, within just a few years, form a dense canopy. This changes the microenvironment because the understory vegetation cover is reduced and surface temperatures and light transmittance through the canopy are diminished. The black willow canopy can provide a regenerating niche for the establishment of bald cypress. Establishment of a black willow canopy as a native nurse species should, therefore, be encouraged on degraded bald cypress sites. The canopy of black willow should, however, be reduced or removed 2 years after bald cypress trees have established because bald cypress regenerate naturally in large forest gaps. Bald cypress grows slowly in partial shade, but optimal growth is only reached in full exposure to sunlight.

A third example of the facilitative effect of a native species to provide a regenerating niche for a target native species can be found in the cedar (*Cedrus libani*) that once formed forests in the mountains of Lebanon (**Figure 11.5**). (Cedar forests can also be found in the Taurus Mountains of Syria and southern

Figure 11.5 Cedar forest in the mountains of Lebanon. (Courtesy of Sigurdur Greipsson, Kennesaw State University.)

Turkey.) The cedar is an evergreen conifer, and individual trees can grow to an enormous size and live up to 3,000 years.

Throughout many periods of history the cedar was highly valued. It was used to build temples, palaces, and ships, and it formed the basis of the wealth and power for the ancient Phoenicians. Ancient logging of the cedar by the Egyptians dates back to 2700 BC. The ancient city of Biblos was the main harbor of the lumber export. The dynamic trading at Biblos resulted in the early Phoenician civilization. Deforestation in the mountains of Lebanon continued throughout early history by civilizations such as the Assyrians, the Romans, and the Ottoman Empire. The cedars were not only used in building ships and palaces but were also wasted tragically as firewood (locomotives of the Ottoman Empire used cedar's wood as energy). The logging was so intensive that today only a few small remnants of the once extensive forests still exist.

The existing fragments of cedar forest are mostly found in inaccessible and isolated places. Today, about 1,700 hectares of fragmented cedar forest are found in Lebanon. Only 12 Lebanese sites still harbor intact cedar forests. These cedar forests represent only a small fraction of the once extensive forests that are estimated to have covered about 81,000 hectares in the mountains of Lebanon before massive logging started. Once the cedar forests were cut they did not regenerate, most likely because of relentless overgrazing by livestock. Scrubby vegetation and grasses typically replaced logged forests. Also, soil erosion usually followed in the wake of forest destruction, and many sites that were once forested are now barren lands. This example demonstrates that compounded disturbances can lead to a permanent alternative state of the ecosystem in question.

Reforestation efforts of the cedar forests in Lebanon began with international support in the 1960s. About 50 hectares were reforested in Ain Zhalta, where cedars were planted directly in stony ground on bulldozed terraces in the mountains. The cedars, however, were planted in monoculture, and today most of these trees are either deformed or severely stunted in growth. The reforestation efforts were halted in 1975, with the onset of the civil war in the Lebanon.

Alternative strategies in the restoration of cedar forests include a delay in the transplanting of cedars until native shrubs, especially oaks (*Quercus infectoria* or *Q. libani*), which grow as large shrubs at high altitudes, have established and grown to a critical size. It might be necessary to facilitate the establishment of the oaks, which provides "facilitative effects" for the establishment of cedars into restoration sites. The cedars can be transplanted on the sheltered site of the oaks that act as native nurse plants. In this case the oaks provide a regenerating niche for the transplanted cedar.

One critical aspect of the restoration of the cedar forest is that cedars do not bear cones until they are 40 or 50 years old. Reforestation must rely on raising nursery-grown plants from seed and transplanting them strategically into restoration sites. Of course, the establishment of local nurseries is needed for this work. Cedars must be raised in such nurseries and allowed to harden outdoors for few years before being transplanted. It is important to use seed of local populations for this purpose because mixing seed of different populations should be avoided. It is also important to sample seed from many trees and avoid using methods of propagation that are based on cloning single or a few genotypes.

Restoration should aim at expanding buffer zones around fragments by using ecological principles such as establishing oak trees as nurse plants before transplanting the cedar, as mentioned above. For example, the 40 hectares of fragmented cedar forest at Barouk could, in an ambitious project, be expanded to about 3,000 hectares. Exclusion from livestock grazing is essential and can only be accomplished by erecting strong fences around restoration sites.

It is important that the local community perceives such reforestation efforts positively. Reforestation of the cedar forests has many potential economical and social benefits. The benefits for the local communities are increased ecotourism benefiting local tourist resorts and a general increase in local revenues. Tourism in the cedar forest is already an important source of income for local villages, but its environmental impact should be carefully examined before being further extended. The hot and dry Mediterranean summer climate will be moderated by the shade of the trees. Wildlife may reestablish in these areas in the long term. Another benefit relates to water conservation, which is a critical issue for the growing population in Lebanon. Today, the distribution of the cedar is mainly confined to places that receive more than 800 mm of annual precipitation. These areas are mainly located high in the mountains between 1,400 and 1,800 m above sea level. Most of the precipitation on these sites comes as snow, and the dense canopy cover of the cedar forest delays snow melting in the spring. The cedar forest can, therefore, improve water retention in this otherwise arid region.

Hardly any precipitation falls during 3 months of intense heat during the summer, when there are great contrasts in the local microenvironment between cedar forests and nearby barren sites that are exposed to intensive sunlight. In the cedar forest a mist can form in the summer during the afternoon and may form drops that precipitate to the ground. The mist also moderates the intensive summer heat. The mist is probably caused by evapotranspiration from the cedar forest. The mist formation might also have brought precipitation across the mountains into much more arid regions north of the Bekaa valley. Deforestation of the cedars, therefore, has probably not only altered hydrology in the mountains but has also changed the regional climate adversely.

Reforestation greatly improves the hydrology in the mountains that feed springs at lower elevations. Currently, most of the water that comes from the snow is lost as runoff during early spring snow-melt. Potable water is becoming one of the most critical commodities in this part of the world. Reforestation of the cedar forests, therefore, has a role in large-scale watershed management.

The final example of the facilitative effect of native species to provide a regenerating niche for target native species can be found in the restoration of birch woodlands in subarctic Iceland. Before transplanting birch on restoration sites, it is important that willow (*Salix* sp.) shrubs have established. The willows act as a native nurse species that facilitates the establishment of birch by providing a regeneration niche. Nursery-raised birch should be planted on the sheltered site of *Salix* shrubs to increase their survival. Such a transplanting strategy also follows random distribution of willow shrubs in the landscape and avoids linear transplanting. Care should be taken in providing container-raised birch seedlings inoculated with appropriate mycorrhizal fungi. Also, strategic transplanting of birch seedlings into restoration sites should consider the prevailing wind direction as a vector for future seed dispersal.

11.3 Tropical Rain Forest

The TRF is located around the equator and has a very uniform, warm temperature (26°C–28°C) throughout the year as well as high and uniform precipitation levels. The TRF once covered about 6% of terrestrial lands, where most of these forests used to be located in Central and South America (63%), then Asia (25%) and Africa (12%). Today, the greatest continuous area of TRF is found in the Amazon basin. The Amazon basin contains about 45% of the remaining TRF in the world.

TRFs have the highest biodiversity of all biomes, and it is estimated that as much as 90% of all species can be found there. As an example of this enormous biodiversity, more tree species have been found in 2 hectares of Costa Rican rain forest than in all of United States and Canada together.

Rain forests have the highest primary productivity of all biomes and, therefore, play a critical role in global CO_2 cycling. In fact, massive destruction of the TRF is responsible for about 30% of anthropogenic CO_2 in the atmosphere. Rain forests are also a critical sink for large amounts of carbon. Restoration of TRFs can be strategically used to reduce net fluxes of CO_2. Carbon sequestration on terrestrial lands may be very effective in the tropics. It is estimated that a maximum feasible worldwide reforestation program with emphasis on the TRF can potentially lower the CO_2 level in the atmosphere by 15 to 30 ppm over 50 years.

Threats to the Rain Forest

It is estimated that the total cover of TRF has declined about 50%. Each year about 15.4 million hectares of TRF are logged. The intensity of TRF logging has been immense in the past, and in just one decade (1959–1968) the loss was about 38% in Central America and about 24% in Africa. Furthermore, about 50% of Costa Rica's TRFs were logged between 1960 and 1986. The distribution of TRFs worldwide has consequently changed due to deforestation. Today, a much higher proportion of TRF is found in Central and South America (80%) followed by Asia (10%) and Africa (10%). In Asia many countries have only fragmented TRFs. In fact, only Indonesia still has some large continuous areas of TRF remaining. In Africa deforestation has been severe, and in West African the only remaining forests are highly fragmented.

The TRFs are currently being cleared at an alarming rate, mainly for agriculture. Shifting cultivation is probably the main threat for TRFs. The senseless destruction of TRF has been paraphrased by Edward O. Wilson in his book, *Biophilia*: "The action [TRF destruction] can be defended (with difficulty) on economic grounds, but it is like burning a Renaissance painting to cook dinner."

After the TRF is cleared, small-scale farming is typically established, but the land is only cropped intensively for 1 to 2 years due to low soil fertility. Substantive areas of TRF (11%) have been cleared to grow cash crops such as bananas, oil palm, rubber, and coffee. Large plantations of eucalyptus and pines (*P. radiata*) are grown as lumber in the tropics. Plantations of cash crops such as sugarcane and soy beans have a drastic impact on the environment and such cultivation has been on the rise. Monocultures of cash crops have negative impacts on biodiversity, soil stability, and hydrology. Brazil is the second largest exporter of soy beans, approaching 55 million tons annually. Most of the soy cultivation takes place in the Amazon basin and has been advancing rapidly, especially as new genetically modified, heat-resistant varieties have been cultivated. The soy cultivation has a detrimental effect on the ecosystem because it alters the soil chemistry. The TRF soil is acidic but is limed before soy is planted. It is unknown what effect this practice has on the soil or potential restoration of the TRF.

The future of the TRF is grim, and it is predicted that by 2035 all lowland TRFs worldwide will be destroyed. About 20% of the total cover of TRF is predicted to be lost over the next two decades. It is also predicted that within the next two decades about 40% of the Amazon TRF will be destroyed. Sadly, almost all the logged timber is wasted. Only about 2% of the cut lumber is marketed. Typically, just the higher priced lumber such as mahogany is marketed. The rest of the forest is cleared, burned, and the land is turned into cash crop cultivation or grassland for cattle ranching.

About 30% of the loss in TRFs is due to cattle ranching. In the Amazon basin cattle ranching can be blamed for roughly 70% of deforestation. Brazil is the world leader in beef export, with about 60 million cattle. The beef is predominantly produced in the Amazon basin.

Intentional fire used in forest clearing is a great threat to the TRF. In just 1 year more than 200,000 hectares of forest have been burned in the Amazon basin. The TRF is permanently damaged by such practices because most tropical tree species are not adapted to fire.

Thousands of hectares of TRF have been destroyed by hydropower dams. Most of the energy derived from these dams is destined for high-consuming international industry such as aluminum smelters. These projects have often been funded in part by the World Bank and other international organizations. There are currently plans to build seven mega-scale hydropower dams within the Amazon basin.

Mining has increasingly impacted the tropics. It results in a direct loss of TRF due to the clearing of land and the severe water and soil pollution that follows. Local inhabitants seldom benefit from such operations. The aluminum industry is especially destructive because large strip mines operate in the TRF. The aluminum industry mines subsoil deposits (bauxite) rich in aluminum. This operation takes place in strip mines where the forest is cleared and the soil is removed before the mines are opened.

In Jamaica 10% of the world's bauxite is produced. Jamaican bauxite mining is responsible for the high rate of deforestation. For the past 50 years the mining operation has cleared more than 5,100 hectares of land, most of which was intact rain forest. The mining operation has also opened an access to the once remote forest by building access roads. Today, the mining industry is required by law to return the land to a productive state after the mining operation ceases. Such work involves grading the mine and spreading a thin layer of topsoil before revegetation starts. More then 3,000 hectares have already been revegetated on the island.

Restoration

Initial steps in restoration of TRFs involve the removal of disturbing agents (e.g., fire, logging, and livestock grazing). This is followed by restoration strategies that increase seed dispersal. In the TRF birds, bats, and other animals typically disperse seed of mid- and late-successional trees. Seed-dispersing animals should be encouraged to move across the deforested landscape. Installing bird perches is an effective strategy in increasing bird-dispersed seed into severely degraded forests.

Using native trees in the restoration of TRFs has many challenges. The lack of studies, for instance, on the biology of TRFs inhibits their use. Restoration efforts would undoubtedly benefit from a more complete set of basic information on the biology of trees targeted for reforestation of TRFs. For instance, seed production of TRF trees is unreliable, and limited information is available on their seed biology. Reliable seed production is often a problem, and high seed predation usually takes place.

In the central Amazon pioneer tree species with small seed size are not appropriate for direct seeding. The ground cover in the Amazon also influences the success of direct seeding. Seedling establishment is, however, most successful on sites lacking vegetation. Although such non-vegetated sites have adverse microclimate conditions, there is no interspecific competition; it is assumed that predation and herbivory is much less than on vegetated sites. Seeding directly on vegetated sites can result in direct fungal attack on the seed, and seedling herbivory by insects is high. In addition, competition with grasses can be severe. Weeds and non-native species can also inhibit establishment of tree seedlings. To control invasion of weeds and non-native species, trees that provide a dense canopy are transplanted. Shading provided by transplanted canopy trees can effectively reduce the vigor of many weeds and non-native plants.

It has been suggested that reforestation of the Amazon basin should be accomplished by the transplanting of pioneer tree species and, at the same time, the seeding of selected nonpioneer species that produce large seed. Such planting strategy is called "stacking succession." This approach maximizes the establishment of trees and simulates secondary succession. When a dense canopy is established, shade-tolerant trees are typically transplanted. Another approach is to simultaneously transplant species of different successional stages, which may accelerate succession. Transplanting native trees in clusters (islands) within a degraded landscape is an option if forest fragments are very scattered. Forest fragments should be used effectively as a seed source.

Obstacles to Restoration

Many developing countries harbor several social obstacles to large-scale restoration of TRFs. These include the exponential growth of the human population and the huge demand for agricultural land. Also, a lack of legislation and governmental infrastructure within many developing countries inhibits TRF protection and restoration. In addition, total lack of funding for large-scale restoration projects of the TRF inhibits any reforestation efforts.

International intervention may be the only option to conserve and restore the TRF worldwide. One strategy that has been suggested is the "debt swap for nature." This strategy encourages countries with huge national debt to work toward protecting its forests and facilitate reforestation. In fact, developing countries owe more than $1 trillion in loans, and it is unlikely that these loans will ever be paid back. The international community can put pressure on these countries to give ownership on degraded TRF in lieu of loan payments. The debt swap for nature program is under the supervision of the Tropical Forest Conservation Act that was enacted in 1998. The debt swap for nature consists of a voluntary agreement between a developing country and its creditors to discontinue debts in exchange for ownership of TRF. This act includes funds for restoration of degraded forests.

Another strategy to save the TRF is to improve the rights of indigenous people. Indigenous people of the Amazon basin are estimated to have been between 2 and 6 million. Most of these people were affected by diseases brought by European settlers, and consequently their population has drastically declined. Today, about 450,000 indigenous people inhabit the Amazon basin. About 25% of the Amazon basin is currently set aside as land for indigenous people. This is an important conservation strategy because the destruction stops where the land of indigenous people starts. Also, the fate of the Amazon TRF and the indigenous people is linked together. The recent expansion of conservation efforts, especially efforts in recognizing the rights of indigenous people in the Amazon basin, has been somewhat successful in slowing down the rate of deforestation.

Summary

Forests are still an important feature of the landscape and provide society with various ecological services. Forests have been reduced drastically worldwide. Currently, the most extensive deforestation is taking place in the TRF. Increases in human populations and demand for products derived from the TRF are the underlying reasons for this destruction. Clear-cutting forests is a common practice. Forests have, however, a great potential to regenerate, especially if large fragments of the original forest still exist within the landscape. Passive restoration is a cost-effective approach where forest fragments exist within the landscape, but this involves long-term protection and, in some cases, weed control and fire management. Another restoration approach is to expand buffer zones around fragments. Dispersal barriers into degraded forests can be overcome by direct seeding or transplanting of tree seedlings. In this regard it is important to establish effective seed dispersal routes by animals, especially birds and bats, by installing perches. If natural recolonization is unacceptably slow, a dense planting strategy using many tree species at a time can be used to facilitate rapid restoration. Another active restoration strategy is to use regenerating niche provided by native nurse plants for transplanting target native trees. TRFs are valued for their high biodiversity. They also play an important role in atmospheric carbon fixation and oxygen production. TRFs are being decimated at an alarming rate, mainly to clear land for agriculture. Restoration of the TRF involves curbing the disturbing agent, especially logging, ranching, and fire. Active restoration of the TRF could involve transplanting pioneer tree species and, at the same time, seeding of selected nonpioneer species with large seed size. Another restoration approach involves simultaneous transplanting of species of different successional stages. Also, forest fragments should be used effectively as seed source for seed-dispersing animals.

11.1 Case Study

Recovery of Forested Ecosystems After Management in Nova Scotia, Canada

Liette Vasseur, Department Biological Science, Brock University, St. Catharines, ON, Canada

In eastern North America forest communities comprise a mosaic of pure and mixed stands of mature deciduous and coniferous species. From a gradient south to north, the temperate forest ecosystem changed to become the boreal forest. Along the eastern coast of Canada and New England, there is a transient forest ecosystem, the Acadian forest, which includes species from both temperate and boreal forests. Historically, these forests have been exploited for timber. With land settlement and forest harvesting most of the regions have seen significant changes in landscape, forest composition, and even genetic makeup. Forest managers have long oriented their goals toward maximizing valued goods and services from the forest ecosystem. With growing pressure from the general public, wildlife scientists, and foreign countries as potential buyers of forest products, however, biodiversity standards as one of the important criteria for sustainability and efficiency in restoration are often emphasized. The maintenance, restoration, and even increase of diversity in our forests represent a concern, which requires a better understanding of forest disturbance on different temporal and spatial scales.

Under natural conditions, the Acadian forests can survive into the late-successional stages of stand development because of low frequency of large-scale disturbance and high longevity of dominant species, especially species such as hemlock and red spruce. Approximately two-thirds of harvested lands in Nova Scotia are regenerating at sufficient densities, with the remainder requiring planting, seeding, and other postharvest silvicultural treatments. Studies to better understand how regeneration can occur under forest management pressures and what best practices in restoration ecology could be developed to mainly ensure optimal regeneration while protecting biodiversity have reported various results from different regions of Atlantic Canada. The current knowledge of the Acadian forest remains limited, with some knowledge on certain types of practices. Previously mentioned studies have focused on specific areas, and few generalizations have been possible. Here the results from studies conducted in Nova Scotia, Canada, are summarized and demonstrate the importance of considering timber management strategies and climatic conditions for restoration.

Study on the Potential of Forest Regeneration

A total of 19 sites representing a chronosequence of forest ages (general three sites of 0, 2, 5, 20, 54, and >100 year old forests) in southwestern Nova Scotia were selected and sampled according to a completely randomized design. Replicates were randomly chosen from a geographical information system database of forest stands, types and age classes provided by Kejimkujik National Park and Historic Site of Canada and forest management units of Bowater Mersey Paper Company and surveyed to ensure similar forest cover, topography, drainage, and silvicultural history. Company records included forest cover before logging, dates of disturbance, and silvicultural history for post–clear-cut stands. Some sites were mechanically harvested and skidded according to industrial forestry protocols, whereas a few originated from hand-cut and logs skidded using horses.

This research project was conducted in two parts: (1) the impacts of anthropogenic and natural canopy disturbance on structure and understory diversity in managed and unmanaged red spruce forests of southwestern Nova Scotia and (2) the influence of seeds and advanced regeneration in the restoration process of these sites after clear-cutting. In each site at least three plots of 20 × 20 m were randomly installed as well as six transects of 80 m originating within the adjacent forests toward the center of the clear-cuts. In each plot plant diversity from the three strata (ground, shrub, and tree vegetation), woody debris, and demographic data for *Gaultheria* were measured. Along the transects seed bank, seed rain, seedling growth, and herbivory as well as plant diversity were measured.

Results suggest that diversity varies in function of the intensity of the disturbance; although most species reestablished after disturbance, some (e.g., *Coptis trifolia, Trillium undulatum*) were restricted to mature/old growth sites. Because of their characteristics some of these species can be considered as indicator species in other research projects on the impacts of forest management on biodiversity or in monitoring in operational conditions. Simple measurements and monitoring of these species can give us some indication of the health of the forest ecosystem and its level of recovery through natural

(continued)

Case Study (*continued*)

processes. The convergence in secondary succession indicates that clear-cutting temporarily increases species richness, but the community rapidly shifts to its initial composition.

Distance from the forest edge has been shown to influence natural regeneration patterns after clear-cutting or fire in other conifer-dominated systems. For example, decreases in seedling abundance and in growth response were observed for balsam fir (*Abies balsamea*) with increasing distance from an intact forest. Similar patterns of decline toward the center of the clear-cuts were documented for red spruce in this study. An immediate drop of roughly 50% in average seed rain per square meter was observed along each transect from within the intact forest into clear-cut areas. The results also show that (1) there is a relatively high and rapid recolonization of red spruce due to advanced regeneration after clear-cutting, (2) seed bank over time and distance from forest edges show high level of diversity, and (3) stems of 50 to 150 cm in height are favored with about 50% of the trees experiencing winter browsing in clear-cuts. White-tailed deer (*Odocoilus virginianus*) (sightings, scat, tracks), porcupines (*Erethezon dorsatum*) (sighting, tracks), and snowshoe hares (*Lepus americanus*) (sighting, scat) have been observed within the harvested and/or adjacent mature stands. Browsing on red spruce seedlings appeared to stop as the availability of herbaceous and shrubby species increased (onset of spring and summer).

Discussion and Recommendations

Concerns exist that current levels of harvesting of softwoods in Nova Scotia are unsustainable. With increasing market demands and the importance of preserving biodiversity a greater concern, ecologically based management strategies must be implemented to ensure a future and healthy supply. This requires a better understanding of the regeneration processes occurring in post–clear-cut sites. Is natural recovery possible or is there a need for active restoration?

Our research in Nova Scotia suggests that even in clear-cuts, the potential for regeneration can remain high if practices reflect the types of disturbance often present in this region (i.e., small-scale blow-down). It is worth noting that in this region oceanic climatic conditions prevail, leading to more humid conditions, favorable for seed germination and seedling survival.

The forest canopy of the oldest clearcut stands has recovered to dominance by red spruce. This finding is in accordance with the classical view of succession, which suggests that forest ecosystems return to the same initial stage after catastrophic event. Convergence in successional outcome toward spruce dominance has been observed by other researchers in stands that were logged before the development of mechanized harvesting methods. The return to predisturbance canopy composition has been attributed to the relatively benign impact that early logging methods, particularly skidding of felled trees with horses, had on the survival of conifer "advanced growth" (i.e., seedlings established on the forest floor before clear-cutting). The technique used by the studies' lands includes small patches, limited compaction by the machinery, and retention of woody debris on site. Combined with favorable climatic conditions, this management strategy has the potential to accelerate or optimize recovery of the ecosystem.

Figure 11.6 Extensive regeneration of conifers occurs from seed within recent clear-cuts as in this deciduous forest in northeastern United States. (© Snehit/ShutterStock, Inc.)

Persistence of advanced growth in stands logged with mechanical harvesting methods is often inadequate due to the poor survival of seedlings after catastrophic canopy removal. This is especially true in the interior of the province where the climate is drier and conifer seedlings often face periods of summer drought stress. In this study because of oceanic conditions leading to a more humid climate, extensive regeneration of conifers within recent clear-cuts from both ex situ seed (**Figure 11.6**) and in situ sources can be observed. Although an emerging hardwood canopy dominated recent clear-cuts at first, by 20 years conifers had returned to predisturbance level. Evidence of an anthropogenically initiated convergence in successional outcome toward mixed softwood composition suggests that the seedling–sapling conifer bank persists after canopy removal with clear-cutting. Under such conditions natural restoration can be promoted, leading to a forest where components are relatively similar to the original ecosystem. Such a restoration approach can be used for much of the otherwise fragmented forests in the United States (**Figure 11.7**).

To ensure long-term, sustained productivity of the wood resource and to better manage forest ecosystems for their nonextractive values and biodiversity, forest resource managers request a better understanding of the sustainability processes taking place in forest stands for the tree species and key indicator species. Increased knowledge of stand-level factors that influence the success of plant reestablishment will help to achieve these goals. Management practices that take into account the effects of these factors can be implemented, if necessary, to help ensure red spruce recovery and to reduce the negative impacts of clear-cutting on the distribution of other native plant species.

(continued)

Case Study (continued)

Figure 11.7 The distribution of fragmented forests in the United States.

Key Terms

Clear-cutting 266
Deforestation 265
Dense transplanting 275
Direct seeding 273
Enrichment planting 274
Reforestation 278
Regeneration niche 275
Tropical rain forest 265

Key Questions

1. What are the main reasons for widescale deforestation worldwide?
2. Describe passive restoration of forests.
3. What ecological services are provided by the TRF?
4. How can seed dispersal into degraded landscape by animals be encouraged?
5. Define the term "regeneration niche."

Further Reading

1. Lamb, D. and Gilmour, D. 2003. *Rehabilitation and restoration of degraded forests.* Gland, Switzerland: IUCN.
2. Laurance, W. F. 1997. *Tropical forest remnants.* Chicago: University of Chicago Press.
3. Mansourian, S., Vallauri, D. and Dudley, N. 1986. *Forest restoration in landscapes: beyond planting trees.* New York: Springer Verlag.
4. Sauer, L. J. 1998. *The once and future forest : a guide to forest restoration strategies.* Philadelphia: Andropogon Associates.
5. Stanturf, J. A. and Madsen, P. (eds.). 2004 *Restoration of boreal and temperate forests.* Boca Raton, FL: CRC Press.
6. Whitmore, T. C. 1998. *An introduction to tropical rain forests.* Oxford, UK: Oxford University Press.
7. Williams, M. 2006. *Deforesting the earth.* Chicago: University of Chicago Press.

12

ENDANGERED ANIMALS

Chapter Outline

12.1 Restoration of Critical Habitats
　　　Area Threshold
　　　Carrying Capacity
　　　GIS Technology
12.2 Captive Breeding
　　　Genetic Structure and Genetic Variation
　　　Giant Panda
　　　Black-footed Ferret
　　　California Condor
　　　Role of Zoos
12.3 Translocation and Reintroduction
　　　Translocation
　　　Reintroduction
　　　Release Strategies
　　　Reintroduction of the Gray Wolf into Yellowstone Park
　　　Complex Reintroduction
Case Study 12.1: Hybridization Between Introduced Walleye and Native Sauger in Montana: Implications for Restoration of Montana Sauger

Numbers of endangered animals on the brink of extinction are on the rise worldwide. Restoration of endangered animals can be accomplished successfully by integrating programs designed to improve critical habitats where these species can survive and by using special techniques that augment the population in question. These techniques include captive breeding, translocation, and reintroduction. These and other integrated restoration programs are certainly a more attractive alternative to the extinction of endangered animals.

Species extinction is mainly due to fragmentation and destruction of habitats, as outlined in Chapter 3. Many endangered species require specific or critical habitats for their survival. Habitat restoration is, therefore, often a prerequisite for successful integrated restoration programs that focus on populations of endangered animals. Habitat restoration is discussed at length in this chapter.

Animals rarely exist as evenly dispersed individuals within a landscape. Instead, they usually thrive among discrete populations in specific or critical habitats. Most populations of animals in nature form a metapopulation structure, as discussed in Chapter 6. The metapopulation generally retains greater genetic diversity than would a single, large population with the same number of individuals. Small and isolated populations typically contain low genetic diversity. It is well known that low genetic diversity is detrimental to the survival of species. Increasing gene flow between populations, therefore, enhances species survival. Restoration efforts should, therefore, aim at establishing metapopulation structures by increasing connectivity between isolated populations. This is especially important for species with specific habitat requirements.

While increasing connectivity between populations, it is important to consider the movement patterns of these endangered animals within a landscape. Such work might involve establishing source populations, establishing satellite populations, or building corridors or stepping stones between populations, as discussed in Chapter 6. In this respect it is important to identify dispersal barriers between isolated populations. It is also valuable to understand how target populations relate to other nearby populations and how metapopulation structure can be enhanced.

Animals are threatened worldwide, especially by the combined effect of habitat loss and degradation. Other contributors to the dilemma of species endangerment include direct hunting, trading of live animals, and competition from non-native species. Non-native species can also alter the gene pool of the native species (see **Case Study 12.1** on page 306). Amphibians, birds, and mammals are particularly threatened. In fact, of the roughly 4,600 known species of mammals found worldwide, 1.8% have become extinct in the last 500 years. Today, about 4% of mammals are listed by the International Union for Conservation of Nature and Natural Resources (IUCN) as "Critically Endangered," about 7.4% as "Endangered," and about 12% as "Vulnerable."

A comprehensive restoration strategy of large animals is certainly needed. This involves the use of "flagship species," which are typically charismatic and emblematic animals. This is especially important to maintain political and financial support of such programs. At the same time, such an approach increases education and awareness.

Restoring endangered mammals requires legal and policy framework of national and international legislation and agreements. One such agreement is the U.N. Convention on Biological Diversity, which especially recognizes the need for species and ecosystem restoration (see Chapter 14).

Different techniques of restoring populations of animals are discussed at length and relevant examples are given below.

12.1 Restoration of Critical Habitats

Restoration of the **critical habitat** is essential for enhancing survival of populations of endangered species. The critical habitat for endangered species is composed of complex factors and is specific for each species. The northern spotted owl (*Strix occidentalis caurina*), for example, has a critical habitat composed of old-growth conifer forests with a multilayered canopy. In this case large-scale logging results in the removal of large old trees that make up the critical habitat.

Selection of a critical habitat by animals is a hierarchical process and must be considered during the planning stage of restoration projects (**Box 12.1**). The hierarchical order used by endangered species in selecting a critical habitat is as follows: the geographical range of the species, the home range of an individual, the use of particular habitat elements within the home range (i.e., foraging, nesting, or rest sites), and the actual food items that an animal selects (**Box 12.2** and **Figure 12.1**).

Destruction of the critical habitat can have detrimental effects on animal populations. For example, the observed decline of terrestrial birds in western North America has been associated with the loss and degradation of riparian habitats. About 95% of these habitats are already lost or have been severely degraded. The degradation is characterized by the loss of willows and cottonwood along rivers. Restoration of these habitats cannot simply be accomplished by outplanting these tree species along rivers. It takes much more effort to restore these critical habitats. First, proper stream functioning must be restored. This involves reinstating the meandering pattern of the river and natural flood regime. This should be followed by stabilizing embankments through revegetation, mainly by using native grasses. This effort protects embankments from

---

Box 12.1   Spatial Scales of Habitat Selection by Animals

- Microhabitat: Microsite
- Macrohabitat: Home range
- Landscape: Dispersal distance
- Regional: Species range

> **Box 12.2** — **Hierarchical View of the Habitat Selection Process for a Migratory Bird**
>
> - Region (North America)
> - Vegetation type (shrubland)
> - Perch site (large shrub)

flood erosion and at the same time filters sediment that can be deposited. After this preparation work, tree species that dominate the critical habitats, such as willows and cottonwoods, can be successfully transplanted.

Restoration programs should provide a natural complexity and disturbance regime for the critical habitat. Restoration of critical habitats must, therefore, consider natural disturbances (fire, floods, etc.) that maintain the quality of the environment over time. It is also important to consider dynamic processes such as succession and shifts between alternative ecosystem states. Throughout the restoration of critical habitats, it is important to implement a long-term monitoring program (discussed further in Chapter 14).

The success of restoration programs for endangered animals depends on the integrity of the critical habitat. If the critical habitat is seriously degraded, a reintroduced population will probably fail to establish. Furthermore, to augment the size of the target population, it is essential that the degraded habitat is restored. During the initial planning process, it is, therefore, important to evaluate available patches of the critical habitat within the landscape of the target population, and identifying dispersal barriers between patches. Such evaluation includes identifying available patches of potential predators, competitors, and invasive species. Restoration efforts should not only focus on the condition of the critical habitat, but also should identify features of the landscape that need to be restored, such as connectivity between populations to enhance metapopulation networks.

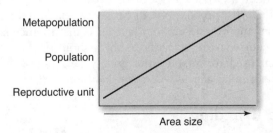

Figure 12.1 Spatial areas of habitat needed to support functionally significant demographic units plotted for one hypothetical animal species.

Area Threshold

The **area threshold** is the minimum area of critical habitat to maintain a viable population. Area thresholds have been estimated for many species. For instance, the area threshold for the ovenbird (*Seiurus aurocapillus*) in New Jersey is 3 hectares of forest patches. Although a species occupies a given habitat, it does not necessarily indicate habitat suitability. Thus, to ensure population growth, habitat patches may have to be much larger than the area threshold.

Restoration of the critical habitat should establish an effective size of the breeding population. Such a restoration effort begins by identifying the area threshold needed to support a breeding population. A population should be self-maintaining and large enough to avoid inbreeding. It is possible to estimate the size of a target population by using the minimum viable population approach, which applies genetic and demographic variables for this purpose.

Carrying Capacity

The **carrying capacity** of the critical habitat is an important factor that always limits restoration efforts. The carrying capacity of a habitat determines how many individuals of a population can exist there. Although this carrying capacity can be augmented temporarily by continuous reintroduction or supplemental feeding, such approaches are not sustainable in the long run. The most agreeable approach to enhance the carrying capacity is, therefore, to expand the critical habitat of the target population through restoration efforts. The goals of such efforts should be restoring the critical habitat and restoring the quality of the habitat (matrix) surrounding the target population in question, which at the same time enhances functioning of any metapopulation network.

GIS Technology

To prioritize restoration efforts of critical habitats, geographical information system (GIS) technologies can be used along with other approaches discussed further in Chapter 14. For instance, a GIS model is used to rank suitable habitats for targeted populations of endangered species. Such ranking is based on various landscape features that have potential for restoring critical habitats. For example, a GIS model was used to prioritize restoration efforts of grizzly bear (*Ursus arctos horribilis*) populations in Montana. Habitat variables such as elevation, vegetation, and human activity (acting as disturbance factor) were used to identify the occurrence of grizzly bears throughout the year. The results of this study were then used to enhance the grizzly bear population by closing roads and restricting human access to critical habitats.

12.2 Captive Breeding

Captive breeding is the first step in the restoration of endangered species when a population reaches a critically low number. A captive breeding program is initiated by catching wild animals and rearing them under surveillance at appropriate facilities. Zoos have played an important role in captive breeding programs. To operate such programs requires skilled personnel with knowledge of the biological requirements of these animals. If a critical population size can be built in captivity, then excess animals can be released into their natural habitat to establish new populations or to augment existing populations.

Genetic Structure and Genetic Variation

Before animals targeted for a captive breeding program are caught in nature, it is important to obtain information on the genetic structure of their populations. Nonrelated individuals of the target population should ideally be selected for a breeding program. Small and isolated populations are more likely to contain low genetic variation than are larger populations or populations that are effectively connected in a metapopulation network. Strategic selection is needed to obtain the greatest genetic variation present in a wild population. Animals that carry desirable traits and animals that are distantly related are given the highest priority for breeding in captivity. It is estimated that the most genetic variation can be captured from a wild population by randomly obtaining at least 20 unrelated animals. If the targeted species has reached a critically low number, however, the only option is to capture all remaining individuals. This approach, however, requires moderate numbers of founding animals for the breeding population. For example, the "50/500 rule" recommends a minimum of 50 individuals for short-term breeding programs and more than 500 individuals for long-term breeding programs. After the founding animals for a captive population are caught, they are allowed to increase in number to the maximum size that can be supported by the restoration program. Random mating in captivity is important to maintain genetic variation of the population. Inbreeding by related animals must be prevented by designing circular mating schemes. An important goal of captive breeding programs is to optimize the genetic diversity of a population.

Giant Panda

An example of captive breeding can be found in studies on the giant panda (*Ailuropaoda melanoleuca*). The panda is probably one of the most charismatic and emblematic endangered species (its image can even be found on the logo of the World Wildlife Foundation). The panda is often the main attraction in zoos where crowds find it adorable, cute, and pet-like. The panda looks and acts peaceful. It eats only bamboo shoots, and this strict vegetarian diet adds to its nonthreatening image. The future of the panda has been jeopardized by low population size in nature due to serious degradation of bamboo forests that form its critical habitat. The panda also faces problems associated with reproduction in captivity. The natural population of the giant panda was recently found to be critically low (roughly 1,000 individuals in 12 populations). In addition, about 100 pandas were kept in captivity, but reproduction in captivity was rare and the survival rate of cubs was very low.

The reproductive biology of pandas makes rearing them in captivity a challenge. Females are only fertile for 1 or 2 days a year. To make matters worse,

296   Chapter 12 ■ Endangered Animals

males in captivity do not show any interest in copulation and are even aggressive toward females during their few fertile days. The only option in this situation was to implement artificial insemination. This technique was improved and especially adapted for the pandas. The reproductive biology of pandas was also studied in detail, and the short timing of female fertility can now be accurately identified. To collect sperm from donor male pandas a breeding technique called electroejaculation was used. To inseminate the females a modified medical technology (laparoscopy) was successfully used.

In addition to these assisted reproductive efforts, the restoration of the bamboo forests is essential for the expansion of the existing panda populations. Such work is already taking place in China and is sponsored by various zoos and organizations.

Black-footed Ferret

The North American black-footed ferret (*Musteles nigripes*) is an example of a species recently rescued from extinction by captive breeding. The ferret was once widespread on the Great Plains, where it occupied almost 42 million hectares (**Figure 12.2**). The population size was estimated to have been between

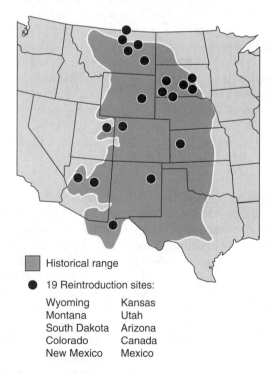

Figure 12.2 Previous distribution for the black-footed ferret. Sites of recent reintroduction are also shown. (Reproduced with permission of the Black-footed Ferrett Conservation Center/USFWS.)

half a million to 1 million individuals. The decline of the ferret population was related to a control program on their main prey, the prairie dog (*Cynomys* sp.). In 1915 a rodent control program aimed at eradicating prairie dogs resulted in a drastic decline and fragmentation of ferret populations. The ferret was listed as endangered in 1967 and was thought soon after to be extinct. In 1981 a population of about 130 animals was discovered near Meeteetse, Wyoming. This discovery initiated a restoration program for the ferret. Throughout this work adaptive management was used successfully in planning restoration efforts (see Chapter 14). This population was monitored intensively using field-based techniques including direct observation, snow tracking, spotlighting, capture–recapturing, radio-tagging, and prey studies. The population contained inbred individuals and probably only a fraction of the total genetic diversity of the original population. In 1985 outbreak of diseases almost wiped out this last population. Between 1985 and 1987, Wyoming Game and Fish caught the last 18 ferrets and started an intensive captive breeding program.

The Captive Breeding Specialist Group of IUCN gave expert advice and direction in establishing the captive breeding program. The captive breeding program for the ferret involved newly developed techniques such as artificial insemination, embryo manipulation, and genetic engineering. Restoration of populations of ferrets involved captive breeding, field pen breeding, and reintroduction into their natural habitat. The last 18 remaining ferrets in nature were caught; however, only 5 animals became the founders of the captive population. This small founder population probably induced a serious bottleneck to the potential genetic diversity of the species. The future consequences of this genetic restriction are not yet known.

The captive breeding program was successful, and the population was gradually increased. In 1987 seven litters were produced in captivity. In 1988 the ferret population grew to 59 individuals. In 1989 the population grew to 67 individuals. In 1990 the population was 180 individuals, and in 1991 the population was 320 individuals. The captive breeding program has produced about 3,500 ferrets, of which 1,500 have been reintroduced into their natural habitat. Local zoos played a critical role in the captive breeding program. Currently, ferrets bred in captivity are kept at different locations to avoid the risk of diseases. Ferrets that are reintroduced require critical habitats for survival. The main component of the critical habitat is large populations of prairie dogs. Population sizes of prairie dogs should not fluctuate drastically to avoid repercussive effects on the ferret population. The critical habitat should also be absent of diseases. Today, however, only few sites meet all these criteria.

The "Habitat Suitability Model" was used to select appropriate sites for the reintroduction of ferrets. Expert opinion was also used to evaluate such sites. Several reintroduction sites were located within Wyoming and adjacent states and Canada. The largest targeted site in Montana was assumed to support more than 450 ferrets. Reintroduction began in the fall of 1991 when 49 ferrets were released at Shirley Basin, Medicine Bow, Wyoming. This 295,000-hectare site can potentially support about 150 ferrets. Initial mortality was mainly due to coyote and badger kills. The goal was to establish more than 10 self-sustaining populations by 2010, with approximately 1,500 animals in their natural habitat.

It is essential in the near future to manage the reintroduced ferret populations as metapopulations. The goal is to establish at least 10 populations with corridors connecting them. Techniques of ferret reintroduction have steadily improved. For instance, using outdoor pens before reintroduction has doubled the survival of ferrets in their natural habitats. Also, the movement of released ferrets was tracked using implanted electronic devices and radio collars. Most monitoring, however, is done either through snow tracking (of footprints) or spotlight surveys.

Successful reproduction was documented in reintroduced populations in South Dakota, Wyoming, Utah, and Montana. In Arizona ferrets successfully reproduce in onsite breeding pens. It is clear that by providing reintroduction sites close to large prairie dog colonies and in absence of diseases, captive-bred ferrets can establish and reproduce in their natural habitats. Restoration of the prairie dog habitat is a prerequisite for successful large-scale reintroductions because the lack of large prairie dog populations remains one of the key limitations to ferret population restoration.

California Condor

Another example of captive breeding can be found in the California condor (*Gymnogyps californianus*). The IUCN Red List puts the California condor as "critically endangered." In the 1930s an estimated 150 California condors lived in nature. By 1987 they had declined to just 27 birds, and soon after that the population went critically low to only 8 individuals. At that time all condors were caught and brought into zoos. Various institutes collaborated on the captive breeding program for this species. In 1992 the first captive-bred condors were reintroduced into their natural habitats in California. In the last two decades the total population has grown to more than 240 birds.

Role of Zoos

Zoos have played an important role in preserving endangered animals. Also, zoos have maintained populations of animals with the goal of reintroducing them into their natural habitats. Today, a number of zoos collaborate in programs for captive breeding of more than 300 endangered animal species. Moreover, research that is undertaken in zoos often complements or forms the basis for captive breeding programs. This includes studies on the behavior, nutrition, reproduction, and physiology of wild animal species. The preserved population of endangered species that is kept in zoos serves as a genetic reservoir of the species. For example, zoos have successfully preserved large species such as the European and American bison (*Bison* sp.) and the Prewalski horse (*Equus prewalski*). Furthermore, improvements in animal management techniques, enclosure design, and biological knowledge of particular species requirements have led to successful breeding rates in most zoos. It should be noted, however, that a number of threatened species, such as the giant panda, the penguin, the killer whale (*Orcinus orca*), and the elephant (*Loxodonta africana*), have proved very difficult to breed in captivity. Improved reproductive techniques are, therefore, needed to ensure the survival of these species in the near future.

12.3 Translocation and Reintroduction

Translocation

Translocation involves relocating animals from a source site to another release site that has not necessarily been occupied by the species before. This process involves both animals caught in nature and animals that are bred in captivity. Translocation can be effectively used in establishing a new subpopulation or a satellite population in a metapopulation network. Translocation can also be used effectively to augment an existing population simply by increasing numbers of individuals in that particular population. The success of translocating animals, to a large extent, depends on the quality or suitability of the release site. Degraded sites can usually not support the translocated animals. Such sites should, therefore, be restored to resemble the critical habitat. Translocation of animals into their original habitat can then take place after that habitat is restored. In extreme cases where habitats have been seriously degraded, permanently destroyed, or converted to other uses, it is possible to preserve the target population temporarily in a surrogate site or in captivity.

The success of translocation depends on many factors, but predation is usually the main cause of mortality of the animals in question. High mortality of translocated animals can lead to a failure of the whole program. The main reason for higher mortality is animals raised in captivity in absence of predators have never developed the necessary behavior and skills to avoid predation. Animals can be trained to avoid predation, however, and survive in the new habitat. Such training is complicated, but it is a critical step toward the survival of the species. This approach has been attempted with tigers reared in Africa and trained to survive in nature with the goal of eventually translocating the animals to their native Asian habitats.

Reintroduction

A **reintroduction** program involves returning animals into an area in which they once inhabited but from which they were extirpated. These animals can be caught in nature, nurtured or raised in captivity, and then reintroduced into the new location. Reintroduction aims to establish viable, self-sustaining populations of animals capable of surviving in their natural habitat. Reintroduction has been used successfully to restore colonies of 49 seabird species in 14 countries. To enhance the success of the reintroduction bird decoys, recorded bird calls, and even mirrors have been used. Reintroduction has been successfully used in restocking populations with new genetic materials and in augmenting the size of declining populations. To increase the success of reintroduction programs, the suitability of the target species for reintroduction should be evaluated. The biological feasibility of the reintroduction program must be assessed where critical needs of the target species should be determined. These include an assessment of the conditions of the critical habitat and presence of diseases, predators, or competitors. Also important are the evaluation of the reintroduction site and the assessment of the success rate of the whole project. Reintroduction is not likely to be successful into degraded habitat that cannot support the target animal population. The need for habitat restoration should be assessed early in the reintroduction program.

The initial step in reintroduction programs is to assess availability of target animals. It is important to select individuals for reintroduction that are (potentially) genetically similar to the original population. Usually, this involves selecting populations that are located close to the target site. At the same time it is critical to maintain genetic variation of the animals involved in the program.

During the initial planning phase of reintroduction programs, it is important to acquire relevant permits that are within the legislative framework. Also, it is important to gain political support and to secure funding for the whole project. This involves effective dissemination of information about the project and effective communication.

After reintroduction, a long-term monitoring system should be installed. This involves setting priorities and objectives and selecting methods to measure success. Reintroduction programs should use adaptive management to assess the effectiveness of the project and intervene, revise, or halt projects as necessary (see Chapter 14).

Reintroductions of animals from captive breeding programs have been attempted for over 120 different species. Most of those reintroduction projects were undertaken over the past few decades. The success of these reintroduction programs has varied, with some failures. However, tremendous experience and information has been gained that will undoubtedly benefit future reintroduction projects. The final success of a reintroduction program depends, of course, on the survival of the released animals and their ability to reproduce and expand their population.

There is a great need for reintroduction programs as a rescue attempt for endangered animals. Reintroduction programs are currently being planned for a variety of species, including the golden lion tamarin (*Leontopithecus rosalia rosalia*), the black and white ruffed lemur (*Varecia variegata variegata*), elk (*Cervus elaphus roosevelti*), and the Florida manatee (*Trichechus manatus latirostris*).

Release Strategies

The release strategy of animals into their new site needs to be determined. Soft release and hard release are two basic techniques. **Soft release** involves placing captured animals from one location into a small enclosure (pen) for a period of time (days to months) to acclimatize them to their new habitat. The enclosure is usually a caged or fenced off pen located at the eventual release site. This allows the animals to become accustomed to the release site. Also, animal conditions can be monitored before their release. Providing shelter for animals is important for their establishment on the release site. In this respect birds can, for instance, be reintroduced by placing bird boxes on restoration sites. **Hard release** involves trapping animals directly from their native habitat, transporting them to a release site, and releasing them without any prior conditioning. This method is usually less successful than soft release, but it eliminates the stress that might be associated with keeping the animals in pens at the new site.

Reintroduction of the Gray Wolf into Yellowstone Park

Reintroduction of keystone animals can have widespread impact on the ecosystem. National parks are actively working on restoring intact habitats by installing ecological control mechanism of different trophic levels. This involves reintroducing top predators, which often act as keystone species, in the ecosystem. Many predators act as top-down trophic level control mechanisms in the ecosystem. One of the best known examples of restoration of top predators is the reintroduction of the gray wolf (*Canis lupus*) into Yellowstone Park. The wolves' primary need is for prey, which is most likely to be elk, deer, and other ungulates in the reintroduced areas. They also prey on smaller animals such as rabbits, opossums (*Didelphis virginiana*), and a variety of rodents.

Yellowstone Park was established in 1872 as the world's first national park. At that time little was known about predator–prey relationships and the role top predators play in ecosystems functioning. This lack of knowledge was shown in the disregard for the gray wolf by the park service and its early management practices.

In 1914 the Predatory Animal and Rodent Control Service was established with its primary objective to eliminate all predators from federal lands. These included coyotes (*Canis latrans*), cougars (*Felinae concolor*), and, of course, wolves. As a result of these actions, the last wolf in Yellowstone Park was killed in 1926. The once top predator in a dynamic ecosystem would remain absent from the park for the next 70 years. In 1933 the National Park Service changed its policy regarding predators and declared that no native predator should be hurt within national parks unless they were threatening the existence of another species. However, for the gray wolf this came too late as the last wolves had already been eradicated from the park. In 1974 the gray wolf was declared endangered in the lower 48 states.

In 1980 the U.S. Fish and Wildlife Service formulated a recovery plan for the gray wolf in the northern Rockies. This plan was later revised in 1987. It proposed a reintroduction of wolves into Glacier National Park, the Bob Marshall Wilderness in Central Idaho, and Yellowstone Park. This led to the 1991 congressional decision to provide funds to the Fish and Wildlife Service to prepare an environmental impact statement on restoring wolves to Yellowstone Park and central Idaho. In 1992 the Secretary of the Interior signed the Record of Decision on the final environmental impact statement for reintroduction of gray wolves. It was decided that the species would be managed as a nonessential experimental population. In the environmental impact statement several studies were conducted using population modeling to predict the future impacts of wolves on the ecosystem of Yellowstone Park. This work acted as a pilot study to the restoration project of the top predator. Today, the National Park Service policy calls for restoring native species when sufficient habitat exists to support a self-perpetuating population and management can prevent serious threats to outside interests. One important criterion for this program is that the restored population should resemble the extirpated population.

Geographic information system (GIS) technology was used to identify large areas of potential habitat for wolf packs. These studies demonstrated that road density was a good predictor of recolonisation by wolves.

The wolves were reintroduced into the park using a soft release technique, which had been used previously to restore red wolves (*Canis rufus*) to the southeastern United States and swift foxes (*Vulpes velox*) to the Great Plains. This involved penning animals temporarily in areas of suitable habitat. Each pen had a small holding area attached to allow a wolf to be separated from the group if a medical treatment was needed. Inside each pen were several plywood boxes to provide shelter for the animals.

Concern was expressed about the wolves becoming habituated to humans or to the captive conditions; however, the temporary holding period was not long in the life-span of a wolf. In Alaska and Canada wolves are seldom known to develop the habituated behaviors that are seen more commonly in other animals such as grizzly bears. Wolves typically avoid human contact and are highly efficient predators with well-developed predatory instincts. Their strong social bonds and pack behavior minimize their need to scavenge food in garbage and other human food sources.

In 1995, 14 gray wolves were reintroduced from Canada into Yellowstone Park. A year later 17 more wolves were reintroduced into the park. At the same time another 30 gray wolves were reintroduced from Canada into the nearby Idaho wilderness area. In 1996 four packs produced 14 pups. In the spring of 1998 nine packs of wolves produced 64 pups. In early 1997, 10 young wolves were orphaned when their parents were killed due to livestock predation on the Rocky Mountain Front, Montana. However, most of the wolves remained primarily within the boundaries of Yellowstone Park. In 2008, 124 gray wolves lived in Yellowstone Park. The first generation of reintroduced wolves into the park lives in several packs and has behaved much as predicted. Today, all wolves living in Yellowstone were born in their natural habitat within the park (**Figure 12.3**).

The greatest concern for the introduction of wolves into Yellowstone Park has been with local ranchers and recreational hunters. In response to concerns about wolves killing livestock, Defenders of Wildlife offered to compensate ranchers for any livestock lost to wolves. (Interestingly, this organization has the image of the wolf in its logo.) The National Park Service had also made a plan for wolves that venture onto private land. Wolves that attack livestock on

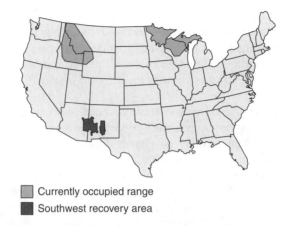

Figure 12.3 Distribution of the gray wolf (*Canis lupus*) in the United States. (Reproduced from USFWS. *Gray Wolf (Canis lupus): Current Population and Range in the United States* [http://www.fws.gov/midwest/wolf/aboutwolves/popandrange.htm]. Accessed April 7, 2010.)

private land are captured, kept for some time in pens, and then released back into the park. Otherwise, wolves suspected of killing livestock are killed. The population is, therefore, not considered essential to the survival of the species. However, in the 4 years after their release, wolves had only killed 8 cows and 83 sheep. This was only 10% of the predicted killing of livestock by wolves. Wolves are responsible for only 1% of sheep loss in the Rockies. In 2005, wolves were suspected of killing about 4,400 cattle in the United States. Domestic dogs, however, killed at least 22,000 cattle within the same time frame.

Speculations about the reintroduction of the gray wolf into the park began in the 1960s, especially to control the large and expanding elk (*Cervus canadensis*) population. The large population of elk were over-browsing and causing vegetation destruction, including riparian aspen and cottonwood, which forms the critical habitat for migrating birds.

The reintroduction of gray wolves into the park was expected to act as a top-down control mechanism (trophic cascade) by predation on the elk population in the park because the absence of wolves in the park led to a tremendous growth of the elk population, which resulted in over-browsing of vegetation. The wolves have, indeed, helped in controlling the elk population. Before the wolves were reintroduced into the park the elk population was about 20,000 animals, but today it is half that size. In fact, wolves act as a keystone species and are capable of changing the entire ecosystem within the park. Today, 16 packs of wolves roam the park, and each pack kills on average one elk a day. As expected, elks are the primary prey for wolves in the park throughout the year, representing 92% of the wolf kills. As a top predator wolves play a critical role in maintaining the healthy state of the elk population, because wolves often remove old, weak, and sick animals and control elk numbers to prevent them from over-browsing the land.

Other factors, however, could also have contributed to the reduction in the elk population, including an increase in hunting permits (quota) for elk. Also, increased numbers of grizzly bears and mountain lions and prolonged drought that reduced elk forage most likely all played a role in reducing the elk population. Winter weather is another great factor in controlling the number of elk and deer populations.

The reintroduced wolves have also indirectly made an impact on the growth of many plant species, especially riparian tree species such as aspen, willow, and cottonwood. The number of beaver dams has increased as the riparian forest regrows, which leads to a more complex landscape. Over-browsing mainly by elk had suppressed the growth of those trees and changed the structure of riparian habitat that is an essential habitat for many bird species.

The wolves in Yellowstone Park have successfully regained their role as top predators. The reintroduction of wolves has also restructured populations of lesser predators. Wolves routinely abandon unfinished elk and moose carcasses. These carcasses provide essential scraps for scavengers such as coyotes, eagles, and ravens. Mountain lions and even grizzly bears have also been found to follow in the wake of hunting wolves. Wolves have, in fact, had beneficial effects on the population size of grizzly bears and mountain lions as well. Wolves have a much more violent relationship with the coyote, however, and 14% of recorded clashes resulted in the death of one or more coyotes. The coyote population has consequently been reduced in size by 50% since the wolf's reintroduction. It is likely that the wolves will drive coyotes out of some areas within the park altogether. During the wolf's long absence from Yellowstone, coyotes reached some of the highest population densities ever observed in North America. Although the coyote reined as top predator, other lesser predators such as red foxes (*Vulpes vulpes*), wolverines (*Gulo gulo*), lynx (*Felis lynx*), and even fishers (*Martes pennanti*) were suppressed.

The gray wolf made a significant population growth in the Northern Rockies. In 2008, about 1645 gray wolves were counted in the Northern Rockies. In 2009 the gray wolf was declared recovered in the Northern Rockies and considered as a game animal. Hunting quotas were established for the gray wolf where 75 animals could be hunted in Idaho and 220 animals in Montana. The return of the wolf to the world's first national park is a remarkable milestone in species recovery and ecological restoration and can be used as a model for other similar situations. The visibility of the gray wolf reintroduction program provided opportunities to educate the public about predator–prey relationships, endangered species restoration, and the importance of restoring intact ecosystems.

Complex Reintroduction

Reintroduction of endangered animals is not always a simple process. An example about complicated interacting factors affecting the success of reintroduction program is derived from studies on the Channel Island fox (*Urocyon littoralis*).

This species is a miniature canine living on small islands owned by the National Park Service that lie off the coast of southern California. This species was reduced from more than 1,500 foxes in 1994 to just 14 foxes in 2000. At this critical point the few remaining foxes on the islands were captured and a captive breeding program initiated. The collapse in population size of the Island fox was mainly due to a direct predation by a golden eagle (*Aquila chrysaetes*) and dogs. The golden eagle had recently colonized these islands due to easy prey provided by feral pigs (*Sus scrofa*). These pigs were introduced to the islands in the 1850s and have since escaped. Their population has since grown out of control. The bald eagle (*Haliacetus leucocephalus*) was once the top predator on these islands, and its presence kept the golden eagle away. It is critical that the bald eagle did not prey on the foxes. The population of the bald eagle began to decline in the 1950s, mainly due to DDT toxicity, and they vanished from the islands in the 1960s, leaving a vacant niche for the golden eagle.

The first step in reestablishing the fox on the Channel Islands required a reduction in the number of golden eagles. This task was implemented indirectly by attempting an eradication of feral pigs from the islands. Also, golden eagles were captured and relocated to places that were far enough away to discourage a return to the islands. Reintroducing the bald eagle on the islands was an important step in this process.

Bald eagles that managed to establish on the islands are closely monitored and their eggs taken away and hatched in the San Francisco Zoo in incubators under controlled conditions to avoid any problems with the notorious egg-thinning effects associated with DDT toxicity. The bald eagle chicks that are hatched in incubators are then transferred back to their nest on the islands via helicopter. Today, more then 15 juvenile bald eagles have established on the islands. The first release of foxes reared in captivity was not successful because most were killed by golden eagles that remained on the islands. It is important, however, to eradicate the golden eagle from the islands before attempting further reintroduction of the island fox. In this case the control mechanisms of native top predators had to be restored before reintroducing the endangered species back into their natural habitat (**Figure 12.4** and **Box 12.3**).

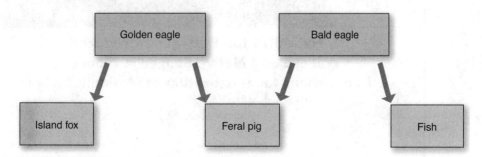

Figure 12.4 Important control effect of top predator on restoration success.

> **Box 12.3 Key Factors for Animal Restoration Projects**
>
> - Is the area threshold large enough to support a viable population?
> - Does the habitat of the release site contain all critical factors to allow establishment of the target population?
> - Are there any ecological constraints associated with land surrounding the release site?
> - Is it possible to manage such constraints to establish a population of the released species?
> - Is it possible to restore metapopulation structure?

Summary

Extinction of animal species can be avoided by establishing integrated captive breeding programs involving translocation or reintroduction of excess animals to selected sites. Such programs usually require restoration of the critical habitat for the target species before translocation or reintroduction. Captive breeding involves rearing the target animals under surveillance and assisting their reproduction using advanced technologies. Captive breeding programs must suit the biological requirements of the target species. The goal of captive breeding programs is to build excess population of target species for reintroduction or translocation. Translocation involves transferring animals from a source site to a release site that they have not occupied before. Reintroduction involves returning animals back to a site they once occupied but for some reason became extirpated. Reintroduction involves two techniques: soft release and hard release. Soft release involves conditioning the animals at the new site before release, whereas hard release does not.

> **12.1 Case Study**
>
> ### Hybridization Between Introduced Walleye and Native Sauger in Montana: Implications for Restoration of Montana Sauger
>
> *Dr. Neil Billington, Department of Biology and Environmental Sciences, Troy University, Troy, AL*
>
> Destruction of habitat is largely responsible for the extinction of native fish species. However, introduced species have been reported to significantly contribute to further loss of native species. A state survey on the use of native and non-native species in

fishing programs reported that 36% of the states had more non-native sport fishes than they did native ones. Additionally, 75% of the non-native fish have become well established and did not need further stocking to thrive. Management agencies, therefore, need to assess native fish populations before implementing the introduction of non-native species in their recreational programs.

In freshwater fisheries, the introduction of a non-native species has led to the decline of native species, by competition, predation, or by altering the gene pool of the native species through hybridization and introgression. Several non-native species have been introduced into most states to support sport fishing programs, in some cases leading to the extinction of native species. In an analysis of 40 extinct North American fish taxa conducted in 1989, introduced species were cited as responsible for 68% of these extinctions, while hybridization and introgression was responsible for 38% (more than 82% of extinctions had multiple causes); the major factor was habitat degradation (73% cases). Natural populations and their supporting ecosystems can be made more vulnerable to the impacts of introduced species when their habitat is degraded.

Hybridization occurs when two (parental) species cross to form an interspecific F_1 (first-generation) hybrid. Interspecific hybridization is a widespread phenomenon in fishes partially due to their use of external fertilization. In some cases hybridization can result in sterility in either one or both of the sexes. Introgression occurs when the F_1 hybrids are fertile and cross with either one or both of the original parental species, causing genetic material from one species to move into the other species, so that the two species involved are no longer genetically pure. In addition, fertile F_1 hybrids may cross with each other to produce F_2 (second-generation) hybrids. In extreme cases introgression leads to the formation of hybrid swarms where the genes of parent species are distributed randomly in a population with no F_1 hybrids or pure parental taxa.

Hybridization and introgression have caused major problems in many fish species, especially in salmonids. For instance, work by Fred Allendorf and Robb Leary of the University of Montana has shown that the westslope cutthroat trout (*Oncorhynchus clarki lewisi*) has greatly declined. This decline has been attributed to introgressive hybridization with the introduced rainbow trout and the Yellowstone cutthroat trout (*Oncorhynchus clarki bouvieri*), resulting in the formation of hybrid swarms. The existence of the remaining westslope cutthroat trout populations continues to be threatened by the migration of rainbow trout.

Montana Sauger

Sauger (*Sander canadensis*, formerly *Stizostedion canadense*) (**Figure 12.5**), a large predatory fish in the family Percidae, is native to Montana. The walleye (*Sander vitreus*, formerly *Stizostedion vitreum*) is not native to Montana but was introduced and stocked across the state to satisfy angler demands into drainages containing sauger since at least 1933 and possibly earlier. Concern has been expressed about the decline in Montana sauger populations since the late 1980s. This decline has been attributed to habitat loss,

(continued)

Case Study (continued)

Figure 12.5 Dr. Neil Billington holding a sauger on the bank of the Missouri River downstream from the Fred Robinson (U.S. Highway 191) Bridge, Montana. (Courtesy of Neil Billington, Troy University.)

low river flows, migratory barriers, over-exploitation, competition with walleye and other species, and hybridization with walleye. A severe drought in the late 1980s was thought to be partially responsible for the decline, but the lack of rebound in sauger abundance despite improved flow conditions raised major concerns for fisheries managers. The role of entrainment in diversion channels on the Yellowstone River in causing nonangling mortality was recently confirmed by studies conducted by Matt Jaeger of Montana State University at Bozeman, although significant over-exploitation was not shown in his study. Sauger is now listed as a "Species of Special Concern" in Montana.

The historical distribution of sauger in Montana included the Missouri River and its major tributaries downstream of the Great Falls as well as the Yellowstone River and its major tributaries downstream. In the past sauger likely occupied about 3,400 km of Montana's riverine habitat, but this has declined by 53% from historical levels, with a reduction of 75% in tributaries. The current distribution of sauger in the Missouri River is limited to the main-stem Missouri and a few sections of the Marias and Milk rivers, and sauger are considered rare or absent in other major tributaries. In the Yellowstone River drainage the present distribution of sauger is confined to the lower main-stem Yellowstone, and they are considered rare or absent in major tributaries such as the Bighorn and Powder rivers. They are found in a small section of the upper Powder River and the Bighorn River system in Wyoming.

Sauger and walleye are known to hybridize naturally and artificially under experimental conditions. Their hybrids, especially saugeye (an F_1 hybrid resulting from a cross between a female walleye and a male sauger), have been produced and extensively stocked in the central United States because they can withstand warm,

Figure 12.6 Mounted specimens of sauger (top) and walleye (bottom) showing morphological differences between the two species in the Montana Department of Fish, Wildlife and Parks office in Lewiston. (Courtesy of Neil Billington, Troy University.)

eutrophic waters with high flushing rates better than walleye, and they have faster growth rates than either of the parental species. However, saugeye have never been stocked in Montana.

Several external morphological characteristics distinguish sauger from walleye (**Figure 12.6**). Sauger have darker skin pigmentation (dark yellow to brown), scaled cheeks, three dark saddles that extend all the way down the sides of their bodies, and a series of dark speckles arranged in a number of lines across their first dorsal fin (Figure 12.6, top). Walleye have lighter skin pigmentation (light yellow to green), unscaled cheeks, 13 short, lightly colored saddles that extend slightly down the sides of their bodies, a dark blotch on the posterior end of the first dorsal fin, and a white patch at end of the lower caudal fin (Figure 12.6, bottom). F_1 hybrids are often intermediate for the characteristics of the parents, but features of both parents are often expressed. Consequently, it is usually very difficult to distinguish backcrosses of these F_1 hybrids to either of the parental species by morphology because they tend to closely resemble one of the parental species. In addition, embryo and larval F_1 hybrids tended to closely resemble the female parent.

Hybrids can be easily detected by genetic screening when diagnostic protein-coding loci between the species involved have been identified. Sauger and walleye show fixed allelic differences at four protein coding loci: ALAT* for alanine aminotransferase and IDDH* for L-iditol 2-dehydrogenase in liver tissue and mMDH-1* for malate dehydrogenase and PGM-1* for phosphoglucomutase in muscle tissue. By using these

(continued)

Case Study (*continued*)

diagnostic loci, it is possible to screen *Sander* specimens by protein electrophoresis to confirm species identification and to detect F_1 hybrids and F_2 hybrids or backcrossed individuals (often jointly referred to as F_x hybrids). The F_1 hybrids are heterozygous at all of the four diagnostic loci, whereas the F_x hybrids are heterozygous at some loci and homozygous at the others; the direction of the backcrossing can often be inferred from the homozygous alleles present. A number of studies have established the natural occurrence of sauger–walleye hybrids, introgression between the two species, and the difficulties of distinguishing sauger, walleye, and their hybrids using morphological characteristics.

Hybridization between native sauger and introduced walleye is among the factors that have been attributed to the decline of Montana sauger. Previous studies have reported hybridization and introgression between sauger and walleye in Montana ranging from 0% to 15%. The extent of hybridization and introgression between native Montana sauger and introduced walleye was examined from 2003 to 2005 by protein electrophoresis at four diagnostic loci for 465 sauger from 18 sites (15 sites in Montana [266 fish from 11 sites in the Missouri River drainage and 107 fish from 4 sites in the Yellowstone River drainage] plus 3 sites from adjacent watersheds in Wyoming [48 fish from 2 sites] and North Dakota [44 fish from 1 site]) in a study carried out with cooperation and financial support from the Montana Department of Fish, Wildlife and Parks (MDFWP).

Sauger containing walleye alleles were found at 11 of 15 sites surveyed in Montana; 4 sites had no hybrids, 4 sites had hybridization rates of <5%, 5 sites had <10% hybridization, and 2 sites had >20% hybridization. Often, multiple loci with walleye alleles were found, suggesting the development of hybrid swarms at some sites. Hybridization rates ranged from 0 to 22% in the Missouri River drainage and from 0 to 4% in the Yellowstone River drainage. All hybrids found in Montana were backcrosses to sauger, except for one F_1 hybrid found in the Fort Peck Reservoir, showing that introgression of walleye alleles into sauger is occurring. The MDFWP personnel tried to collect fish that looked like sauger, so this might explain why most of the individuals with walleye alleles were backcrosses to sauger and only a single F_1 hybrid was found.

All hybridization found in the Yellowstone River drainage occurred in Montana; to date, no walleye alleles have been found in sauger from two sites on the Bighorn River system in adjacent Wyoming. Previous studies had reported values ranging from 0% to 10% in the Missouri River system and from 0% to 15% in the Yellowstone River system. The lower hybridization values recorded in the Yellowstone River drainage likely indicate that sauger–walleye hybrids might have been underestimated in this study due to the small sample sizes for some sites. In a survey of 175 potential sauger brood stock from the Yellowstone River conducted spring 2003 to 2005, 4% to 10% of the fish contained walleye alleles, and these values were likely underestimates because only the two muscle loci were screened (see below). A 20% hybridization rate of was found in sauger collected from Lake Sakakawea, North Dakota, an impoundment on the Missouri River downstream from the confluence of the Missouri and Yellowstone rivers.

No physical barriers prevent sauger from migrating from the Missouri River system below the Fort Peck Reservoir in Montana and the Yellowstone River in Montana to Lake Sakakawea.

Various studies have reported cases of hybridization between sauger and walleye and have shown that morphological analysis is usually inferior compared with protein electrophoresis for identification of sauger–walleye hybrids. This study provides further evidence that morphology alone is insufficient for identifying sauger, walleye, and their hybrids. Twenty-three fish identified as sauger in Montana plus nine fish from Lake Sakakawea were hybrid or introgressed individuals.

Hybridization can occur when the spawning periods of the species involved overlap, where there is a shortage of spawning sites, or when one species is more abundant than another so that individuals find it difficult to find conspecific mates with which to spawn. Sauger and walleye have not coexisted for long in Montana (<80 years), so they likely have not developed adequate reproductive-isolating mechanisms and their spawning periods overlap, leading to hybridization. In addition, in the Missouri River above Fort Peck Reservoir sauger are threatened by walleye migrating downstream from the Canyon Ferry Reservoir, where they were illegally introduced. Montana sauger are declining so it is possible that there are fewer conspecific individuals with which to spawn, forcing some sauger to spawn with walleye.

This study does not report any significant increase in hybridization between sauger and walleye in most Montana populations, but two sites on the Missouri River system showed an increase of more than double, from 10% to 22%, from previously reported values. However, continued stocking and range expansion of walleye may lead to the formation of more hybrid swarms contributing further to the decline of sauger. If such populations arise, there will likely be no pure sauger remaining, presenting serious problems in sauger conservation efforts in Montana.

In Montana the construction of dams and water diversion structures on rivers has likely affected sauger populations, because sauger are highly migratory and depend heavily on unimpeded habitats provided by large rivers, perhaps making them more susceptible to the effects of introduced walleye. Recent studies by Matt Jaegar of the Montana State University have confirmed that distances traveled by Montana sauger range from 5 to 350 km in the Yellowstone River. In California extensive water projects combined with introduced species that can better tolerate the degraded habitats have resulted in the decline of many species, and again introduced species were cited as a primary factor in the status of 49% of species in that state that are now extinct, endangered, or require protection.

Restoration Implications

The MDFWP recently amended its fishing regulations from a daily limit of 5 walleye/sauger and a possession limit of 10 walleye/sauger to note that only 1 of the 5 fish in the daily limit and 2 in the possession limit are allowed to be sauger over most of the range of the sauger in an effort to reduce relative numbers of walleye to sauger. Hopefully, this will reduce the relative numbers of walleye to sauger where they co-occur and thus, reduce the likelihood of hybridization.

(continued)

Case Study (*continued*)

The MDFWP have used supplemental stocking to maintain and restore sauger populations for a number of years. They collect brood fish from the wild, spawn them, and raise fingerlings for stocking. They may, however, inadvertently propagate individuals that possess walleye alleles. The potential exists to seriously impact the genetic integrity of recipient sauger populations after stocking, because a few hybrid or backcrossed individuals accidentally included as brood fish can result in the production of many hundreds of thousands of fry and fingerlings containing foreign alleles. Every effort, therefore, should be made to prevent the inadvertent culture and stocking of hybrid or introgressed individuals by fisheries management agencies.

Genetic screening of potential brood fish is an important precaution that can be taken to ensure correct species identification and to prevent the inadvertent inclusion of F_1 or F_x hybrids in fish culture operations. One concern with screening potential sauger brood fish is that it is normally only possible to collect muscle samples from live fish either by a fin clip technique or with a biopsy needle. With only two diagnostic loci that can be scored in muscle (mMDH-1* and PGM-1*) it will be possible to confirm species identification and to eliminate all of the F_1 hybrids, but $(½)^n = (½)^2$ in this case = 25% of the backcrosses would likely be missed. The use of the two additional diagnostic loci in liver could improve the chances of detecting backcrosses to 6.25%, but MDFWP personnel were concerned about performing surgery on fish that were about to be spawned in the field to collect liver samples such as has been used for largemouth bass (*Micropterus salmoides*).

Data collected between 1999 and 2005 in surveys of potential sauger brood stock from the lower reach of the middle Missouri River and Yellowstone River middle reach showed the percentage of fish that contained walleye alleles ranged from 2.7% to 10.0%; these fish were either not used as brood fish or, if they were used, the fertilized egg batches were destroyed. Note that only the two muscle loci were screened, so a possible 25% of backcrossed individuals would have been missed. An alternative approach is to kill the adults once they have been spawned, screen all four diagnostic loci (because the two liver loci could now be included) by electrophoresis, and then destroy any batches of fertilized eggs that derive from hybrid or introgressed fishes. Managers were reluctant to kill mature fish that might spawn for many years, however, on the slight chance of them possibly being a hybrid or backcrossed individual. They also believed it would have been difficult to conduct liver biopsies in the field.

Concern over the potential inclusion of hybrids in sauger brood stock and difficulties in being able to adequately screen them by protein electrophoresis led the MDFWP to suspend their sauger spawning program in 2006, hampering sauger restoration efforts. If wild sauger are to be used as brood fish to produce fry and fingerlings for supplemental stocking in the future, it will be important to search for additional protein-coding loci that are diagnostic between sauger and walleye that can be scored in muscle, but seven diagnostic loci would be needed to reduce the likelihood of missing backcrosses individuals to <1%. Also, it might be possible to use another genetic analysis method such as microsatellite DNA, if suitable markers can be developed.

Key Terms

Area threshold   294
Captive breeding   294
Carrying capacity   294
Critical habitat   292

Hard release   301
Reintroduction   299
Soft release   300
Translocation   299

Key Questions

1. How are flagship species used in restoration of endangered populations?
2. Define area thresholds.
3. Describe essential steps in a captive breeding program.
4. What are the main differences between translocation and reintroduction?
5. What are the main differences between hard release and soft release techniques?

Further Reading

1. Bowles, M. L. and Whelan, C. J. 1994. *Restoration of endangered species*. Cambridge, UK: Cambridge University Press.
2. Morrison, M. L. 2002. *Wildlife restoration*. Washington, DC: Island Press.

13

AQUATIC ECOSYSTEMS: WETLANDS, LAKES, AND RIVERS

Chapter Outline

13.1 Degradation of Aquatic Ecosystems
  Degradation of Wetlands
  Degradation of Lakes
  Degradation of Rivers
13.2 Restoration of Wetlands
  Passive Restoration
  Active Restoration
  Restoration of Florida's Everglades
13.3 Restoration of Lakes
  Restoration of the Great Lakes
  Restoration of the Broads in Norfolk, United Kingdom
13.4 Restoration of Rivers
  Active Restoration
  Restoration of the North Atlantic Salmon
Case Study 13.1: Restoration of Coastal Salt Marshes in Brazil Using Native Salt Marsh Plants
Case Study 13.2: Biomanipulation as a Tool for Shallow Lake Restoration in the Norfolk Broads, United Kingdom
Case Study 13.3: Restoration of the Kissimmee River, Florida

Aquatic terrestrial ecosystems, including wetlands, rivers, and lakes, form an integrated part of the landscape. These ecosystems can be connected within the landscape, forming one continuous water body often referred to as **watershed**. A watershed can cover enormous areas of land. For example, the watershed of the Mississippi River covers more than 3.2 million km². The integrity and function of aquatic ecosystems is important in determining the quality of the watershed. In addition, aquatic ecosystems provide an important service to society in delivering and storing valuable potable water. Potable water will probably become one of the most critical commodities for many societies in the near future. Watershed management plays an important role in meeting societies' demand for potable water. The reality, however, is that aquatic terrestrial ecosystems are being degraded worldwide. Therefore, today conserving, managing, and restoring these ecosystems is emphasized.

Wetlands are characterized by soil that is water inundated for at least some part of the year and are modified by local landscape, regional climate, water chemistry, and regional flora. Wetlands are classified into marshes, swamps, bogs, and fens. The main characteristics of marshes are the cover of herbaceous vegetation. Swamps harbor trees and shrubs and are inundated by water throughout the year. Bogs contain peat (organic matter) and receive water only by precipitation; they are dominated by a canopy of evergreen trees and shrubs and a moss-dominating understory. Fens receive only groundwater and have waterlogged soils that form peat; they are characterized by grasses, sedges, and wildflowers.

Wetlands are among the world's most productive ecosystems. They form an important sink for atmospheric CO_2 by accumulation of organic matter such as peat in temperate climate. They are also important for maintaining biodiversity by providing wildlife sanctuaries. For example, an estimated 30% of all plant species live in wetlands. Wetlands provide important habitats for local and migrant birds. More than half of North American bird species either nest or feed in wetlands.

Wetlands perform many important ecosystem functions such as water storage, flood mitigation, erosion control, and even stabilization of local climate. Wetlands can store excess water derived from torrential precipitation and floods. Wetlands found along riverbanks, lakes, and coastlines are important to the adjacent environment. These wetlands can absorb excess floodwater and reduce flood damage.

Wetlands also provide society with various ecological services including potable water supply through groundwater regulation, aquifer recharge, and water purification. Wetlands improve water quality by filtering and storing contaminated sediments and excess nutrients that form deposits. It is estimated that the water quality protection provided by wetlands in the United States is worth at least $1.6 billion. In addition, because they form important spawning grounds for many fish species, more than two-thirds of the world's fisheries depend on the integrity of wetlands. Wetlands also serve in wildlife resources, recreation, and tourism. They further provide agriculture with soil and are important in the maintenance of water tables. In total, ecosystem services attributed to wetlands are estimated to have worldwide value of $4.9 trillion.

Lakes play an important role in storing potable freshwater; more than 92% of all liquid surface water is stored in lakes. Lakes provide additional ecological services for society, such as fishing (commercial and recreational) and transportation of freight. In addition, water from lakes is used for industry and agriculture (irrigation).

A simple process is used to estimate the water quality of lakes. This water quality is determined by several factors, such as pH, dissolved oxygen, biological oxygen demand, chemical oxygen demand, transparency, chlorophyll (free-floating algae) content, and nitrogen and phosphorus concentrations.

Degradation of lakes has taken place worldwide. Increased human population, intensified agriculture, and industry are the main factors behind this degradation. Restoration of lakes aims at the water quality and maintaining diversity of fish species. This is particularly important because about half of all animals in the United States that are on the threatened and endangered lists are freshwater species. Several cases of lake restoration are discussed at length in this chapter.

Rivers play an important role in transporting fresh water that can be used for human consumption, agriculture (irrigation), and industry. Large rivers are critical in generating hydropower and for navigation. To improve the functions of rivers for navigation, a number of engineering structures have been installed, such as dams, floodgates, and levees. Channeled rivers have lost almost all their integrity and characteristics.

Rivers are modified by the landscape they run through. River morphology is defined by the slope, discharge, and flow regime. Other factors that characterize rivers include features of embankments that can change as rivers run through the landscape. The flow pattern of a river includes volume and velocity of the water. The water flow can vary on a daily or seasonal basis. Variation in the water flow pattern can affect water turbidity, sediment load, and chemical characteristics. Processes leading to river degradation and restoration efforts are discussed below.

13.1 Degradation of Aquatic Ecosystems

Degradation of Wetlands

Wetlands are being degraded worldwide, mainly due to ongoing drainage, conversion to other uses, pollution, and over-exploitation. These and other mistreatments of wetlands have been occurring for centuries. Today, less than 50% (about 40 million hectares) of the original wetlands remain in the United States. The problem is worse in some states than in others (90% of wetlands have been lost in Ohio and California). Current loss of wetlands is the greatest, however, in Florida (about 9,400 hectares per year) (**Figure 13.1**). This is, undoubtedly, an international problem. For instance, much of Europe has lost more than 50% of wetlands. The loss of wetlands in New Zealand is estimated to be about 90%.

Degradation of Aquatic Ecosystems 317

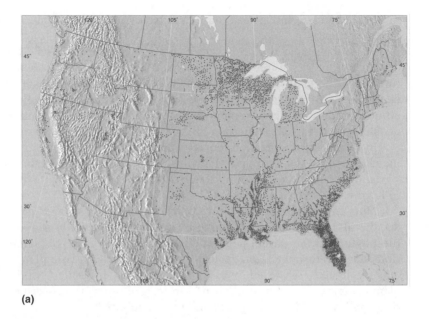
(a)

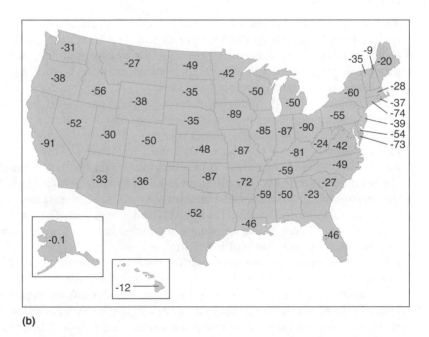
(b)

Figure 13.1 Maps showing wetland destruction in the United States according to states. These maps show (a) the remaining wetlands in the lower 48 states and (b) the percentage lost in each state.

In Asia, Thailand, Indonesia, and China have also experienced extensive areas of lost wetlands.

Anthropogenic activities have been a direct cause of wetland degradation and destruction. In the United States wetland losses are mainly attributed to agriculture (30%), forestry (23%), and rural development (21%). Other contributing factors include urbanization, mining, hydrological alterations, pollution, sewage effluent, industrial development, and road construction. Wetland degradation usually results in transformation of the original hydrological system.

In particular, urbanization can have direct impact on wetlands mainly through increased runoff and increased sediment loads, which have adverse effects on the wetland habitat. Toxic pollutants derived from urban runoff may even accumulate in nearby wetlands.

Nutrient enrichment (i.e., eutrophication) of wetlands generally results in a decline in aquatic plants, a decline in invertebrate density, and an increase in the phytoplankton (algae) populations. An increase in phytoplankton populations due to eutrophication typically out-shades the aquatic plants and consequently depletes dissolved oxygen concentrations. In particular, loss of aquatic plants is critical because of their role in oxygen production, sediment stabilization, and as a habitat for aquatic and terrestrial life.

Degradation of Lakes

Shallow freshwater lakes are commonly subjected to anthropogenic degradation including drainage, eutrophication, overfishing, sedimentation, and introduction of non-native species. Lakes act generally as sinks for sediment and pollutants generated by agriculture, industry, and sewage effluent. In addition, recreation pressure has increased its impact on lakes.

Eutrophication of shallow freshwater lakes can result in a cascade of events leading to general ecosystem degradation and reduction in biodiversity. After eutrophication, aquatic plants increase in size and abundance. Also, the **phytoplankton** (algae) population increases in density. This increase in the phytoplankton leads to a gradual shading-out of aquatic plants. An increase in the sediment load leads to more turbidity and shading, which further degrades the environment of aquatic plants. Eutrophication of lakes can also result in shifts in the fish population from **piscivorous** (fish-eating) to **planktivorous** (algae-feeding) fish. The loss of aquatic plants leads to more open waters, favoring fish that remove **zooplankton**, which are important grazers of phytoplankton. This exemplifies how perturbation such as eutrophication can lead to a trophic cascade and an alternative state of the ecosystem.

Excessive accumulation of organic matter can result in **hypoxia** of lakes. Hypoxia refers to conditions where dissolved oxygen in water is so low that it can no longer sustain aquatic life. Effects of hypoxia include fish kills, unattractive or smelly water, and decrease in biodiversity.

Excessive use of water from lakes for irrigation purposes can be disastrous. As an example, the Aral Sea has been reduced considerably in size due to large-scale irrigation practices, and this has contributed to extinction of local fish populations. Similarly, Lake Chad in central Africa has been reduced drastically, and a recent drought has exacerbated the situation. The dry climate of sub-Sahara (Sahel) Africa makes the lake vulnerable to such events. About 37 million people rely on the water of Lake Chad. Also, fisheries in Lake Chad provide an important protein source for the local human population. Fishing from the lake has declined drastically, and adjacent wetlands have dried. These events have brought instability to the human population surrounding Lake Chad.

Degradation of Rivers

Rivers have been impacted by humans worldwide. Anthropogenic activities such as industry and agriculture have seriously degraded many rivers. Activities within the watershed of a river can affect a large stretch of that river. This includes any changes in land use within a river's watershed (i.e., deforestation, agriculture, and overgrazed pastures). Such activities can increase eutrophication and **sedimentation** of rivers. Excessive sedimentation, one of the greatest degrading factors for rivers, is usually caused by soil erosion of adjacent agricultural lands. Sedimentation adversely affects the riverbed, which forms a critical habitat for spawning fish. Excessive sedimentation of riverbeds excludes many fish species from spawning.

More than 850,000 dams have been built in rivers around the world. Most impoundments in rivers are, however, just earth-filled structures. These dams obstruct more than two-thirds of all rivers worldwide. In Europe and North America water flow of about 70% of large rivers is regulated. About 50,000 large dams (above 30 m) have been built worldwide. On average, these dams create about 23 km^2 of reservoir. Most of these dams were constructed in China (35%) between 1940 and 1970. About 45,000 medium-size dams (above 15 m) have been constructed worldwide. Most of these dams serve irrigation and hydropower purposes. Although it can be argued that using hydropower is a better option than burning fossil fuel in regard to CO_2 emission, methane (another greenhouse gas) emission is considerably high in shallow tropical reservoirs. Dams can result in increased erosion of embankments and the amount of suspended material in the water. Also, dams can lead to alterations in water discharge and water temperature. Moreover, dams create a barrier that prevents migration of spawning fish.

13.2 Restoration of Wetlands

The initial step in restoration efforts of wetlands usually focuses on reestablishing the hydrological functioning. This work typically aims at building the original hydrological conditions of the degraded site by using engineering solutions such as building levees or dams or filling in existing drainage systems. For this work heavy machinery is needed. This effort can effectively restore important components of the natural hydrology of the wetland in question.

Restoring only the hydrology, however, does usually not provide complete restoration of wetlands. Unfortunately, drainage of wetland soils can lead to irreversible chemical and physical changes such as affecting the soil chemistry and intensifying nitrogen mineralization and acidification. Also, increasing the water table through restoration of wetlands hydrology is usually followed by substantial losses of nitrogen and phosphorus. In addition, draining wetland soil can lead to irreversible physical conditions such as excessive compaction.

Wetland restoration projects take place under different environmental conditions. These projects are often connected to management of larger landscapes such as watersheds, lakes, rivers, and even aquifers. Wetland restoration is important along the Mississippi River to decrease agricultural runoff into the river. For example, about 3,600 hectares of wetlands are being restored along the Raccoon River in Iowa. This river is one of the tributaries of the great Mississippi River. The Raccoon River produces one of the highest nutrient loads of tributaries feeding the Mississippi River, which already contains high levels of pollution. Restoration projects of wetlands along the Mississippi River will not only reduce sediment and nutrient load into the Mississippi River, but they will also be important for the waters of the Delta and the Gulf of Mexico. The Mississippi River Delta comprises about 1.2 million hectares of coastal wetlands. In the Gulf of Mexico large-scale offshore hypoxia (referred to as a dead zone) resulting in massive fish death has been linked directly to the high nutrient load of the Mississippi River.

Passive Restoration

Passive restoration is a practical approach where degraded wetlands are located adjacent to pristine or less-degraded wetlands. Seed dispersal into degraded wetlands can be achieved by allowing free flow of surface water where seed can drift between the degraded and the intact wetlands. Also, rhizome pieces can be dispersed in floodwater as such propagules can be effective in establishing new populations.

Active Restoration

Active restoration involves introduction of native plants and may be an option if intact wetland sites do not exist adjacent to the degraded wetland in question. This can be achieved by strategic seeding, outplanting of seedlings, mature plants, rhizomes, or cuttings of native plants. Sowing native seed is usually a cost-effective method compared with other options in wetland restoration. Using native seed, however, requires initiating a program of seed harvesting and storing the seed. Another active restoration option is to transplant whole turfs of wetland plants strategically into degraded wetlands to increase species richness. Restoration of coastal marshes is discussed in **Case Study 13.1** on page 333.

Nurse crops (commercial species) can be used to stabilize the ground temporarily. As the cover of nurse crops declines, natural colonization of native plants is usually facilitated. The cover of nurse crops should ideally decline within a few years and be replaced by a diversity of native plants. Careful selection of nurse crops is important to avoid the introduction of invasive species.

Wetlands that were previously used in agriculture usually have high potential for restoration. This is especially the case if intact wetlands, which can act as a seed source, are found nearby. These restoration efforts, however, can be challenging. The first step in such work is to deal with the high nutrient availability in the soils. This may require soil nutrient exhaustion. This can be accomplished, for instance, by removing the topsoil. Such methods, however, might be too destructive. Otherwise, continuous biomass removal through livestock grazing, mowing, or simply cutting vegetation and removing it eventually exhausts the soil of excess nutrients. It is also critical to reduce the biomass of dominating plants. This approach usually facilitates restoration of species richness. Cultivation of wetlands can lead to impoverished seed banks of native species. Therefore, introduction of native species is often needed. Also, the use of agrochemicals may have resulted in buildup of pollutants in the soil, which may need to be remediated using bioremediation or phytoremediation techniques, as discussed in Chapter 10.

Often, cultivation of wetlands is followed by an excessive increase in propagules of weeds and non-native species. Control of invasive or other undesirable plants is critical and can be achieved using a combination of methods (see Chapter 7).

Restoration of Florida's Everglades

An example of a large-scale restoration of wetlands can be found in ongoing restoration efforts of the Everglades in Florida. The Everglades comprise a unique wetland ecosystem that spans most of south Florida, from Lake Okeechobee in the north to the Keys in the south. In fact, the Everglades are the largest remaining subtropical wilderness in the United States (that once covered more than 16,000 km^2). The Everglades is a World Heritage Site and an International Biosphere Reserve. They are based on a shallow braided river that is roughly 180 km long and 100 km wide. The water flows slowly, or only a few hundred meters per day, from Lake Okeechobee to the sea (**Figure 13.2**). The wetlands around Lake Okeechobee play an important role in regulating the flow of this water.

The Everglades contain heterogeneous landscapes; these include saw grass prairies, lakes, ponds, marshes, tree islands, and mangrove forests. The saw grass prairies are a dominating feature of the Everglades and were phrased by Marjory Stoneman Douglas as the "river of grass." Tree islands are scattered in the Everglades and comprise about 8% of the area. They provide critical habitats for animals and act as nesting sites for birds. The steady flow of fresh water confines a border of the mangrove forests to the coast. The heterogeneous landscapes of the Everglades support high biodiversity, which includes 25 mammal species, some of which are endangered (including the manatee [*Trichechus manatus latirostrus*] and the Florida panther [*Puma concolor coryi*]). Currently, populations of almost 70 endangered species living in the Everglades are being restored.

Figure 13.2 Maps of the Everglades showing (left to right) historical water flow, current water flow, and planned water flow after large-scale restoration. (Graphic by Stephen Rountree, www.stephenrountree.com.)

The first human impact on this ecosystem was through poaching; several species were almost decimated due to over-hunting. For example, the alligator (*Alligator mississippiensis*) population has been estimated to be reduced to only 1% of its original size by 1960, though it is now recovering. The Florida panther, however, is still on the verge of extinction.

The greatest damage to the Everglades ecosystem came from a drainage operation that interfered with the natural flow of the river. In the late 1800s northern parts of the Everglades were drained; this work concentrated on the land south of Lake Okeechobee. The natural flow of the river through the Everglades was changed in 1948 by a large engineering project named Central and Southern Florida Project for Flood Control and Other Purposes. The aim of this project was to provide flood control, especially during hurricanes, and to provide a water supply system for residents in the area. In addition, the goal was to reclaim more land for agriculture and residential development. Over about 20 years more than 3,000 km of levees and canals were built. In addition, 150 water control structures (spillways and floodgates) and 16 major pumping stations were built. The largest rivers were dredged and straightened from their natural meandering patterns. As a result, about 70% less water flows through the Everglades, much of this reduction occurring because fresh water was expelled into the sea via constructed canals. Installing these structures changed some parts of the Everglades from a largely uncontrolled, uninhabitable wetland into an agricultural and residential area.

Networks of channels were dredged to drain the peatland for agricultural purposes. The peatland formed the basis for agriculture, but topsoil is eroding at an alarming rate of 2 cm per year. Because of pressure for land for residences and farming, much of south Florida was drained for agriculture and residential development. The peak of this operation took place between 1905 and 1927.

The human population has increased drastically because of the land drainage program. Before this land was drained about 30,000 people lived in south Florida. Today, the number of residents in this part of Florida is close to 7 million. This figure is predicted to reach 12 million by 2050.

Farmlands adjacent to the Everglades leach agrochemicals (fertilizers and pesticides) into the surface waters. Eutrophication has resulted in abnormal growth of algae, and its overextended areas have turned the saw grass prairie into cattail plain. Eutrophication has also resulted in abnormal growth of algae, leading to red tides in some tributaries that adversely affect native species such as manatees, dolphins, and oysters. Eutrophication of the water has increased the abundance of invasive non-native plants, such as the Brazilian pepper (*Schinus terebinthifolius*) and cattail. Also, offshore coral reefs have been adversely affected by the massive eutrophication. To reduce the nutrient loading into the Everglades, the Everglades Nutrient Removal Project was initiated. This project constitutes one of the largest constructed wetlands in the United States. Cattails are mainly used in this project to purify flowing water. The cost of this restoration effort could reach $465 million.

The lack of free-flowing water is the most critical factor for the recovery of the Everglades. The loss of water changed the wetlands drastically. Plant and wildlife supported by this water flow were drastically affected. For example, populations of wading birds such as wood storks (*Mycteria americana*), herons, and egrets have been reduced by at least 90%. Saltwater intrusion into wetlands adjacent to the sea has occurred due to the drastically reduced flow of inland fresh water.

The Everglades form a critical habitat for several vulnerable and endangered species. One of the most emblematic is the Florida panther. The Florida panther is a subspecies of the cougar (*Puma concolor*). It is a big cat, weighing up to 70 kg and reaching more than 2 m in length. The Florida panther mainly preys on deer and wild hogs. Before the arrival of the European settlers, the panther ranged throughout the southeastern United States from Texas to the Carolinas. They were, however, relentlessly hunted by the settlers, and by 1900 the last remaining population was found in southwest Florida. Even this population was overhunted, until the panther finally gained federally protected status in 1973. The population size reached a critically low number in the late 1980s when only about 30 animals were left.

The small surviving population of panthers suffered from serious inbreeding. A controversial plan to save this population was to introduce breeding females from another subpopulation of the cougar. Most likely a gene flow existed between this subpopulation of the cougars and the Florida panther in the past. Therefore, eight female cougars were brought in 1995 to southwest Florida from west Texas. Consequently, the population size increased to about 100 animals. Wildlife experts predict, however, that the minimum viable number is about 240 animals to ensure long-term survival of the Florida panther. To protect the panther, the Florida Panther National Wildlife Refuge (10,680 hectares) was established within the Everglades. The Florida panther needs a large area to survive because it roams over large areas to hunt. A single panther might roam over a territory exceeding 50,000 hectares. A strong demand for more land for developers in the south of Florida will obviously restrict the potential population size of the panthers in the future.

In 2000, the Comprehensive Everglades Restoration Plan began with a budget of $7.8 billion. In fact, this restoration project is one of the most expensive and ambitious project of its kind in the world. It aims at returning a natural water flow by converting straight canals into braided streams. At the same time this project ensures potable water for the rapidly growing population in south Florida. The restoration plan involves 971,000 hectares of the Everglades over a period of 20 years. If realized, it will remove more than 390 km of levees and canals. The goal of the project is to restore the hydrology of the Everglades. This involves restoration of wetlands and the construction of 18 reservoirs with the goal of recharging deep aquifers. In addition, wells will be built to inject water into deep aquifers. This water can then be used in the future when needed. The plan is to increase water flow through the Everglades by 20% within 30 years. The goal is to redirect the flow of about 6.4 billion liters of fresh water into the Everglades. This water is currently expelled through canals into the sea. Restoring the water flow is the key to the restoration plan; at the same time water supply and flood protection will be provided for local residences.

The restoration of the Everglades can serve as a model for other future megarestoration projects such as Louisiana's coastal wetlands, San Francisco Bay, and wetlands adjacent to the Great Lakes.

13.3 Restoration of Lakes

Initial restoration efforts of freshwater lakes target the main disturbing agent. For instance, such efforts to restore eutrophic shallow lakes focus on curtailing nutrient loading. Engineering methods are effective in reducing excess nutrient loads and involve sediment dredging and removal. Although dredging is effective in reducing nutrients, such operations are expensive as well as detrimental for any bottom-dwelling organism. However, nutrient levels in shallow lakes can also be lowered by removing fish from the lakes. In deep stratified lakes, on the other hand, excessive nutrients are reduced by chemical treatments (using alum and iron salts). In these deeper lakes injecting oxygen has even been used in an attempt to reduce the nutrient load.

Nutrient-rich water is often associated with high concentrations of phytoplankton populations. This in turn gives the water its murky appearance. The murky state of the water inhibits photosynthesis by bottom-dwelling aquatic plants, which, consequently, decline in population size. Unfortunately, the result here is often a reduced oxygen level of the water. Restoration of murky waters to clear waters, therefore, involves efforts to increase the zooplankton population. An increase in zooplankton populations can result in lower algal populations. Restoration of the zooplankton populations is achieved by reducing or removing temporarily resident fish populations. This type of lake management is called **biomanipulation**. Biomanipulation can also be achieved by selective fishing of zooplanktonvorous fish and careful introduction of piscivorous fish. Biomanipulation involves top-down trophic control of algal populations and has been successfully used in restoration of shallow lakes. Biomanipulation is discussed further in **Case Study 13.2** on page 338.

After the restoration and clearing of murky waters, aquatic plants are reestablished. To increase their chance of establishment, these aquatic plants can be seeded or transplanted into lakes. Small sites can be permanently protected to build populations of aquatic plants for seed production. After the establishment of aquatic plants, fish populations are usually reintroduced as well.

Widespread pollution can impact aquatic ecosystems. For instance, atmospheric pollution, especially the burning of coal, can result in acid rain that can adversely affect aquatic ecosystems. Acidification of lakes leads to clear waters and reduction in biodiversity. Hundreds of acidified lakes in Scandinavia and Canada have been affected by acid rain. Some of these lakes, however, have been restored. The success of restoration efforts depends on the water retention time of the lake in question. Initial restoration efforts involve chemical treatments to raise pH of the water. For this purpose a sludge containing calcium carbonate and/or calcium hydroxide is generally spread on lake surfaces from boats. Chemical treatment on large or remote lakes has been accomplished by the use of aircraft. This method has been used successfully to increase water pH. Chemical treatment usually needs to be repeated several times to ensure restoration success.

Restoration of the Great Lakes

An example of restoration efforts on deep lakes is provided by studies on the Great Lakes in North America. The Great Lakes comprise an enormous, interconnected inland water body. They contain the largest volume of liquid fresh water in the world. In total, the Great Lakes cover about 244,000 km^2 and contain about 22.7 km^3 of water. The largest is Lake Superior (82,367 km^2), with a maximum depth of 406 m and a total water volume of 12.2 km^3. Most of the Great Lakes are deep oligotrophic (nutrient-poor) waters. The exception is Lake Erie, which is relatively shallow (maximum depth of 64 m) and tends to be eutrophic. The water retention time in Lake Superior is about 190 years. Therefore, any nonvolatile pollutants that enter the lake will probably stay in the water for almost two centuries. Concentrations of toxic chemicals have been shown to increase as the water flows through the lakes to Lake Ontario. Release of toxic chemicals into the lakes is now strictly controlled.

The total watershed of the Great Lakes covers about 295,000 km^2. The total watershed of Lake Superior is about 125,000 km^2. Most of the watershed around Lake Superior is primary forests (94%), with agricultural and residential making up the remainder (6%). The total human population in the watershed of Lake

Superior is just above 500,000 people. Therefore, anthropogenic factors that cause degradation, such as eutrophication, acidification, and wastewater pollution, are not a serious problem.

Although the human population in the watershed of the Lake Superior is relatively low, the total human population around all the Great Lakes has increased drastically during the last two centuries to about 38 million. Huge metropolitan areas such as Chicago, Detroit, Cleveland, and Toronto have extended practically on the shores of the Great Lakes. Most resident around the Great Lakes rely on the water from the lakes as a source of potable water.

Agriculture and industry around the Great Lakes have correspondingly increased with the human population, and the lakes have experienced excessive anthropogenic eutrophication. Eutrophication, generally, results in a reduced biodiversity of fish species. In addition, eutrophication increases the growth of littoral aquatic plants, which facilitates snail populations that lead to a higher incidence of parasitism on fish by trematodes. To limit eutrophication, restoration efforts have focused on phosphate stripping. In a continuous effort, restriction has been put on the use of phosphates in detergents.

Overfishing has resulted in a major decline in fish populations in the Great Lakes. The larger catch such as whitefish (*Coregonus clupeaformis*) and lake trout (*Salvelinus namaycush*) were among the first species to be heavily overfished. The fisheries, however, have gradually moved to a smaller catch, such as lake herring (*Coregonus artedii*).

Before this eventual switch to smaller catch, the Atlantic salmon became extinct from Lake Ontario as a result of overfishing and habitat degradation. Streams were degraded by sawmill dams and sedimentation resulting from logging and farming. Excessive sedimentation in the streams prevented spawning and, eventually, contributed to the extinction of the Atlantic salmon from Lake Ontario. Restoration efforts to reintroduce this species back into Lake Ontario involved restocking. Such restocking of native fish populations back into the Great Lakes has generally been successful. These efforts involve raising native fish in hatcheries and then releasing them back into their native habitat. During the work, however, great care must be taken in maintaining genetic diversity within populations of hatchery-raised native fish species.

Introduced non-native species have also played a dramatic role in the decline of native fish populations in the Great Lakes. The Great Lakes harbor about 160 invasive non-native species, including 25 fish species. Several of these invasive species have caused considerable ecological and economical harm.

One of the most notorious non-native species is the sea lamprey (*Petromyzon marinus*). The sea lampreys are native to the Atlantic coast of the Gulf of Mexico, the Mediterranean Sea, and most coastal areas of Europe. This species first entered Lake Ontario around 1860 via the Welland ship canal. Although the Niagara Falls forms a natural barrier for fish migration between Lake Ontario and the upper Great Lakes, construction of the Welland Canal (opened in 1829) between Lake Ontario and Lake Erie enabled the sea lamprey, along with other invasive species, to bypass the Niagara Falls and invade the upper Great Lakes.

Consequently, sea lamprey can now be found in all the lakes. The sea lamprey is parasitic on larger fish. It simply attaches itself to its prey and makes a permanent wound where it sucks blood until the host fish is dead. The invasion of sea lamprey has devastated populations of fish of economic importance, such as lake trout, burbot (*Lota lota*), and whitefish. In addition, the sea lamprey has caused a great decline in trout catches.

To control the problems associated with sea lamprey, the Great Lakes Fishery Commission was established in 1955. Early control efforts involved installing electrical and mechanical barriers to block access of sea lamprey to spawning grounds. This method was later deemed unsuccessful and was, therefore, abandoned. The use of a specific pesticide (i.e., lampricide: 3-trifluromethyl-4-nitrophenol) has, however, been used successfully in reducing populations of sea lamprey in the Great Lakes. For example, in Lake Superior the use of lampricide resulted in a 92% decline in sea lamprey catches.

Other invasive non-native fish species in the Great Lakes include common carp (*Cyprinus carpio*), which can damage shoreline aquatic plants, and the invasive non-native zebra mussel (*Dreissena polymorpha*). Native to Russia and first discovered in the Great Lakes in 1988, zebra mussels have had a great impact on the Great Lakes ecosystems. They were probably introduced by ballast water. Invasion of zebra mussels has resulted in drastic declines in populations of native mussels and decreased density of phytoplankton (algae), leading to increased water clarity. Zebra mussels can form high-density colonies (700,000 individuals per m^2). This high density can cause blockages of water intake pipes. Currently, control efforts for zebra mussels exceed $1 billion a year.

More recently, other non-native species have invaded the Great Lakes. These include the ruffle (*Gymnocephalus cernuus*), an invasive member of the perch family that was found in the Great Lakes in 1986. Also, in 1990 round gobies (*Apollonia melanostoma*), another aggressive invader, were found in the Great Lakes. In 2002 their total population was estimated to be about 9.9 million individuals in Lake Erie alone.

Restoration of the Broads in Norfolk, United Kingdom

An example of restoration of shallow lakes can be found in studies on the Broads in Norfolk, United Kingdom (**Figure 13.3**). The Broads is a commonly used name for a series of small, artificial shallow lakes. These lakes were dug over the centuries as peat was mined for fuel. They are typically shallow (on average just 1 m deep) and are naturally eutrophic. Dikes and rivers that are deep enough to be navigated by small boats connect many of these lakes. Today, these lakes are of importance for nature conservation because they harbor a number of species not found anywhere else in the United Kingdom.

Management of these lakes throughout the centuries has maintained characteristic vegetation of the Broads. For example, the ancient practice of cutting the dominating wetland grass reed (*Phragmites australis*) inhibited succession of adjacent wetlands to woodlands (*Alnus* and *Salix* species). Today, reed cutting

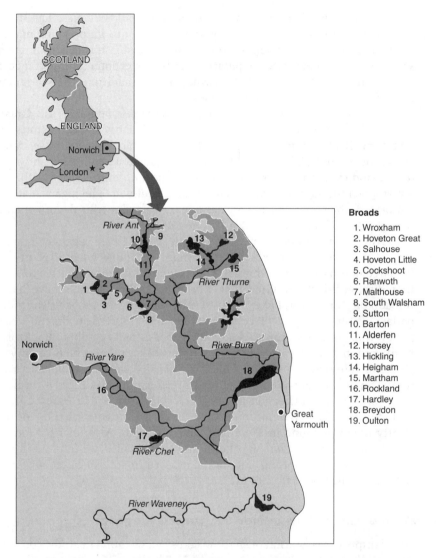

Figure 13.3 Map of the Norfolk Broads. The dark gray indicates wetlands, and the black indicates rivers and small lakes. (Adapted from the Broads Authority [http://www.broads-authority.gov.uk/visiting/getting-to-the-broads/location-maps.html] and data from the Microsoft Corporation [http://www.microsoft.com/maps/developers/]. Accessed May 21, 2010.)

has been mechanized on these wetlands to arrest succession and maintain the landscape characteristics of the Broads.

Increased eutrophication of the Broads eventually reached a critical threshold, changing the appearance of the lakes and resulting in an abrupt change in the state of the water from clear to murky. Eutrophication of the Broads was mainly due to phosphorus loading and, to a lesser extent, nitrogen from sewage effluent and agriculture. Sedimentation due to erosion of nearby agricultural lands has also added to this problem. The original state of these lakes was clear waters dominated by submerged aquatic plants. However, eutrophication resulted in high phytoplankton (algae) populations, which led to the murky state of the water. Fish had free access to the lakes, and lakes had low zooplankton (*Daphnia* sp.) populations in the summer. Eutrophication has also resulted in anaerobic state of the waters, which has been connected to outbreaks of pathogens. For example, birds as well as fish have been killed by a notorious flesh-eating anaerobic bacterium (*Clostridium botulinum*).

The dikes and lakes have been navigated by small boats, and over the last two centuries pressure from the tourist industry has escalated. Unfortunately, increased boat traffic on these lakes has had an adverse effect on shoreline vegetation. For example, water lilies have been damaged extensively by boat traffic. Restriction on the boat traffic was, therefore, necessary in an effort to restore shoreline vegetation.

Restoration of the Broads has consisted of both engineering and ecological approaches. The first step in the restoration of these lakes aimed at reducing the nutrient load from sewage effluent. Two methods were effectively used. First, streams containing sewage effluent were diverted away from the lakes. This work involved isolation of lakes from dikes and rivers so they received only drainage water from adjacent land. As result, phosphorus and nitrogen concentrations were dramatically reduced in the water. The combined effect of this work resulted in a clear water state in the lakes and facilitated the recolonization of aquatic plants. Second, phosphate stripping was used by dredging and removing nearly a meter of phosphorus-rich sediment from the bottom of selected lakes.

During the initial restoration work, interesting trophic-level interactions were observed. After isolation of lakes from dikes and sediment removal by dredging, algal populations were reduced; however, zooplankton populations were dense. Fish populations declined temporarily because of the dredging operation, but recovered quickly. The increase in the fish population resulted in a decrease in the zooplankton population. Consequently, the algal populations increased, maintaining the murky state of the water, although phosphorus and nitrogen concentrations did not change.

The next step in the restoration work involved biomanipulation where fish populations were removed. After the fish removal, zooplankton populations increased in the lakes, resulting in declined algal populations and subsequently a clear water state in the summer and continued colonization of aquatic plants. However, nitrogen and total phosphorus concentrations in the water remained steady. Surprisingly, populations of aquatic plant collapsed after 8 years of high biomass production. Following this event the cover of aquatic plant populations were low for about 2 years and then recovered gradually. The zooplankton and algal populations have fluctuated in a typical "prey–predator cycle". Some of these changes in the ecosystem may be cyclic, and restoration efforts should, therefore, not aim at the state of an ecosystem.

The restoration of the Broads has demonstrated how eutrophication of shallow lakes can result in regime shift (clear water state vs. murky water state) (see Chapter 5). Also, restoration efforts involving a decrease in nutrient loading and biomanipulation have had cascading effects at different trophic levels. Ecological surprises have taken place, but it is not known if these will have permanent effects on the lakes.

13.4 Restoration of Rivers

Successful restoration programs require information on the geomorphic, hydrological, and ecological functioning of the river in question. In this context it is important to restore the functional integrity of rivers, especially if the water flow has been affected by impoundments such as dams or if water has been diverted to other uses. Regulating the flow pattern of rivers by installing dams generally leads to degradation and fragmentation of their environments. Reservoirs created by large dams can also affect the ecological integrity of a river.

Regulating the water flow of rivers by building dams and reservoirs can result in decreased capacity of the river to carry sediment, which leads to siltation of the riverbed. Soil erosion of agricultural lands is one of the main factors behind siltation of rivers. Increasing sediment in rivers can lead to a loss of critical spawning habitat, especially gravel bottom. Therefore, initial restoration efforts of rivers with excess siltation aim at introducing irregular flow patterns to increase the hydrological capacity to carry or "flush" the sediment. Restoration of these functions can be achieved by allowing natural but variable water flow regime. Destruction of dams and other impoundments that inhibit natural water flow is often necessary in this process. In efforts to restore rivers, more than 650 dams have been removed in the United States. Usually, such restoration strategies increase the complexity and heterogeneity of the river environment. Restoration of the Kissimmee River in Florida is discussed in **Case Study 13.3** on page 347.

Restoration of water flow regime is usually followed by a passive restoration approach where the river's streams reshape the river geometry in meandering or braided patterns.

Active Restoration

An active restoration approach involves specific measures that rapidly modify geomorphology of the river. This may involve channel bed raising, replacement of riffle pools, creation of gravel beds to facilitate fish spawning, and the general removal of any obstacles to fish migration (**Figure 13.4**). The impact of impoundments can be improved. For example, to facilitate fish migration, dams can include fish ladders. This is especially important for commercial or recreational fish species. Active restoration may involve restoration of adjacent wetlands, floodplains, and embankments to improve river habitats. Restoration of such buffer zones is important to reduce sediment loading and leaching of

Figure 13.4 Restoration of riffle pool and gravel bed to avoid erosion in a small stream. (Courtesy of Sigurdur Greipsson, Kennesaw State University.)

agrochemicals into rivers. In addition, management of land that is adjacent to rivers is critical (including measures to prevent livestock from entering rivers).

Restoring the flow pattern and geomorphology of a river is usually followed by natural recolonization of native organisms or translocation of selected native fish species. River restoration often centers on recreational and commercial fisheries. Native trout species have been seriously overfished in the United States. Native trout species have also hybridized with introduced non-native trout species, especially brown trout (*Salmo trutta*). The brown trout was introduced in 1883 to improve fishing in streams in the United States. To restore streams with native trout species, invasive species including brown trout are eliminated by using specific pesticides (i.e., piscicides: antimycin or rotenone). After such elimination of invasive species, native species are reintroduced into the streams.

Use of piscicides, when followed by reintroduction of native species, has been used successfully in the recovery of Gila trout in New Mexico and Arizona. Also, populations of cutthroat trout have been restored using this method in California, Nevada, and the Great Basin. Conservation groups such as Trout Unlimited have emphasized the importance of genetic diversity of native trout species along with habitat restoration. Similar emphasis has been set by the National Fish Habitat Action Plan of the U.S. Fish and Wildlife Service. In addition, much less emphasis is now put on releasing hatchery-raised fish into the rivers.

Funding for river restoration has, in part, been provided by tourist- (recreational-fishing) related money. For instance, rivers in Montana attract large numbers of tourists, which account for $422 million each year. In Montana part of this tourist-related money has been used in restoration efforts to prevent erosion of river embankments. This work has involved planting riparian forests, including willows and cottonwoods, along river embankments. The river restoration in Montana started in 1990 and has involved some 45 tributaries. Also, access for migratory fish has been opened to 740 km of rivers that were previously blocked. At the same time efforts were made to restore and conserve land adjacent to rivers. This work has involved efforts to prevent livestock from entering rivers. In total more than 1,050 hectares of wetlands and 930 hectares of grassland have been conserved for this purpose. Also, more than 37,600 hectares of private land has been placed in perpetual conservation easements. Tax credits have been used as compensation to private land owners that participate in this program.

Restoration of the North Atlantic Salmon

Restoration of migrating fish populations can be complicated and may need an international effort. An example on such an effort is the restoration of the North Atlantic salmon (*Salmo salar*). The North Atlantic salmon migrates long distances from the ocean to its home river for spawning. An extensive study has shown that most rivers in Iceland have genetically distinct populations of native salmon. Hybridization of native salmon with farmed-raised salmon is a serious problem. Farm-raised salmon can escape and enter rivers. Gene flow from farm-raised salmon can reduce the adaptation of native populations. Therefore, a special precaution must be taken to avoid gene flow between native salmon populations and farm-raised salmon. Another problem is that the number of wild salmon returning to their home rivers has declined sharply. The reason for this decline is complex but may include commercial fishing in the sea, degraded or blocked (by dams) rivers, and gene flow from fish farms. Fish farms can, however, be improved by using sterile salmon or only local populations.

In 1989 it was recognized that populations of the wild Atlantic salmon could be seriously damaged and even decimated over large regions. To combat this trend, the North Atlantic Salmon Fund (NASF) was founded. The NASF is an international conservation organization that aims at restoring populations of the wild Atlantic salmon to their previous abundance. Its main effort has been to reduce commercial salmon fishing in the sea. It was realized that protecting the salmon during their oceanic migration was an important step in the restoration of the total population size. For this purpose, NASF has negotiated and funded agreements with the fishing industry in various countries to reduce salmon fishing in the sea. The NASF has aimed at reducing commercial fishing on the salmon's ocean feeding grounds and migrating routes in the North Atlantic by providing financial compensation to the fishing industry. More than $150 million have been raised for this purpose.

For example, compensation for fishing salmon in nets in the sea in Norway was about $3 million paid over 5 years. Drift nets are especially damaging to migrating salmon and have been the primary target of NASF. It has been estimated that more than half a million salmon were killed by drift nets over just

3 years in the North Sea. In 2003 NASF compensated fishermen for the total use of drift nets in the North Sea. Consequently, the number of salmon has increased in rivers that flow to the North Sea. Along with restoration of habitats in rivers, restriction on direct fishing in the ocean has resulted in recent increase in total population of the Atlantic salmon.

Summary

Aquatic ecosystems, including wetlands, rivers, and lakes, are an important compartment of the landscape. Wetland ecosystems have high primary productivity. Wetlands play an important role in groundwater hydrology and recharge of aquifers. Wetlands, however, are seriously degraded and transformed to other uses worldwide. Wetland restoration involves restoring the proper hydrological functioning and strategically establishing cover of native plants. Restoration of lakes involves curtailing the main disturbing factor. For instance, restoration of eutrophic shallow lakes involves inhibiting the nutrient loading. This can be accomplished by dredging and removal of nutrient-rich sediment. After eutrophication of lakes, high populations of algae (phytoplankton) build up, often leading to murky waters. Biomanipulation is an effective method to restore such lakes. In this case resident fish populations are reduced to increase zooplankton populations, which reduce the populations of algae. River restoration usually focuses on commercial or recreational fish species. River restoration aims at establishing functional integrity of this ecosystem. This often involves hydrological restoration and regulation of the flow pattern of rivers. If native fish species do not establish themselves, restocking might be necessary. Maintaining genetic diversity in the hatchery-raised population is, however, essential.

13.1 Case Study

Restoration of Coastal Salt Marshes in Brazil Using Native Salt Marsh Plants

César S. B. Costa, Oceanographic Institute, University of Rio Grande, Rio Grande, Brazil

The artisanal transplantation of cordgrass (*Spartina alterniflora*) clumps on wave-eroded margins of the Patos Lagoon estuary was already practiced by some farmers of southern Brazil at the beginning of the twentieth century. The scientific work on ecological engineering with coastal halophytes, however, is very recent in Brazil.

Creation and restoration of coastal habitats is today a critical conservation issue facing the cumulative loss of wetlands and degradation of remaining environments. All coastal salt marshes in Brazil are protected by the Forestry Law of 1965 that

(continued)

Case Study (*continued*)

defines salt marshes as a permanent preservation area and restricts any activity leading to their conversion and any nontraditional uses. Over the last two centuries land filling for urban and industrial development has destroyed hundreds of hectares of salt marshes in the main estuarine areas of the southwest Atlantic, and extensive livestock grazing induced the short vegetation physiographical aspect of several marshes. Most urban centers along the Brazilian coast either lack or have inadequate sewage treatment facilities, which tend to overflow during heavy rains and discharge raw sewage directly into bays, lagoons, or estuaries. Recently, an increasing amount of floating organic and industrial debris has been observed, resulting in extensive drift-line deposits and environmental degradation, with loss of ecosystem value for human populations. Additionally, seawater shrimp farming is uprooting mangroves and polluting the tropical coast of Brazil at an exponential rate. Evidence of erosive processes on coastal wetlands suggests a long-term trend of relative sea-level rise along the Brazilian coast.

This short review presents the latest developments on coastal habitat restoration in Brazil using native salt marsh plants. Most of the propagation and trial plantings of salt marsh plants has been done under supervision of the Department of Oceanography of the Federal University of Rio Grande (FURG) located in the city of Rio Grande (RS, southern Brazil). Several restoration projects of mangroves surrounding metropolitan areas occurred in Brazil during the last decade involving about 10,000 hectares, mainly in the northeast and southeast coasts, but they are not the focus of this review. We focus on the study cases of the amelioration of shrimp farm effluent pollution in the tropical coast of northeast Brazil by *Salicornia* crop-filtering systems and habitat creation and environmental quality improvement of an estuary in southern Brazil by *Spartina* species transplantation.

Reduction of Shrimp Farm Effluent by **Salicornia Gaudichaudiana** *Crop*

The application of aquaculture effluent to field crops has been successfully used as a method of biological treatment, and it could be extended to saline effluent streams if suitable, salt-tolerant crops were available. The perennial glasswort (*Salicornia gaudichaudiana*) occupies intertidal salt flats above the mean high water line along the entire Atlantic coast of South America, and it has being tested as a seawater crop suitable to treat shrimp farm effluent in the northeast of Brazil.

During the last decade thousands of hectares of mangrove forests and tropical salt marshes associated with salt flats were destroyed and converted to shrimp tanks. Brazilian shrimp farming extended over 15,000 hectares and produced 90,000 tons of shrimps during 2003 (mainly the exotic *Litopenaeus vannamei*). However, 90% of the tanks were located in the northeast region of the country, and each ton of produced shrimp released 50,000 to 60,000 m^3 of effluents rich in organic particulate matter, microorganisms, and macronutrients into estuarine and coastal waters. Shrimp farm pollution caused marked events of fish and invertebrate mortality associated with low oxygen concentration in the water as well as outbreaks of viral, fungus, and bacterial diseases in shrimp farms.

Salt crops such as *Salicornia* present no harm to farm profitability but do provide a way to remove a significant fraction of the nutrients in the irrigation stream and provide an economic return to the grower. Elsewhere, *Salicornia* species already showed the potential to produce biomass, forage, and oilseed crops using saltwater irrigation.

Field tests of *Salicornia* crops took place in the northeast Brazilian state of Ceará (4° S) during the 2005 drought season when no rainfall occurred and effluent salinity averages 40 g $NaCl^{-1}$. Seedlings were grown for 30 to 45 days in nurseries before being planted into a 10 × 9 m plot irrigated with shrimp farm effluent throughout shallow ditches. Plants were spaced 25 cm apart and watered by filling up the ditches once a day with 1,350 liters of effluent. Salt crust was removed by overflowing the plot and leaving the effluent to drain for 5 minutes in the low side of the plot.

Plants grew vigorously and started to flower after 100 days when they were harvested at ground level (**Figure 13.5**). *Salicornia* crop yield per harvest was a mean (± standard error) of 553 ± 148 kg dry weight $hectare^{-1}$, corresponding to 3,109 ± 715 kg fresh weight $hectare^{-1}$. Considering shoot mean concentrations of protein (7.6%) and ashes (26.9%) and the completion of three harvests per year, every hectare of *Salicornia* crop can produce annually 126 kg of plant protein and 446 kg of biosalt rich in potassium, calcium, magnesium, and micronutrients, such as iron, zinc, and selenium.

During cultivation the shrimp effluent concentrations pumping into *Salicornia* crop averages 0.902 mg NH_4-N L^{-1}, 0.008 mg NO_2-N L^{-1}, 0.705 mg NO_3-N L^{-1}, and 0.189 mg PO_4-P L^{-1}. *Salicornia* crop reduces effluent water volume through plant evapotranspiration, and a projected 1- hectare plot should remove annually 52 kg of NH_4-N, 41 kg of NO_3-N, and 11 kg of PO_4-P. Even if continuous flow is used to irrigate *Salicornia* crop in place of once daily filling of the ditches, field tests showed

Figure 13.5 Experimental crop of *Salicornia gaudichaudiana* irrigated with shrimp farm effluent in the semiarid northeast coast of Brazil. Detail: Shoots with fertile segments. (Courtesy of César S. B. Costa, Universidade Federal do Rio Grande.)

(continued)

Case Study (*continued*)

that crop system effluent can be consistently lower in ammonium (−43%), nitrate (−54%), and phosphate (−67%) concentrations than the inputted shrimp effluent.

Spartina *Marsh Creation*

Well-developed *Spartina* salt marshes are found at tropical and subtropical latitudes of South America in coasts where intertidal zone is advancing seaward, mainly due to sedimentation processes. Commonly, the cordgrass *S. alterniflora* forms a characteristic few meters wide fringe extending at the seaward of the mangrove forest. Over 2,000 km^2 of salt marshes occur along the hot temperate coast of southern Brazil, Uruguay, and northeastern Argentina. The coastal brackish water marshes of southern Brazil and Río de la Plata estuaries (Argentinean and Uruguayan coasts) are dominated by mixed and monospecific stands of the South American cordgrass *S. densiflora*, whereas in the southern part of southwest Atlantic seasonally hypersaline semiarid bays are dominated by cordgrass and perennial glassworts.

Salt marsh creation was proposed as an environmentally friendly and cost-effective alternative to urban runoff interception and dredged material disposal. In Brazil, propagation and trial plantings of native *Spartina* species started during the 1990s. Work on the practical aspects of using *S. alterniflora* in ecological engineering projects has been in progress only after 2004 (**Figure 13.6**).

The first pilot transplantation of both *S. alterniflora* and *S. densiflora* in southern Brazil took place during the winter of 1995. A mixed stand of 100 individual tillers was successfully planted on a tidal mud flat, spaced 1 m apart, in the Pólvora Island

Figure 13.6 Dr. César S. B. Costa next to *Spartina alterniflora*. (Courtesy of César S. B. Costa, Universidade Federal do Rio Grande.)

(Rio Grande, RS, 32° S). Tillers with rhizomes and roots of each species were excavated from adjacent monospecific natural stands. Mixed stand transplant was attempted because these species dominate areas between low and mid-marshes in southern Brazil with a high degree of overlap, and monospecific and mixed stands of these species can often be found at the same tidal level.

Thirty-five percent of the *S. alterniflora* clumps survived, produced 25 ± 2 tillers, and grew to an average 95 ± 8 cm radius (± standard error) in 10 months. After 18 months they had reached 303 ± 7 cm, and the clumps had coalesced. The main causes of deaths of *S. alterniflora* tillers were grazing by the Grapsidae burrowing crab *Chasmagnathus granulata* (56% of deaths) and tidal removal (31%). At the end of the first growing season, *S. densiflora* vegetative propagules showed better survivorship (50%), produced twice as many tillers (40 ± 4), and were less predated by crabs (40% of the deaths) than *S. alterniflora* transplants. Lower sensitivity of *S. densiflora* to grazing seems to be explained by its clumped growth form and a capacity for rapid tillering in response to perturbation of its aerial parts. On the other hand, *S. densiflora* growth form resulted in short horizontal extension of its rhizomes that reached 9 ± 1 cm in 10 months. This rate of horizontal extension was kept for 18 months when *S. densiflora* clumps were surrounded by *S. alterniflora* runners; the former continued to grow and reached 35 ± 2 cm radius after 7 years of transplantation. By the end of 2003, even affected by intense crab grazing, the plantings had spread and reached an area of 0.1 hectares.

Tillers of *S. alterniflora* were transplanted together with propagules of the mangrove *Avicennia schaueriana* to protect bank erosion of a new highway built over a landfill in Santa Catarina Island (29° S). Clumps were collected in a natural stand during the spring of 1997 and placed just below the mean tidal level along the embankment. After 3 months plants produced several clones and improved the establishment success of *Avicennia* propagules. No further information for this transplant is available.

In 2002 the Rio Grande do Sul state environmental agency contracted a construction firm to recuperate 0.15 hectare of estuarine margin of Patos Lagoon irregularly filled with gravel in the city of Rio Grande. Before transplantation the gravel was covered with 25-cm-thick layer of fine sand, and the planting was done in two stages, because there was a 1-m deep cliff separating the upper and lower sides of the intertidal zone. In September 2003 the upper cliff was cover with geotextile and 175 clumps of *S. densiflora*, ranging from 10 to 20 tillers per clump, were planted along two parallel 30-m-long rows just above the mean high water level, at 35 cm apart. After 12 months all clumps survived and multiplied to an average 139 ± 74 individuals per clump, a 10-fold increase. For the recovering of the low side of the site, 400 clumps of *S. alterniflora* (5–10 tillers per clump) were planted in January 2005, forming a fringe marsh of 0.1 hectares. The survival rate was 90%, and plants showed vigorous growth and horizontal spreading.

Recent Developments

Since 2005 the Inter-American Development Bank is supporting the Costa Sul Program, a technical cooperation for integrated management of Patos Lagoon estuary. Among its actions the program includes the creation and restoration of salt marshes to

(continued)

Case Study (continued)

overturn the historical depletion and improve the public valorization of these vital habitats. Eight 0.5-hectares marsh units were established between 2006 and 2007 with *S. alterniflora* clumps, spaced 1 m apart, along slightly erosive and stable margins and across urban runoff drains. The marsh design addresses the creation of unvegetated salt pans, which are a major component of mid-low marsh habitat used extensively by migratory shorebirds such as plovers and sandpipers and migratory waterfowl such as waders. *Spartina* marshes will be planted with the collaboration of Rio Grande Prefecture and Rio Grande Harbour Administration.

Two 80-m^2 unheated greenhouses have been established at FURG for germoplasm conservation, ecophysiological studies, and plant propagation by seedlings and vegetative growth. A 0.75-hectares tideland nursery was established in a proposed intertidal flat by discharging dredged material. Clumps of *S. alterniflora* from native donation sites and from the unheated greenhouse nursery will be established in the tideland nursery for further propagation spaced 1 m apart. Elsewhere, practice showed that *Spartina* plants multiplied more rapidly and grew more vigorously in a tideland nursery than in paddy nursery or open gardens. Additionally, the tideland nursery site is itself a recovering site, having for 80 years received the disposal of raw domestic sewage, and became available for restoration after the construction of a sewage treatment plant in 2005.

13.2 Case Study

Biomanipulation as a Tool for Shallow Lake Restoration in the Norfolk Broads, United Kingdom

Martin Perrow, Ecological Consultancy Limited, Norwich, UK / Environmental Change Research Centre, Department of Geography, University College London, London, UK

The impacts of a human-related increase in the rate of nutrient supply (anthropogenic eutrophication) have been felt in lakes worldwide, with serious declines in water quality, fisheries and conservation, and recreational value. In simple terms, clear waters with submerged vegetation and high biodiversity have been replaced by turbid, often algal-dominated water with low biodiversity. The reduction of nutrients to prompt a return to clear-water conditions, although desirable, is no easy task. As well as reducing the external nutrient supply, the legacy of previous enrichment stored within the lake sediments typically has to be controlled. Chemical means of binding phosphorus or physical means of removing nutrient-rich sediments (e.g., by dredging) have often been used. Even where nutrients have been reduced, however, a clear-water state is not guaranteed. This is because nutrient levels are typically only reduced to a level where either the clear or turbid state may exist as an alternative stable state and the nature of the ecological community has great influence on which ecological state is maintained.

Fish are a particularly important stabilizing (or destabilizing) influence through a range of top-down and bottom-up interactions, which cascade through the system via different trophic levels. Important trophic interactions include the selective predation of large-bodied zooplankton, which would otherwise control (by grazing) algal populations by small zooplanktivorous fish (e.g., roach *Rutilus rutilus*), and the turnover of bottom sediments by large benthivorous fish (e.g., bream *Abramis brama*) as they forage, thereby releasing nutrients available for algal uptake, resuspending fine lake sediments, or uprooting submerged plants. Biomanipulation, which is generally defined as the removal of undesirable zooplanktivorous and/or benthivorous species or the introduction of piscivorous species to control the other groups, is now seen as the most cost-effective means of restoring shallow lakes, in particular where the density per lake volume and thus impact of fish is at its highest.

The Norfolk Broads is a series of around 50 shallow (mean depth ~ 1 m) lakes of high landscape and ecological value in the east of England. Early writings and photographs from at least the turn of the twentieth century suggest that all these lakes, created from the flooding of medieval peat diggings (to gather fuel material), were dominated by a diverse submerged and emergent flora. However, by the 1960s only a handful of sites maintained this condition. Attempts to restore the Broads over the last 25 years initially focused on reducing nutrient supply, generally through stripping of nutrients from human effluent at the treatment works, although only now has this started to reap rewards after a long recovery time. Over the last 15 years or so, biomanipulation has often been used in a stepwise program of restoration following attempts at external and internal nutrient control (e.g., at Cockshoot Broad), although it has also been used as the primary restoration tool (e.g., at Ormesby Broad) in whole lakes (from 5 to 55 ha) as well as in experimental exclosures from a few square meters to around 2 hectares.

What follows here is a personal view of the successes and failures of biomanipulation in the Broads with a view to the application of the technique elsewhere. I include work from Cockshoot (5.5 ha, 1989 to ongoing), Pound End (1990–1997), Hoveton Great fish exclosure (1 ha, 1992–1995), Alderfen (5.6 ha, 1994–1999), Ormesby (55 ha, 1995 to ongoing), and Barton Broad (two fish exclosures to 2.4 ha from 2000 onward, with a further two exclosures to 0.8 ha from 2003 onward). The work described was undertaken by staff at ECON Ecological Consultancy (notably Mark Tomlinson and Adrian Jowitt) and funded (partly through the EC LIFE Programme) and supported by the Broads Authority, the statutory authority responsible for the Broads National Park, in partnership with The Environment Agency, English Nature, and Essex & Suffolk Water in the case of Ormesby Broad. A potable supply of water to local people is drawn from the latter, and thus biomanipulation of the lake had commercial as well as conservation implications.

Preparation for Biomanipulation

Because many of the Broads are interconnected with other lakes and rivers, isolation of the lake or area (exclosure) has to be undertaken before biomanipulation. Many innovative designs of barrier—preventing the passage of fish but ideally allowing water

(continued)

Case Study (*continued*)

exchange (to prevent stagnation)—have been developed by the Broads Authority and their consultants over the years. Some have inevitably been more successful than others, with common problems being undercutting of barriers in soft sediments and banks. Moreover, many Broads are tidally influenced and very low lying, and large areas may flood in winter. This inevitably influences the success of biomanipulation, with incursion of invading fish requiring further removal effort.

Many fish, especially the smaller ones, are known to leave the open waters of the connected Broads to spend the winter in sheltered locations along the main rivers (especially boatyards and marinas). This means the timing of the introduction of a barrier may have been an effective means of biomanipulation in its own right. Although this was not specifically undertaken, the introduction of barriers to create exclosures in Barton and Hoveton Great Broads did not trap many fish because the habitat present was rather poor (**Figure 13.7**). In contrast, at Ormesby fish from the wider system of connecting lakes, known as the Trinity system, had already become concentrated in connecting streams or small ditches (known locally as dikes). A similar situation occurred in Pound End. Such aggregations of fish inevitably became the focus of removal effort in both cases (see below).

At least in the case of Ormesby, a thorough survey of the fish populations and their distribution was undertaken before manipulation. This survey deliberately chose a method of sampling—point abundance sampling by electric fishing—that allowed representative sampling in all habitat types (littoral margins and dikes as well as open

Figure 13.7 Biomanipulated fish exclosure at Barton Broad. (Courtesy of Mike Page Aerial Photography.)

water) and could still be used to monitor fish populations once the lake had changed its habitat structure and become dominated by submerged macrophytes. Not only did this initial survey provide a baseline against which future change could be monitored (1995 to the present day), but it also proved invaluable in determining the locations of aggregations of fish and thus where biomanipulation should be concentrated.

Because the policy was to translocate all fish live to other water bodies (at the discretion of the Environment Agency), this has demanded large-scale transportation where large numbers of fish are expected (e.g., 302,258 individuals weighing 9,088 kg were removed in the first winter/spring); in the case of Ormesby Broad the receiving water body was not directly adjacent to the biomanipulated area/lake.

Fish Removal in Practice

In the absence of a thorough understanding of which components of the fish stock were responsible for the undesirable conditions in each situation, and accepting that fish readily shift diet and foraging strategy, creating a problem where they did not previously, the goal of removal has been to remove the entire population of all species and age groups that conceivably contribute to the undesirable algal-dominated state. Only the principal predatory species, such as Northern pike (*Esox lucius*) and European eel (*Anguilla anguilla*), and those species that would hopefully predominate in the future community of the restored lake, such as tench (*Tinca tinca*) and rudd (*Scardinius erythrophthalmus*), have generally been left behind. Eurasian perch (*Perca fluviatilis*) have been problematic in that although young fish are often zooplanktivorous, older fish are piscivorous and through cannibalism may control recruitment of young fish. The realization that perch may be functionally very important, particularly as an "open-water chasing piscivore" (as opposed to the "sit-and wait" ambush tactic of pike, restricting them to submerged vegetation and littoral margins) led to a change of policy for this species, and, now, not even young fish are removed in the hope of building an age-structured population of predatory perch.

Overall, the removal strategy is a combination of (1) a massive initial removal, (2) manipulation of the remaining potential spawners, and (3) subsequent "top-up" removals, especially of young-of-the-year recruits or invaders. Manipulation of undesirable species was then to continue until self-sustaining populations of appropriate fish species, probably at equivalent biomass to the populations removed, had developed. Because this has proved difficult to attain (see below), some form of manipulation has either continued longer than initially intended or even resumed where a clear-water state with submerged macrophytes has not been realized.

A range of methods have been used to capture fish, including standard survey methods such as electric fishing and seine netting as well as more traditional (local) means of fish capture such as fyke nets—in essence comprised of a length of netting (a leader) along which fish swim passively into a long cylinder of net containing a series of hoops connected by increasingly smaller funnels of netting that fish push through but cannot return, ultimately becoming trapped in the net bag at the end of the cylinder. As a general principle, whichever method offers the best return of number of fish per unit effort has been used. Consequently, concentrations of fish have been specifically

(continued)

Case Study (*continued*)

targeted. Aggregations tend to develop in winter, which has the added advantage of colder temperatures, reducing swimming speed and escape ability and enhancing survival during capture and transportation. Capture of fish during spawning may be undertaken but with caution because fish are easily stressed and subject to additional mortality at this time.

Further concentration of fish to increase removal efficacy has been achieved in a number of ways. For example, electric fishing has been used to drive small fish, already concentrated in their winter quarters in ditches and streams outside of the main lake, into seine nets set at convenient locations for subsequent transport. Further, even in the open water zone of the lake, large fish such as bream have been successfully driven by scare lines—lines of weighted, colored lines attached to ropes strung between boats—again into waiting seine nets. Just three runs of this in the summer of 1995 provided 55% of the 690 large bream ever removed.

Knowledge of the distribution and ecology of the fish in the lake in question is especially important in large lakes where it has not been cost effective to systematically fish the whole system. In the 1-hectare exclosure at Hoveton Great Broad and in the 5.5-hectare Cockshoot Broad, however, systematic electric fishing of "lanes" of water created between stop-nets across the entire width of the area or lake has been undertaken. "Leap-frogging" of the back net created a new lane of water, which was then fished several times until no fish are captured, and so on, until the whole area/lake was covered. Such a method yielded over 73,949 fish, mainly small (<10 cm) individuals from Cockshoot Broad.

Success of Removal

The initial removal exercise was generally highly successful in significantly reducing the fish population. In Ormesby, surveys indicated that 95% of fish had been removed in the first year. Similarly, in Cockshoot, after the initial removal of 1.35 individuals m^{-2} equating to 15.3 g m^{-2}, subsequent removal efforts over the next few years suggested a residual stock of around 1 g m^{-2} was likely to remain, thereby suggesting removal of at least 94% of fish. Counterintuitively, smaller exclosures have been more problematic as a result of being dogged by undercutting of barriers, allowing fish to invade. At Barton, high densities of 3.4 individuals m^{-2} and 1.5 individuals m^{-2} were removed from the two exclosures in 2000. However, in the former, 1.9 individuals m^{-2} were again present in the spring of 2001. After the removal of as many as possible over the summer, the pattern was again repeated over the following winter, and 2.4 individuals m^{-2} were present in 2002, again prompting removal all over again. After each removal, clear-water conditions were attained within a matter of weeks (see below), suggesting this was effective each time.

The success of subsequent spawning operations has been more difficult to judge. In Ormesby, despite the removal of large numbers of eggs on artificial spawning media, young-of-the-year bream have recruited on a biennial basis, with no obvious explanation. In good years for recruitment, densities reach 0.3 to 0.6 individuals m^{-2}. In the intervening years spawning intervention seems to have the effect of suppressing recruitment at least one order of magnitude to below 0.02 individuals m^{-2}. However, despite setbacks, experiences at Cockshoot illustrate that it is dangerous to stop manipulation completely.

Here, roach populations recovered after the discontinuation of removals in 1993 and by 2000 had reached an equivalent density (~1.5 individuals m^{-2}) to that present before biomanipulation in 1989. Renewed effort, particularly spawning intervention, saw this maintained at about 0.3 individuals m^{-2} from 2001 onward.

Recolonization of Submerged Macrophytes

Despite the risks of subsequent incursion and recruitment, the initial response to biomanipulation is often extremely dramatic and rapid. A clear-water phase has been virtually guaranteed in all circumstances within the first season, with the rapid regeneration of large populations of large-bodied zooplankton (*Daphnia* sp.) including species not seen for some time. Evidence from unmanipulated Broads, such as the 2-hectare Cromes Broad, suggested macrophytes could recolonize quickly once a suitable light climate had been achieved. Here, after isolation from the nutrient-rich River Ant, one macrophyte species, hornwort (*Ceratophyllum demersum*), became dominant within a year and has more or less remained so over the subsequent 14 years.

In contrast, manipulated Broads have shown a lag response of variable length. This may be explained by the need for a source of colonists or a preexisting inoculum of propagules such as seeds, tubers, or vegetative segments (turions) in the case of asexually reproducing species. Ormesby was partly selected on the basis of its relatively good condition before biomanipulation, which promoted a very short response time. Indeed, the long-term (>20 years) monitoring of macrophytes in the system showed an obvious increase in rooted vegetation in the first season. After 6 years pondweeds (*Potamogeton* sp.) were completely dominant, especially in the early season (**Figure 13.8**). Further succession of the community to include a greater range of species occurred subsequently, and the lake remains largely clear and dominated by macrophytes.

Recolonization was slower at Cockshoot, mainly limited to an approximately 0.25-hectare area of dike until 6 years after manipulation. In 1995, however, spontaneous generation of up to nine species of macrophytes occurred, including rare species such

Figure 13.8 Recolonization and dominance of fine-leaved pondweeds *Potamogeton* sp. in Ormesby Broad in the interconnected Trinity Broads 6 years after biomanipulation. (Courtesy of Martin Perrow, ECON Ecological Consultancy Limited.)

(continued)

Case Study (continued)

as holly-leaved naiad (*Najas marina*). The reasons behind this remains unclear, and although some regeneration may have occurred from recently introduced propagules carried in the guts of wildfowl (known to be of importance for *Najas*, for example), it seems likely that most species were generating from dormant reproductive material. As the site had been suction-dredged 13 years previously, this introduced the possibility that long-buried seeds exposed by dredging had remained viable, to break dormancy and germinate in the spring of 1995. Incredibly, a dramatic reverse occurred the following year, and, in what is remembered as a cold spring, virtually all species failed to germinate. Macrophytes then fluctuated at a low level over the next 6 years coincident with the reversion in the fish community back to dominance of zooplanktivorous fish, partly as a result of the lack of effort to control them. With renewed biomanipulation effort in 2000 (see above), the conditions for macrophytes improved, and there are now promising signs of recovery, with *Najas* the most numerous species.

The Barton exclosures show a wide range in the success of macrophyte recolonization within the same lake. Those in one area of the Broad (the Neatishead Arm) have shown a much better response than those in another (the area known as Turkey Broad). The limited area over which macrophytes have responded, the speed and magnitude (achieving virtually 100% cover in 2 years) of the response, and the fact that one of the principal species involved, fan-leaved water crowfoot (*Ranunculus circinatus*) (**Figure 13.9**), is now rare in Barton (as well as much of the Broads area) again point to the exposure of a dormant seed bank after suction dredging and thus possible true restoration of a formerly existing community as an important mechanism in recolonization. Unfortunately, the barrier behind which plants have recolonized is now in an advanced state of decay, and a "biomanipulated" status cannot be maintained. With incursion of fish and waters rich in algae from the main Broad, it seems unlikely that clear water and dense macrophytes will be self-sustaining maintained, unless the mechanisms stabilizing the macrophyte-dominated state (see below) prove to be particularly resilient.

In contrast to the other sites, Alderfen, Hoveton Great Broad, and especially Pound End showed limited recovery of macrophytes. At Alderfen, hornwort appears to have been undergoing a population cycle with a return period of around 7 years, since the 1980s. This cycle appeared to be largely unaffected by the biomanipulation of the lake, which had been partly achieved by natural fish-kill of the population of large benthivorous bream some 5 years earlier. Although both Hoveton Great and especially Pound End became clear, it was filamentous macroalgae, especially *Enteromorpha* sp., and not submerged macrophytes that became dominant. Unlike at Cockshoot, where the occurrence of filamentous algae mostly seemed to be a transitional phase before recolonization of macrophytes, the latter did not occur despite transplantation of a range of species.

Because grazing by birds, especially coot (*Fulica atra*), damaged transplants, these were also protected within bird-proof exclosures. The subsequent success of protected plants led to the belief that grazing by birds was limiting recolonization. Experiments involving simply protecting areas of sediment, however, showed clearly that germination

Figure 13.9 Two submerged macrophytes, *Ranunculus circinatus*, small flowers that have yellow buds and white petals, blooming from underwater in a pond. (Courtesy of Martin Perrow, ECON Ecological Consultancy Limited.)

was not occurring naturally even in the absence of grazing. Pound End in particular had been a severely degraded site, which had been filled with nutrient-rich fine sediment. Although suction dredging removed large quantities, there was no evidence that the ancestral peat bed was reached, and thus there was little potential for exposure of a former seed bank. In Hoveton Great Broad, where suction dredging was not undertaken, in retrospect there seems to have been little chance of rapid recolonization of macrophytes from a system that retained very few anyway.

Fluctuating Fish Communities

Achieving the desired fish community has been seen as the cornerstone of restoration, as a system dominated by piscivorous species should mean that the potential negative impacts of zooplanktivorous or benthivorous species are effectively suppressed. Over the short to medium term, there has been a great deal of fluctuation in the fish communities of all biomanipulated sites, as may be expected from a manipulated fish

(continued)

Case Study (continued)

community also subject to recruitment or incursion events of some species and not others. In all cases, however, large benthivorous bream have either been virtually eliminated or considerably reduced. Recent research points to the particular importance of benthivores and suggests that the positive aspects of restoration may owe much to their relative demise. Although tench, the benthic equivalent of bream in macrophyte-dominated systems, have generally recruited well in manipulated sites, they have barely begun to occupy the portion of biomass vacated by bream because they are particularly slow growing. In contrast, the replacement of roach by rudd (both smaller, shorter-lived species) is well underway in Ormesby and Alderfen but not in Cockshoot. Although rudd is typically associated with and favored by macrophytes, it is both zooplanktivorous and herbivorous in different life stages. Consequently, self-sustained control of this and other small species by piscivores is seen to be essential in the longer-term.

Unfortunately, self-sustaining populations of piscivores have not yet been attained, although there are promising signs of the emergence of perch at Cockshoot, some 15 years after the initial biomanipulation. Moreover, in Ormesby populations of both perch and pike seem to have declined in the 10 years since biomanipulation. In the case of pike this, rather perversely, may be because of the decline of prey species, perhaps ultimately limiting recruitment by reducing fitness of individual females and reducing the number of eggs laid or promoting cannibalism of any young fish by older conspecifics. Perch, on the other hand, were expected to benefit from the lack of roach with which they compete in their initial zooplanktivorous stage. Although growth of perch in the first year is extremely rapid, suggesting competitive release is indeed occurring, subsequent survival appears to be poor. Possibilities for this include predation of perch by the remaining pike, competition with pike for the available fish prey, or perhaps even competition with the remaining bream and tench for benthic invertebrates, which are an important prey resource for larger perch, especially where small prey fish are unavailable. Perhaps some combination of all these factors is the most likely cause of poor perch survival. Only further research holds the answers.

In an attempt to speed up the development of piscivore populations, an attempt was made to stock perch from another lake into Alderfen in 2001, coupled with the introduction of artificial macrophytes ("brushes"), which monitoring by Mike Jackson revealed could provide a suitable habitat and a source of suitable invertebrate prey. Unfortunately, the introduction failed, probably as a result of poor survival after transportation. Given the success of this technique in other similar circumstances, especially in similar lakes in Denmark, a further attempt seems warranted.

The Future

Multiple lake studies in Denmark, Germany, and recently in the United Kingdom show clearly that if lake nutrients can be reduced to low levels (<50 µg total phosphorus), then restoration of appropriate fish and macrophyte populations is probably guaranteed, with macrophytes self-perpetuating a clear-water state with low algal populations through several mechanisms: (1) effective monopolization of available nutrients, (2) production of algal-inhibiting substances by some species, (3) supporting grazing invertebrates such as molluscs to control epiphytic algae, (4) harboring zooplankton grazers of planktonic algae from the attentions of zooplanktivorous fish, (5) reducing resuspension of fine sediments, (6) reducing access to the lake bed for benthivorous fish, (7) shifting competitive interactions between fish species and providing habitat for piscivorous species, and (8) promoting conditions (especially uptake of oxygen at night) that kill nonadapted fish.

In the Norfolk Broads, despite the successful and considerable reduction of nutrient levels in the main rivers as a result of removal of nutrients within sewage treatment works, and even if in-lake control of nutrients is attempted and successful, background levels of nutrients are still likely to be too high for the clear-water state to be readily restored without further intervention. Biomanipulation is, therefore, still likely to be required to force the switch to the clear-water state. While clear water itself is virtually guaranteed within a short time scale, considerable patience may be required before submerged macrophytes take hold and dominate (>5 years). In fact, rather than simply trusting that this will happen, the likely type and sources of colonists and thus the potential speed and pathways of the restoration of macrophytes should be undertaken before biomanipulation is even attempted. Where natural recovery is predicted to be slow, further investment in the means of speeding it up, for example, by translocating propagules, should be attempted. Otherwise, repeated or ongoing biomanipulation to maintain or regenerate clear-water conditions will be necessary over a long time scale of 6 to 10 years or even more. Although this can be achieved, this leaves the system vulnerable to undesirable indirect effects (from the incursion of fish to the buildup of invertebrate predators of zooplankton where fish are absent). Moreover, the evidence to date suggests that it takes some time to establish an appropriate fish community (perhaps >15 years) through natural recovery, in keeping with the long generation time of many species. Indeed, it seems this final step has yet to be achieved. As with macrophytes, a more interventionist approach to shorten and speed up the restoration trajectory through stocking of appropriate species, especially piscivores, would seem to be appropriate and ultimately more cost effective than continuous management.

13.3 Case Study

Restoration of the Kissimmee River, Florida

Louise A. Toth, Department of Research, South Florida Water Management District, West Palm Beach, FL

The ongoing reconstruction of the Kissimmee River in central Florida, United States, may be the largest and most ambitious river restoration project in the world today. The project's goal of restoring "ecological integrity" has a consummate, unparalleled

(continued)

Case Study (continued)

scope and confers an opportunity to recapture a unique piece (ecosystem) of North America's natural heritage. Although its outcome is still unfolding, the history of the Kissimmee River effort provides an insightful chronicle of the scientific, social, and political intricacies of large restoration projects.

The Kissimmee River saga began in the late 1940s when several hurricanes struck the Kissimmee River Valley and impelled the state to appeal for federal assistance for flood protection measures for the increasing urban and agricultural development in the river's headwaters (i.e., the Orlando region). With congressional authorization, the U.S. Army Corps of Engineers channelized the entire 166 km of river and floodplain with a 9-m deep, 100-m wide canal and 6 dam-like water control structures. Although the U.S. Fish and Wildlife Service voiced concerns for potential impacts on natural resource values, including the river's renowned largemouth bass fishery, the channelization occurred at a time (1962–1971) when an emerging environmental consciousness had little influence on water policy. I presume that channelization projects of this scale and associated environmental impacts would not be acceptable within the morals and standards of today's society.

An alarmed public outcry was first heard near the end of the channelization when the murky, sediment-laden water and barren spoil mounds covering former floodplain wetlands and portions of the river channel provided a vivid glimpse of the magnitude of environmental destruction. The restoration initiative was launched in 1971 when a report revealed that the river's receiving body of water and second largest lake in the continental United States, Lake Okeechobee, was undergoing accelerated eutrophication as a result of high inputs of nutrients (phosphorus). Environmentalists reasoned that channelization had eliminated the nutrient filtration processes that formerly occurred as the river flowed through its floodplain wetlands and that the flood control canal was a conduit for rapid downstream transport of nutrient loads from the developing upper basin watershed.

During ensuing years, this initial concern for water quality impacts in the downstream lake was exacerbated by documented evidence of resource degradation within the river/floodplain ecosystem, including a tremendous loss of wetlands, virtual elimination of wading birds and waterfowl, and a chronic decline in the river's sport fishery. With increasing environmental activism and a "fill in the ditch" rallying cry, the restoration movement gained enough political support by 1976 to garner state and federal legislation for evaluating river restoration alternatives. These "mandates" for restoration, however, would lead to lengthy studies that soon would become a source of frustration for the environmental activists, while biding time for a contingent of restoration opponents, particularly landowners who had benefited from the drainage provided by the river's channelization. Political directives for studies have perhaps often been misused as an expedient way to avoid decisions on restoration and other controversial environmental issues.

The next milestone in the restoration movement occurred in 1983 when Florida's Coordinating Council on the Restoration of the Kissimmee River Valley endorsed a restoration plan (the "Partial Backfill Plan") that called for dechannelization of approximately 35% of the river by filling the central portion of the flood control canal. Although this recommended plan was based on 7 preceding years of public input and technical and scientific studies, restoration opponents continued to voice concerns, particularly about the stability of a backfilled canal, loss of flood protection on privately owned land, and the likelihood of recovering lost fish and wildlife resources. These remaining questions regarding the feasibility of the dechannelization plan were addressed in a demonstration project that included field tests of restoration measures and related physical and mathematical modeling studies.

My initial involvement with the Kissimmee was to monitor ecological responses during the demonstration project through which we provided compelling evidence of the potential for restoration of the river and floodplain ecosystem. Reintroduction of flow through the remnant river channel enhanced diversity and quality of river habitat and led to changes in fish and invertebrate communities that were indicative of a natural riverine system. Wetland plant communities quickly reestablished on reflooded floodplain and were accompanied by dramatic increases in wading bird and waterfowl utilization. The physical model demonstrated that a backilled canal could be stabilized and thereby alleviated concerns for catastrophic washouts and transport of sediments downstream (i.e., to Lake Okeechobee). Most importantly, the modeling studies indicated that the central portion of the river could be dechannelized without compromising flood control in the most developed portion of the river basin (i.e., the headwater lakes), although land acquisition would be needed to compensate for loss of flood protection along the river corridor. Thus, environmental impacts of the flood control project could have been reduced greatly if channelization was limited to the upstream and downstream reaches of the river.

Results of the demonstration project also emphasized what the Corps of Engineers would soon reveal in their independent congressionally authorized feasibility study—*River restoration could be accomplished only with reestablishment of natural hydrological characteristics and required modification of the managed inflow characteristics from the upper basin watershed, which had been regulated since the flood control project was constructed.* The need to incorporate a headwaters component to reestablish inflow regimes presented a new obstacle for river restoration planning as increased development since channelization had amplified flood control requirements in the upper basin watershed. Water availability in the headwater lakes and delivery to the river was constrained by flood control requirements, so how much water was needed for restoration of the river?

The addition of this water management requirement for restoration imposed a more quantitative (objective?) basis for ecological aspects of project planning. In recognition of the multiple objectives (e.g., water quality, wetlands, sport fishery, endangered species, and wading birds) that had been espoused during the course of the restoration initiative, engineers suggested they could use hydrological requirements for individual components to model the plan such that it would optimize restoration benefits—*if the project biologists could only provide these criteria!* To this end, in 1988 a symposium was organized to review available information. During the symposium I and others pointed out that optimal requirements of individual components are often

(continued)

Case Study (continued)

conflicting and that the lost natural resource values of the Kissimmee River were dependent on physical, chemical, and biological processes and interactions, including intricate food webs and an array of river and floodplain habitat characteristics. As a result, participants, including representatives of state and federal agencies that traditionally focus on different resource interests (e.g., sport fishes vs. endangered species), agreed that Kissimmee River restoration planning could not be based on optimization of benefits and needed to embrace a holistic ecosystem perspective. This pivotal event led to the adoption of a restoration goal of reestablishing *ecological integrity*—a naturally functioning river with a similar complement of resource values as occurred before channelization. In lieu of a maze of conflicting, if not impossible to define, requirements for individual components, I suggested that historically based hydrological (e.g., stage and discharge) characteristics and physical form guidelines such as lateral and longitudinal connectivity of river and floodplain habitat would provide the best criteria for restoration of the river.

The hydrological criteria were used to craft a new water management plan for the headwaters that would not only provide the necessary flow regime for restoration of the river but also increase littoral habitat around the lakes in the river's upper basin. With incorporation of this headwaters revitalization component, a federal–state partnership for restoration of over 100 km^2 of the Kissimmee River ecosystem was authorized in the Water Resources Development Act of 1992. This $372 million authorization was a precedent-setting use of Water Resources Development Act legislation for environmental restoration and provided a prototype for funding even larger projects, such as the $7.8 billion Comprehensive Everglades Restoration Plan.

Dechannelization of the Kissimmee began in 1994 with a pilot project that filled a 300-m section of the flood control canal. Between 1999 and 2001, 30 years after the restoration movement was initiated, reconstruction of the river finally began with the backfilling of 12 km of canal (**Figure 13.10**) and demolition of one of the water

Figure 13.10 An aerial view of a meandering stream restored in the channelized Kissimmee River. (Courtesy of Louis A. Toth, South Florida Water Management.)

control structures (dam). This first of three phases of the restoration project reestablished flow through 24 km of river channel and reflooded over 4,000 hectares of floodplain. With reestablishment of river and wetland habitat, the reconstructed area has seen increased use by wading birds and waterfowl and sustained optimistic expectations for rapid recovery of lost resource values. However, restoration of the Kissimmee River remains impeded by headwater inflow regimes, which have not yet been rectified.

The first phase of reconstruction of the Kissimmee River was a monumental step in a restoration project of unprecedented scale and scope, which is intended to eventually span over 70 km of river channel and 11,000 hectares of floodplain wetlands and benefit more than 320 fish and wildlife species. Its ultimate success and potential significance among environmental restoration initiatives remains dependent on modification of water management in the river's headwaters and completion of the other two phases of reconstruction.

Key Terms

Biomanipulation 325
Eutrophication 318
Hypoxia 318
Phytoplankton 318
Piscivorous 318
Planktivorous 318
Sedimentation 319
Watershed 314
Wetlands 315
Zooplankton 318

Key Questions

1. List ecosystem services that are provided by wetlands.
2. Describe processes that contribute to river degradation.
3. How does eutrophication affect ecosystems of shallow lakes?
4. What options are available in restoration of wetlands that have received excess amounts of nutrients?
5. Describe the process of biomanipulation.

Further Reading

1. Loucks, D. P. (ed.). 1998. *Restoration of degraded rivers: challenges, issues and experiences*. Dordrecht, Netherlands: Kluwer Academic.
2. Middleton, B. A. 1999. *Wetland restoration, flood pulsing, and disturbance dynamics*. New York: John Wiley & Sons.
3. Moss, B. 1980. *Ecology of fresh waters*. Malden, MA: Blackwell.
4. Stoneman Douglas, M. 1997. *The everglades: river of grass*. 50th anniversary edition. Sarasota, FL: Pineapple Press.
5. Westcoat, J. L, Jr. and White, G. F. 2003. *Water for life. Water management and environmental policy*. Cambridge, UK: Cambridge University Press.

14

MANAGEMENT OF RESTORATION PROJECTS

Chapter Outline

14.1 Setting Goals
14.2 Planning
14.3 Action Plan
14.4 Adaptive Management
14.5 Monitoring
14.6 Aftercare and Final Assessment
14.7 Legal Framework and International Agreements
      Environmental Legislation
      Wetlands
      Mined Lands
      Invasive Non-Native Species
      Soil Erosion
      International Conventions
      Human Population Overgrowth
      Desertification
      Biodiversity
Case Study 14.1: Integrated Restoration Efforts of Severely Degraded Subarctic Heathland Ecosystem
Case Study 14.2: Decision Making in Ecological Restoration

A restoration project should follow a comprehensive, science-based plan with clear goals. Projects should be guided by pragmatic, realistic, and attainable goals. Restoration projects should be flexible, however, allowing for adjustments or continuous improvements on strategies as needed. Such an approach, called adaptive management, is outlined in this chapter. It is important that restoration practitioners spend a reasonable amount of time defining the aims of projects.

Monitoring systems are an essential part of restoration programs and are critical in following up on and improving restoration practices. Monitoring systems allow early detection of ecosystem regression as well as possible intervention in the form of appropriate aftercare. Final assessment of restoration programs compares attributes of a selected reference site with the restored ecosystem.

Restoration activities should be supported by a legislative framework. Both regional environmental laws and international agreements can aid restoration programs. The involvement and acceptance of the public is also important for the success of restoration programs. This involvement and acceptance should be long lasting. Public relations are critical, and the flow of information should not be restricted. A limited financial budget is often the main hindrance of restoration programs, and pragmatic solutions are required to sustain the work in the long term. The process of planning, executing, and monitoring restoration projects is discussed in details in this chapter. Also, environmental legislation (both national and international) is discussed in this chapter, and examples relating to ecosystem restoration are provided.

14.1 Setting Goals

The initial phase of ecosystem restoration includes identifying specific goals of the project by defining the structure and function of the final ecosystem and outlining restoration strategies to meet these goals. Goals, therefore, define the vision of the restoration project. These goals vary in focus. They can, for example, focus on the success of a single species (a rare or endangered one) whose population is being restored or focus on the integrity and functioning of an entire ecosystem. Goals should be realistic and attainable within a reasonable timeframe and should also adhere to a sensible financial budget. Clearly set goals are not only the beacon of restoration projects but also an aid in evaluating the success of the work.

Interests of all stakeholders should be represented when goals are defined. Stakeholders include people concerned with a restoration project, such as landowners, land-user groups, local authorities, politicians at local and federal levels, technical and environmental consultants, and nongovernmental organizations. Participation of all stakeholders in defining goals reduces the risk of conflicting interests at later stages in the restoration program.

Goals should be flexible, and they should be reevaluated throughout the restoration process. This is especially applicable if the final ecosystem is considered to be a "moving target." Also, flexible goals are more realistic for ecosystems or populations that go through cyclic changes, especially under today's global climate change scenario.

Goals should address the causes of ecosystem degradation. These causes could include ecological degradation caused by, for instance, fire suppression, invasion of non-native species, nutrient deficiencies, or eutrophication.

Restoration goals generally focus on the return of an ecosystem to a state approximating predegradation. The restoration process can be guided by several complementary short-term or interim goals that act as subsets (or "milestones") of the final goal. More attainable, short-term goals pertaining to a particular ecosystem service or function can therefore be reached within a short timeframe. (Such short-term goals can act as milestones) or "benchmarks" in the restoration process. Such milestones usually focus on specific aspects of the restoration process, for instance, revegetation (i.e., establishment of surface cover), nutrient cycles, or establishment of keystone species. The use of milestones in restoration aids in the evaluation of the project at regular intervals and can, therefore, act as "quality control."

The use of milestones can be beneficial in achieving political acceptance of the entire restoration project. This is especially true if it takes decades to reach the final restoration aim, because early in the project it might seem almost impossible to get to the end point. For example, in restoring drastically disturbed ecosystems, the first goal might focus on ground stabilization by the use of commercial cover plants that initiate subsequent succession. Revegetation is usually achieved within years, but subsequent succession could take decades. Goals that relate to succession management then follow. In this context it is important to set realistic short-term goals (i.e., establishment of cover species), followed by interim goals that relate to succession management, and a final goal that relates to functioning and species composition of the final state of the restored ecosystem.

In defining the final state of the restored ecosystem in question, a suitable reference ecosystem should be selected. Careful selection of a reference ecosystem is essential to evaluate restoration success. A **reference site** should ideally be located close to the restoration site in question and should represent a predisturbance state or should be in almost "pristine" condition. It might be challenging, however, to find a nearby ecosystem in pristine or predisturbance state to serve as a reference site because widespread pollution, anthropogenic disturbances, and global climate change have affected almost all ecosystems worldwide. Reference sites should include similar landscapes and should be exposed to a similar regime of natural disturbances as the restored ecosystem. Replicated reference sites should be selected to account for a variation in ecosystems within a landscape. Baseline information of the reference site can be established by examining its structure (taxonomical richness) and function (physiological activities). Such information can be used when comparing with the restored ecosystem in question.

If acceptable reference sites cannot be located, then historical reference data should be collected. The historical state of an ecosystem can be assessed using information describing sites before human impact occurred. These include such historical data as written accounts, oral histories, anecdotes, maps, old photographs, and archeological information. Also, preserved pollens (in sediment or soil) can be used to reconstruct ancient communities. However, such data are often not available or lack necessary details on biodiversity and ecosystem functioning. It should, however, be considered that an historical description of an ecosystem is most likely to be subjective.

14.2 Planning

Restoration plans outline predetermined specific actions to be implemented as well as measurable outcomes of restoration projects. It includes a timeline and a realistic budget. Restoration plans are based on scientific principles and technological expertise. Restoration plans also incorporate ecosystem processes such as succession and disturbance regime. Ecosystem surprises such as invasion of non-native species and regression must be incorporated in the plan as well.

Succession models have traditionally provided restoration practitioners with theoretical framework, as discussed in Chapter 4. However, more recently, "assembly rules" have gained considerable attention (see Chapter 5). This is, in part, because succession can follow various trajectories due to unpredictable patterns of species colonization. A plan of a logical sequence of assisted colonization and establishment of nexus species that are critical for restoration success is, therefore, provided by assembly rules.

Planning a restoration project is a team effort and requires participation of all stakeholders. It is common during the planning phase that stakeholders form a task force to bring technical expertise into the project. Planning should avoid any conflict of interest among stakeholders. Such conflict may include different vision or goals of the project, property right issues, and compatibility of the project with end users. It would be unrealistic to expect that no conflicts of interest will show up during the planning process. Any disagreement that arises between stakeholders must be addressed immediately through conflict resolution. A professional facilitator or mediator could lead such a process by considering all viewpoints and proposing a solution.

Coordinated resource management and planning (CRMP) can serve as a model for planning restoration projects. CRMP uses an interdisciplinary team effort to solve management problems that relate to natural resources. CRMP allows all stakeholders to actively participate in the planning process. It assumes that planning requires input from people with different backgrounds. Involvement in CRMP is usually voluntary. The CRMP model encourages communication and technology exchange among stakeholders. One benefit of this process is the decreased risk of a conflict among participants. This model has been used successfully in natural resource management in California and is rapidly gaining further acceptance as an essential tool in planning ecological restoration.

The initial planning should focus on an **ecological management unit** that is a well-defined but limited area. The extent of degradation of the ecological management unit must be evaluated. Such evaluation should determine the restoration effort needed for the project. The first step is to delineate (using maps) the ecological management unit and evaluate how it links with other ecosystems within the landscape. In this context it is valuable to assess connectivity between fragments and potential colonization patterns. This is followed by an assessment on the biodiversity and genetic diversity of populations that are the focus of the restoration project. Also, natural and human disturbances must be outlined.

Communication networks should be established early in the planning process to keep stakeholders informed and updated throughout the restoration project. This could involve networks of electronic communications (e-mails and blogs), interactive home pages, or regular publication of an ordinary newsletter that is sent to all stakeholders. Extensive public outreach through local media and involvement of volunteer groups and local schools are usually effective. Favorable public relations are essential for the political acceptance of restoration projects.

Improper planning can result in serious problems at later stages, such as unclear or unrealistic goals, lack of scientific information on the ecosystem in question, lack of technical expertise, or conflict about final land use. Improper planning can even result in lack of funding for the whole project. If planning is done improperly, it can result in declining interest of the public and abandonment in the project.

During the planning phase appropriate legislative framework in support of the restoration work and provision for final land use must be considered. In addition, appropriate permits must be obtained, and some phases of the planned restoration project (especially involving engineering structures) should be assessed.

Environmental impact assessment is often required for the entire restoration project or some part of it. Such assessment can agree on the project, suggest mitigation, or recommend that the project be rejected. As more mega-restoration projects emerge, their global impact should also be considered.

Securing adequate funds for the entire restoration project is an important part of the planning process. Planning restoration projects involves careful budgeting to guide the restoration project. Careful budgeting is cost effective in the long run and should include a long-term financial commitment to the entire restoration program, which should be carefully considered. The budget should also include a provision on the benefits of ecosystem services provided by restoration. Such a provision should emphasize the cost of continuous ecosystem degradation or loss of critical ecosystem services. It is, therefore, wise to include in the budget the benefit of restoring essential ecosystem functioning.

The scale of restoration projects dictates different types of budgets. For example, volunteers often accomplish small-scale projects successfully on low budgets, whereas the cost of mega-restoration projects involving ecological engineering can run into the billions of dollars. Budgets should distribute costs between local, state, and federal levels. Also, private and public organizations should share the cost. Funding allocation into restoration projects varies but can be distributed, for instance, in the following way: 10% in baseline studies to define goals, 10% in planning the project, 50% in project implementation, 20% in monitoring (including aftercare), and 10% in the final assessment. The budget should be reviewed by stakeholders and experts. Budgeting should also prioritize restoration activities to optimize financial resources. Careful budgeting of restoration projects ensures political acceptance as well as committment to the entire project. It is also critical to gain commitment of organizations for the restoration project to ensure its momentum.

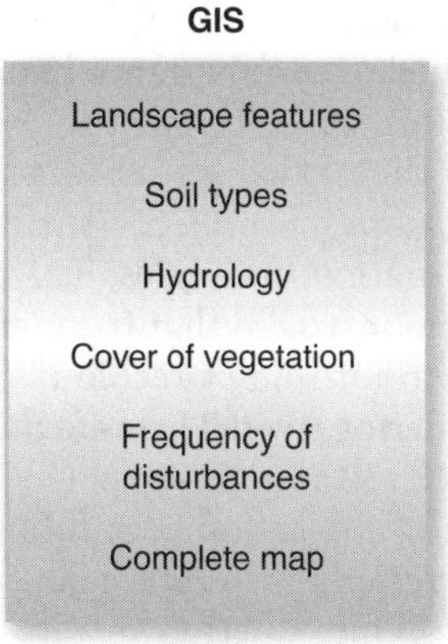

GIS

Landscape features

Soil types

Hydrology

Cover of vegetation

Frequency of disturbances

Complete map

Figure 14.1 Information included in the GIS for planning restoration projects.

Planning restoration projects also involves the use of various expert systems. For example, geographic information system (GIS) is an invaluable tool for planning restoration projects (**Figure 14.1**). The GIS technology is used in manipulating large quantities of data at the landscape level, and it allows restoration planning according to features in the landscape. It is especially practical in the restoration of fragmented landscapes. The GIS can, for instance, be used early in the planning phase to identify clusters of fragmented habitats, which have the greatest potential for restoration based on their size and location within the landscape. These fragments can, therefore, act as foci for expansion of fragmented native habitats. Also, effective integration of restoration sites within the landscape is essential. In addition, GIS is commonly used to store information on particular restoration practices for future uses and to keep track of restoration efforts and progress.

Spatial analysis using GIS combined with the "analytical hierarchy process" (AHP) is particularly powerful in assessing habitat restoration at the landscape level. The AHP provides ranking of important criteria (usually based on expert opinion) that relate to the critical habitat and target species in question. Basically this method identifies clusters of sites (that receive high ranking) in the landscape that have, therefore, high potential for restoration. This method has been used successfully in, for instance, identifying sites for elk (*Cervus elaphus*) restoration in Arkansas, assessing suitability of olive plantations in Andalusia, Spain, for wildlife habitat restoration, and for planning restoration of urban estuaries.

14.3 Action Plan

A **model project** is commonly used to demonstrate, on a small scale, how a restoration project would work if it was fully implemented. Such an approach helps in adjusting restoration methods to local conditions and also in gaining political acceptance for the entire project. For example, before restoring the meandering river pattern of the 170-km Kissimmee River in Florida, a model project was established over only a small portion of the river. The model project provided critical information before the entire restoration project was implemented. The cost of the model project was only a fraction of the $8 billion total cost of the entire restoration project. During the model project it was possible to test various restoration strategies. For example, the model project of the Kissimmee River demonstrated that fragments of wetlands along the channeled river played an important role in recolonization of embankments of the newly reconstructed meandering river. This information alone improved the cost effectiveness of the whole restoration project.

Implementation of the action plan follows a **pilot study**. This type of study involves initiating ordinary restoration actions such as seeding or transplanting of keystone species, reintroduction of rare and endangered species, enhancing genetic diversity, and eradication of non-native species. Initiating succession on restoration sites is often the focus of the restoration implementation. Implementing restoration plans can follow predetermined strategies, such as the "assembly rules" described in Chapter 5 or various succession models described in Chapter 4. Implementing restoration projects also involves managing human resources (hiring and training skilled labor, coordinating work with volunteer groups, or subcontracting some of the work). Members of the community who are involved, either as volunteers, paid workers, or professionals, should be encouraged to participate at all stages of the planned restoration work. This allows more community participation into restoration projects and increases general awareness of the work.

The first step in implementing restoration projects is usually to curtail the main disturbing agent and/or to establish essential natural disturbance regime. This may involve restriction or prevention on livestock grazing. Also, public access to restoration sites is often restricted, especially vehicular traffic. It is, however, more successful to guide public access on restoration sites rather than inhibiting it. Natural disturbance regimes, including fire regime or natural flow patterns in rivers, are reinstalled during this phase.

Severely damaged sites need immediate active restoration efforts. For example, saline soils must be desalinized and contaminated areas need soil remediation efforts (phytoremediation or bioremediation), as discussed in Chapter 10. Soil erosion must be halted by rapid revegetation and ground stabilization. Also, severe lack of nutrients must be overcome by addition of organic matter or judicious application of chemical fertilizers or by strategic use of N_2-fixing legumes.

Implementing restoration plans involves establishing keystone species and maintaining them on site. Stress should be minimized on keystone species by transplanting keystone species according to landscape patterns and especially by using regenerating niches if possible. Also, eradication and control of non-native species should be implemented early, and continuous efforts and monitoring are required, as discussed in Chapter 7. Implementing active restoration of critical habitats is important in the restoration of rare or endangered species (see Chapter 3).

14.4 Adaptive Management

Although restoration ecology has a strong theoretical background, restoration practitioners may prefer strategies that are based on empirical findings or even intuitive experience. Traditionally, restoration projects have been carried out on a **trial-and-error** basis where one restoration technique is implemented at a time; however, this approach should be avoided. The trial-and-error approach typically does not include a feedback system to correct mistakes or optimize the restoration processes. Instead, the whole project is implemented full scale at once, and it either works or fails. This approach is often used because funding agencies and institutes prefer to implement restoration projects rather than to fund long-term monitoring on restoration sites. Also, reporting setbacks of restoration projects or providing contradicting results is usually avoided. Such information could, however, be of value to restoration practitioners in the long run.

An alternative approach is to test restoration methods in small-scale field trials within the degraded ecosystem of concern before implementing on a large-scale. Such small-scale field trials can yield important baseline information that benefits the entire restoration project. These small-scale trials can, for instance, reveal constraints to restoration (as discussed in Chapter 5) of the degraded ecosystem. Some of the most serious biotic constraints that impede restoration are, for instance, lack of nutrient cycling and lack of available propagules of native species. These constraints can be overcome by using appropriate restoration methods. It is important to prioritize restoration efforts by using small-scale trials to solve any constraints. High priority should be given to groups of constraints that can be dealt with simultaneously. These could, for instance, include factors that inhibit establishment of native species on degraded sites or factors that promote nutrient cycling.

To identify constraints in a degraded ecosystem, a sequence of small-scale trials are installed. This approach is called **adaptive management** (**Figure 14.2**).

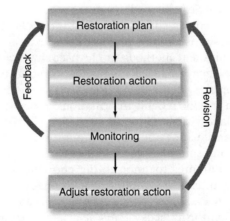

Figure 14.2 Model of adaptive management. Restoration methods are adjusted (through feedback and revision) according to results from small-scale field trials.

Adaptive management establishes an effective feedback system based on small-scale field trials where the goal is to adjust restoration methods to local conditions. Restoration methods that work for one degraded ecosystem may not work for another. The "fine tuning" or "adjusting" of restoration methods is, therefore, necessary even within one degraded ecosystem, due to heterogeneous environmental conditions. Such methods could, for instance, include adjustment of fertilizers application including type of fertilizer and timing and dose of fertilizer application. This information can be used to optimize fertilizer application onsite and then be implemented in the restoration plan. Similarly, methods of seeding of a particular species can be tested involving seed rate, depth of sowing, and timing of sowing. These are just a few factors that need to be adjusted to local conditions.

Adaptive management takes into consideration site-specific problems and suggests adjustments of restoration methods to solve these problems. Adaptive management allows beneficial adjustments or modifications of methods along the restoration project in response to the results of the small-scale trials. Adaptive management is not a trial-and-error process but rather involves a continuous feedback system that adjusts restoration methods to local conditions. Another strength of adaptive management is that it incorporates unpredictable events such as fire, invasive plants, or outbreak of pests into the restoration plan. Adaptive management, therefore, allows adjusting and conforming to a plan in an unpredictable environment.

Generally, adaptive management involves small-scale testing at various stages in the restoration project. It is essential that a feedback mechanism is established to allow the restoration plan to adjust methods rapidly to the results of such trials.

14.5 Monitoring

Monitoring systems act as quality control for restoration programs; however, they are primarily established to assess the progress of restoration projects. Also, monitoring systems evaluates how well the goals of the project have been met. Monitoring systems should elucidate both restoration success and failures. For this purpose different methods can be used. Innovative site-specific monitoring programs are always needed for each restoration project. Information derived from monitoring systems made from replicated "permanent plots" on restoration sites can provide reliable information on the progress of restoration projects. Monitoring involves sampling or surveying permanent plots for high levels of details where data are gathered at certain time intervals. Such data may include hydrological and soil analysis as well as evaluation on ecosystem structure and function. Timeframes for monitoring permanent plots typically run for 1 to 10 years with intensive sampling at least once a year.

Other more extant or "proxy techniques" can be used in monitoring restoration sites. Monitoring ecosystems at a large spatial scale involves the use of aerial photographs and/or GIS maps. False-color aerial photographs can, for instance, be used in a time series to infer vegetation dynamics on restoration sites. Such studies are usually calibrated by intensive studies on permanent plots. Aerial photographs are especially useful in identifying landscape features that are not easily identified on the ground. Colonization patterns can also be followed by the use of aerial photographs in a time series.

Short-term monitoring only covers such goals as establishment of ground cover vegetation. Monitoring efforts require long-term commitments. Restored ecosystems should be monitored until they are functional and do not require any further restoration efforts, such as addition of fertilizers, introduction of native species, or eradication of non-native species. Such monitoring efforts might, for instance, involve assessing vegetation cover in permanent plots every 1 to 5 years for 25 to 50 years. Also, monitoring efforts must include unpredictable events that may appear over a long time scale. Theses include outbreak of fire, climatic fluctuations, and outbreak of pests. Cyclic population fluctuations (e.g., prey–predator cycle) must also be included in the monitoring system.

A monitoring system acts as an "early warning system" for restoration efforts. Such a warning system should indicate if restoration efforts are proceeding as planned. For such a monitoring system it is important to set up appropriate "action thresholds." For example, slow growth of keystone species can result from a lack of available nutrients. Intervention in the form of fertilizer application or weed management is, therefore, needed. If intervention is not applied early enough, regression due to lack of nutrients or competition with weed could lead to a serious decline or a collapse of the degraded ecosystem, thereby increasing enormously the cost of the restoration project.

In particular, it is important to monitor invasive species because their control is often the primary focus of restoration efforts, as discussed in Chapter 7. Lack of proper onsite monitoring can lead to massive invasion of weedy or non-native species. Also, monitoring of invasive species is cost effective and usually critical for the success of restoration projects.

General parameters to be monitored must be selected carefully. Such parameters can vary from molecular to landscape levels. These could, for instance, include ecosystem diversity within a landscape or genetic diversity of a target population. Genetic diversity is especially important in restoration of animal populations (see Chapter 12). At the population level, parameters describing population size and growth (including immigration and emigration) are of interest. These parameters are also important when assessing metapopulation networks. At the ecosystem level, parameters assessing function including nutrient cycling and biomass production should be targeted. At the community level, parameters involving keystone species are of interest, as are parameters assessing biodiversity. Finally, at the landscape level, parameters assessing fragmentation and patch sizes are of interest.

Appropriate "indicator species" for adverse ecosystem conditions should be targeted for monitoring purposes. These include, for instance, species that indicate adverse nutrient levels in the environment or those that are characteristic of certain successional stages. Indicator species could be plants, animals, or soil microbes that are easily surveyed. Monitoring indicator species is critical in assessing restoration progress.

14.6 Aftercare and Final Assessment

Aftercare involves the implementation of restoration efforts in response to long-term monitoring. Aftercare is implemented to correct adverse **ecosystem regression**. Regression on a restoration site can be caused by several factors: decline in soil pH, increase in metal solubility, lack of nutrient cycling, decline in nitrogen fixation, over-grazing of vegetation, invasion of weeds or non-native species, and water logging. A proper monitoring system should detect regression early enough to avoid serious ecological damage. In this context it is important to identify the underlying cause(s) of regression and to evaluate if these can be corrected to enhance restoration efforts. For example, a progressive decline in soil acidity may be halted by periodic applications of lime. Nutrient deficiency in vegetation can be corrected by continuous application of fertilizer or by strategic use of legumes. Also, invasion of weeds and non-native plants can be eradicated by implementing an appropriate weed management program.

Public relations should focus on restoration success. Both restoration success and setback should, however, be documented. Negative results should be addressed and appropriate plans made to deal with them. Results should be published for meetings with stakeholders and in scientific journals as well.

Final **assessment** of restoration projects should focus on ecosystem attributes including ecosystem function and structure. Providing assessments on these attributes can give a decent indication of the success of restoration efforts. Successful restoration of ecosystem structure including biodiversity is important for the long-term stability and resilience of restoration sites in response to natural disturbances (see Chapter 2). Successful restoration of ecosystem function including nutrient cycling is important for the stability of the restoration site in question.

Assessing ecological structure including biodiversity involves measuring the richness and abundance of organisms on a particular site. Plant and arthropod richness are commonly used to measure biodiversity recovery in restoration projects. In addition, determining the diversity of species within different functional groups of an ecosystem provides an indirect measure of ecosystem resilience, as discussed in Chapter 2. Assessing vegetation structure is accomplished by measuring vegetation cover, plant density, biomass, or height of plants.

Assessing ecological functions includes nutrient cycling, which is usually measured indirectly by estimating nutrient availability at different tropic levels. The recovery of symbiotic associations (e.g., mycorrhizae, *Rhizobium*, and *Frankia*) is also critical for the long-term functioning of a restored ecosystem. Assessing final restoration success is, therefore, accomplished by comparing the attributes mentioned above with those of a reference site.

14.7 Legal Framework and International Agreements

Environmental Legislation

Legislative frameworks can enhance the success of restoration programs. Certain challenges exist with legislating ecosystem restoration, however, such as the fact that minimum efforts such as revegetation or reclamation are usually the main requirements rather than a complete restoration of a self-sustainable ecosystem that resembles an original ecosystem. Short-term goals are, therefore, usually the target of legislation. Restoration legislation should ideally focus on long-term goals in building self-sustaining ecosystems that resemble original ecosystems.

Several pieces of environmental legislation in the United States implement direct restoration efforts. These have typically aimed at preventive measures. These legislative frameworks include pollution control measures such as the Clean Air Act of 1963, which, in fact, was the first national air pollution control effort, and the 1990 revision of the Clean Air Act, which focused on pollutants that cause acid rain. The goal of the Clean Air Act is to reduce the emission of sulfur dioxide (SO_2) (the main culprit behind acid rain) in the United States from 23.5 million tons to 16 million tons by 2010. Also, the target is to reduce emissions of nitrogen oxides (NO_x) from 7.5 million tons to 5 million tons by 2010. The 1990 revision has market-based approaches especially designed to lower sulfur dioxide pollution levels and lead (Pb) emission. Under this scheme the government gives the industry permits for emitting certain amount of sulfur dioxide each year. The industry is then allowed to buy and sell their emission quota. Such a flexible marked-based approach will encourage the industry to install efficient pollution-control devices. At the same time it will probably stimulate inventions of pollution-control technologies.

Acid rain is an international problem because SO_2 and NO_x typically cross over national borders. International agreements are, therefore, needed to curb this problem. The U.N.'s Long-Range Transboundary Air Pollution Agreement involved 25 developed countries in emitting less amount of NO_x than 1987 levels. International agreements have resulted in drastic reduction in SO_2 emission. In Europe SO_2 emissions decreased by 40% from 1980 to 1994. To improve this even further, in 1994, 12 European nations signed an agreement to reduce SO_2 emissions by almost 90% by 2010. Reductions in lead (Pb) emission have been particularly successful (they have been reduced by 95% recently) in the United States mainly by curbing the use of leaded gasoline.

Wetlands

Restoration of degraded wetlands was outlined in Chapter 13. Several federal legislations have been developed to protect and restore wetlands. The National Environmental Policy Act of 1969 was the first attempt to protect wetlands in the United States. Today, the Clean Water Act of 1972 is the primary legislative framework for wetlands. The North American Waterfowl Management Plan of 1986 was signed by the United States and Canada but only endorsed by Mexico. The plan focuses on restoring several priority habitat areas. The Food Security Act of 1985 contains the "swamp buster" provisions that discourage the destruction of wetlands. The Emergency Wetlands Resources Act of 1986 advanced wetland protection even further. Farms that destroy wetlands risk losing federal program benefits. However, farmers can regain lost federal benefits if they restore damaged wetlands.

In 1998 a coastal restoration plan was approved, entitled "Coast 2050: Towards a Sustainable Coastal Louisiana." This plan aims mainly at restoring sustainable coastal wetlands of Louisiana. The North American Wetlands Conservation Act of 1989 provides grants for the restoration of wetlands for migratory birds, fish, and wildlife. This Act was implemented by matching funds to state and private landowners. The U.S. Environmental Protection Agency's (EPA) Five Star Restoration Program supports community-based wetland restoration projects. The program provides grants, technological assistance, information, and support for wetland restoration projects. The EPA is currently working to restore wetlands in about 500 watersheds throughout the United States. By targeting grassroot efforts this program has improved public understanding of the importance of wetlands and streams.

Mined Lands

Restoration of mined lands was already discussed in Chapter 10. The Surface Mine Control and Reclamation Act of 1978 regulates coal surface mining activities in the United States. This Act established an Abandoned Mine Reclamation Fund for use in reclaiming and restoring land and water resources adversely affected by the coal-mining industry. The Act requires companies to post bond and a detailed reclamation plan before mining operations can commence. The Act considers different land provision such as forestry, wildlife reserves, and even golf courses. Restoration success is assessed after 5 or 10 years, and bond is released if the project is termed successful. Primary criteria include restored topography, sufficient vegetation ground cover, density of trees, and water quality.

The 1980 Comprehensive Environmental Response, Compensation, and Liability Act, also known as "Superfund," provided federal agencies with the power to respond directly to releases or threatened releases of hazardous waste that may pollute the environment. Over 5 years a $1.6 billion budget was used for cleaning up abandoned or uncontrolled hazardous waste sites. Under the Superfund provision remedial actions can be conducted only at sites listed on the EPA's National Priorities List. This list helps the EPA to prioritize sites for remediation projects. The EPA currently lists about 1,604 sites, many of which have already received remedial action. Lack of sufficient funds has, however, impeded the remediation of most sites that are found on the National Priorities List.

Invasive Non-Native Species

The ecological harm of invasive non-native species was outlined in Chapter 7. Restrictions on introducing non-native species to the United States are found in several pieces of legislation. This includes President Carter's Executive Order 11987 of 1977, which simply forbids the import of any exotic species. Under the Lacey Act of 1900, the U.S. Fish and Wildlife Service is ordered to restrict the entry of fish or wildlife that threatens humans, agriculture, horticulture, forestry, or wildlife. The Non-Indigenous Aquatic Nuisance Prevention and Control Act of 1990 authorizes the Fish and Wildlife Service and the National Oceanic and Atmospheric Administration to regulate introductions and control infestations of aquatic nuisance species such as the zebra mussel. The Fish and Wildlife Service, however, regulates the trade in wild animals. The Animal and Plant Health Inspection Service (APHIS) of the U.S. Department of Agriculture prohibits the introduction of several vertebrate species because they may carry pathogens that affect livestock. The protection role of APHIS also includes wildlife and ecosystems that are vulnerable to invasive pests and pathogens.

The U.S. Public Health Service prohibits importation of vertebrate species because they may carry pathogens that threaten human health. Under the Federal Noxious Weed Act of 1974 and the Federal Seed Act of 1939, APHIS also has the authority to prohibit the entry of plants. However, the U.S. Department of Agriculture keeps the noxious-weed list that includes many invasive non-native species. These actions have not all been preventive, and between 1935 and 1942 the U.S. Soil Conservation Service grew 85 million kudzu seedlings, promoted them for erosion control, and paid farmers to plant them, thus facilitating the widespread introduction of this invasive plant. The National Invasive Species Council was established in 1999 and targets not only weeds but also non-native invasive species.

Soil Erosion

The serious issues of soil erosion and soil conservation were outlined in Chapter 8. The Farm Bill of 1985 aims at conserving agricultural lands. Under this Act farmers were rewarded for efforts to prevent soil erosion and to conserve water and wildlife habitats. The Farm Bill formed the Conservation Reserve Program, which aims at stopping cultivation on agricultural land that is prone to soil erosion. Under this Act more than 15 million hectares were taken out of agricultural production. In addition, the "Sodbuster" program required highly erodible farmlands to implement soil conservation practices or lose federal subsidies.

The Farm Bill has played an important role in providing federal funding for conservation on private land in the United States. This is important because about 61% of land in the United States is privately owned. The Farm Bill is renewed every 5 years. In 2002, the Farm Bill received $74 billion from Congress to be distributed over 5 years. In this budget $17 billion were allocated to conservation alone. Funding from the Farm Bill has been targeted toward landscapes with high conservation value. The Farm Bill has resulted in better protections for various ecosystems and endangered species.

International Conventions

International conventions that deal with environmental issues and restoration have been on the rise. Several conventions and international agencies address the root causes of current environmental problems. The International Union for Conservation of Nature (IUCN), founded in 1948 in France, is one such agency. The IUCN consists of members from 140 nations and numerous nongovernmental agencies. A major priority of the IUCN Programme has been to identify and build awareness on how the life standard of indigenous people in developing countries depends on the management of sustainable resources.

Human Population Overgrowth

Environmental problems associated with the fast-growing human population were already discussed in Chapter 1. To deal with this problem, the U.N. International Conference on Population and Development took place in Cairo, Egypt in 1994. The main resolutions of this conference emphasized the empowerment of women in developing countries through education, improved health services, employment, and involvement in policymaking and decision making in their countries. The conference also included goals of family planning and reducing levels of infant, child, and maternal mortality. The ultimate goal was to stabilize the growth of the human population in developing countries.

Desertification

Desertification is a major economic, social, and environmental problem of concern to many countries (see Chapter 8). In 1977 the U.N. Conference on Desertification adopted a Plan of Action to Combat Desertification. However, the problem of land degradation in arid, semiarid, and dry sub-humid areas has gradually intensified. Desertification was still a concern at the U.N. Earth Summit, which was held in Rio de Janeiro in 1992. The conference supported integrated approaches to the problem, emphasizing action to promote sustainable development at the community level. It also called on citizens to establish an intergovernmental negotiating committee to prepare the 1994 Convention to Combat Desertification, with particular emphasis on the problems in Africa.

In the implementation of U.N. Conference on Desertification, the key instruments are national action programs. They are strengthened by action programs on subregional and regional levels. National action programs are developed in the framework of a participative approach involving the local communities. They outline the practical steps and measures to combat desertification in specific ecosystems. In many African countries combating desertification and promoting sustainable development go together due to social and economic importance of natural resources and agriculture. Desertification forces these people into internal and cross-border migrations, which can lead to social and political tensions and conflicts.

Many African countries have organized national awareness to launch the process of formulation of their national action programs to implement action for the people most affected by desertification. The preparation of action programs is a dynamic ongoing process, and the status of each country is subject to change over time. It is important, if the U.N. Conference on Desertification is to succeed, that countries facing desertification adopt appropriate legal, political, economic, financial, and social measures to meet the challenges.

Biodiversity

The importance of biodiversity and the trend of declining global biodiversity were outlined in Chapter 3. The U.N. Convention on Biological Diversity was adopted at the 1992 Earth Summit in Rio de Janeiro. This convention aims at conserving biodiversity. Also, it encourages sustainable use and fair and equitable share of benefits of biodiversity. In 2002 in Johannesburg, South Africa, this convention was extended to include a plan to slow down significantly the rate of biodiversity loss by 2010. Unfortunately, this target has not been met according to the U.N.'s Global Biodiversity Outlook 3 report. However, the U.N. has declared 2010 as the international year of biodiversity in its continuous effort to build awareness of the value of biodiversity and to slow down environmental destruction.

Summary

Restoration projects should be guided by realistic goals. Properly set goals also allow evaluation of restoration projects. In this respect it is important to select nearby reference sites that can serve as a model ecosystem for the restoration project. Planning restoration projects must consider the needs of all stakeholders, including local residents. A restoration plan should be based on scientific principles and establish achievable timetables. Budgeting of restoration projects that considers all phases of restoration is essential. Adjustments to the restoration plan should be made through pilot projects and performance-based evaluation. The use of GIS technology is important in targeting restoration efforts according to landscape features and also to keep track of restoration efforts and progress of the work. Implementation involves curtailing degradation factors, installing natural disturbance regime, establishing keystone and nexus species, introducing target species, and controlling weeds and non-native invasive species.

Installing monitoring system is essential for adaptive management where restoration efforts are adjusted to local conditions. Monitoring systems also provide warning signals if the restoration is not following a desirable trajectory. Aftercare involves correcting adverse environmental conditions (regression) that can arise. Assessment of restoration projects usually targets ecosystem structures and function of a particular restoration site. Restoration efforts should be guided by legal framework such as the Surface Mine Control and Reclamation Act or the EPA's Five Star Restoration Program for wetlands. International agreements also deal with restoration of particular ecosystems, such as the U.N. Action Plan to Combat Desertification.

14.1 Case Study

Integrated Restoration Efforts of Severely Degraded Subarctic Heathland Ecosystem

S. Greipsson, Department of Biology and Physics,
Kennesaw University, Kennesaw, GA

Severely disturbed ecosystems have usually lost their integrity and function and require major restoration efforts. In such cases critical components of the ecosystem function and structure need to be restored in several complementary phases. This often involves engineering solutions such as reinstalling proper hydrology, long-term protection from livestock, soil biotechnology, and relentless innovative restoration efforts.

An example is given in this case study on severe degradation and subsequent restoration efforts on the heathland ecosystem Haukadalsheidi, located on the highland plateau of south Iceland. The heathland was important for livestock grazing by the early settlers in the region. Catastrophic soil erosion that took place for 300 years (1660–1960) has resulted in almost complete loss of the original ecosystem. Consequently, less than 20% of the original ecosystem still exist as fragments.

Most of the heathland is located just below 300 m of altitude, and, therefore, it lies between the subarctic and arctic region. The growing season is only 3 months. The average summer temperature is about 8.5°C, and annual precipitation is about 1,000 mm. The landscape was characterized by heterogenous topography where rocky hills stand above the willow dominated heathland and wetlands in depressions were dominated by few species of sedges (*Carex* sp.). The original soil is classified as Andisol, mainly derived from volcanic debris and loess (wind-blown) material. On the other hand, soil on the degraded land is mainly sand and gravel (glacier moraine) with poor water-holding capacity and very limited macronutrient levels.

The heathland community was assembled shortly after the Ice Age finished, and the first plants to colonize barren moraine and sands probably survived on nearby high mountain tops and represent the arctic flora. Later on, boreal plants from Europe immigrated mainly by gradual colonization of islands in the North Atlantic ocean, which served as stepping stones for plant dispersion. More recently, humans introduced non-native plants, and about 25% of the island's flora is anthropogenic in origin. This sequence of plant introduction over a long time represents historical contingency and has influenced the composition of heathland communities. It is not know if heathland communities can today reassemble from scratch to their original state; most likely new communities will be dominated by grasses (*Agrostis*, *Poa*, and *Festuca* sp.). Dense cover of grasses can especially have negative impacts on colonization of birch (*Betula pubescence*) and dwarf birch (*Betula nana*), which is characteristic of the pristine heathland community.

The heathland community contains only about 50 species of vascular plants. The heathland is characterized by few dominating species such as dwarf birch, willow species (*Salix* sp.), fescue grasses (*Festuca richardsonii* and *F. vivipara*), woodruff

(*Gallium boreale* and *G. verum*), and horsetails (*Equisetum pretense*). The nitrogen-fixing mountain avens (*Dryas octopetala*) is found in low cover mainly on rocky hills. Nitrogen-fixing lichens such as *Peltigera canina* are widespread in the heathland understory and probably play an important role in the nitrogen budget of the ecosystem.

The low diversity of plants on the heathland has resulted in low resistance of this ecosystem to compounded disturbances (natural and anthropogenic). The low diversity of vascular plants makes this ecosystem particularly vulnerable for disturbances because only few species are represented in each functional group.

The heathland forms the natural habitat of the bird ptarmigan (*Lagopus mutus*), field mouse (*Apodemus sylvaticus*), and top predators such as arctic fox (*Alopex lagopus*), gyrfalcon (*Falco rusticolus*), raven (*Corvus corax*), and snow owl (*Nyctea scandiaca*). Mountain char (*Salvelinus alpinus*) can be found in streams and rivers. About a handful of migrant bird species nest in the heathland where the most obvious ones are golden plover (*Plurialis apricaria*), parasitic jaeger (*Stercorarius parasiticus*), whimbrel (*Numenius phaeopus*), whooper swan (*Cygnus cygnus*), gray lag goose (*Anser anser*), and winter wren (*Troglodytes troglodytes*). Occasionally, ducks make nests along streams, but unfortunately the non-native mink (*Mustela vison*) has probably eliminated most of the ducks and reduced the number of other nesting birds. The mink is currently under a program of eradication.

The land use in the past has undoubtedly contributed to the degradation of this ecosystem. The land use was inferred by written and anecdotal evidences. Also, the grazing pressure was estimated from the number and location of grazing huts and shiels (used to house livestock) in the region. Furthermore, the chronicle of the catastrophic soil erosion that took place on Haukadalsheidi has been traced by the use of tephrochronology (use of volcanic ash layers to date soil profiles). This method estimates the rate of soil erosion by measuring soil thickness between known thephra markers. Intensive grazing practices on fragile heathland ecosystems in the subarctic is not sustainable in the long run and has facilitated widespread soil erosion on the island that has left in its wake vast desertified landscape.

The intensive land use was influenced by traditional use of the heathlands by the settlers. This included use of grazing huts and shiels, which were small huts kept at a distance from the farm but widespread over the landscape. The grazing huts usually only housed small number of sheep (20–50 animals), but they were located strategically on the heathland. Intensive grazing pressure was, therefore, maintained throughout the year. Usually, sheep housed in grazing huts were only hay-fed during days of continuous snow cover; otherwise, the animals were free-grazing. Nearby birch trees were cut and used to build the roof of these huts that were otherwise made of stones and turf. Also, birch was used as a fuelwood to process the sheep milk during the summer. During the 19th century several grazing huts and sheep shelters were located on Haukadalsheidi on land that is now desertified. The grazing huts in this area were gradually abandoned due to catastrophic soil erosion. In 1929 few grazing huts were still in use, and their ruins can still be located. The practice of shieling was generally abandoned in the 19th century, and the use of grazing huts was abandoned in the 20th century.

(continued)

Case Study (continued)

Intensive grazing pressure on the heathland resulted in changes in plant community structure where willows disappeared and were often displaced by low-lying herbs (*Armeria maritima, Rumex acetocella, Cardaminopsis petraea,* and *Silena uniflora*) and thick cover of mosses. The resistance of the ecosystem to soil erosion is reduced when the dense willow canopy disappears. Changes in plant communities are often the first visible indication that ecosystem degradation is taking place. However, the action of such degradation is usually expressed in damage to physical, chemical, and biological properties of the soil. Intensive grazing also significantly reduced covers of live plants and litter on the ground. In turn, more exposed ground leads to rill erosion, which may develop into gullies by water erosion (**Figure 14.3**).

Soil erosion was enhanced by multiple factors. Catastrophic soil erosion was initiated by compounded natural disturbances including frequent volcanic tephra (ash) fallout and sand encroachment associated with floods of glacier rivers. Enormous amount of sediment was brought with glacier rivers and deposited on sand plains where it provided a constant source of drifting sand. Sand encroachment onto vegetated lands fastened the process of soil erosion. The eroding front when sustained by sand encroachment moved rapidly (as much as 30 m per year) over vegetated land and typically left sand and gravel terrain in its wake. The eroding front only slows down or stops if certain landscape features such as rivers, rivulets, wetlands, or lee sides of hills exist where the eroding power of the wind is significantly slowed down. Soil erosion is, therefore, a landscape phenomenon but not only a simple physical movement of soil by wind and water. The catastrophic soil erosion was very intensive and resulted in desertification of most of the landscape.

Figure 14.3 Deep gully formation on Haukadalsheidi. (Courtesy of Sigurdur Greipsson, Kennesaw State University.)

Case Study

Figure 14.4 Sand-stabilizing fences and lymegrass dunes on Haukadalsheidi. (Courtesy of Sigurdur Greipsson, Kennesaw State University.)

Restoration efforts on Haukadalsheidi have involved ecological engineering and active and passive restoration approaches. The restoration effort has so far aimed at protecting the areas from livestock grazing, stabilizing drifting sand, stabilizing exposed soil, halting destruction of vegetated remnants, reclaiming barren lands, and improving the hydrology of the area.

Restoration efforts have aimed at the following goals. The first goal is protecting the whole area from livestock grazing. This was accomplished by erecting barbed wire fences that enclosed the area. The entire area (about 11,000 ha) was protected in 1964. Second, this work was followed by stabilizing drifting sand by erecting low timber fences perpendicular on the main (northeast) sand-drifting wind direction. This was followed by sowing (harrowing) the sand-stabilizing grass Leymus arenarius (**Figure 14.4**). Subsequently, chemical fertilizer was applied for a few years to ensure successful establishment of the grass. The initial sand-stabilizing effort was followed by a large-scale mechanized seeding of *L. arenarius* to stabilize drifting sand. In total about 700 hectares of *L. arenarius* were seeded in the area. In addition, *L. arenarius* produces viable seed in favorable summers and extends its cover by natural seeding (**Figure 14.5**). Leymus arenarius is a native plant in Iceland but is not a natural component of the heathland ecosystem. This species, however, is a keystone plant in sand stabilization. Its use is a prerequisite for further restoration efforts. Sand drifting was, therefore, a serious ecological threshold that could be overcome by building sand fences and seeding of *L. arenarius* (**Figure 14.6**). Natural colonization of native heathland plants such as willows (*Salix* sp.) has already taken place on some of the oldest stabilized sand dunes where *L. arenarius* is rapidly being replaced by several willow species. Seeds of willows spread from vegetated remnants into the dunes. Natural succession is undoubtedly taking place on stabilized lymegrass dunes, which are being replaced gradually by willow shrubs and other grasses.

(continued)

Case Study (continued)

Figure 14.5 Self-seeding of lymegrass on Haukadalsheidi. (Courtesy of Sigurdur Greipsson, Kennesaw State University.)

Figure 14.6 Stabilizing drifting sand on Haukadalsheidi by using lymegrass. This barren land was vegetated about 250 years ago. (Courtesy of Sigurdur Greipsson, Kennesaw State University.)

Third, halting destruction of vegetated lands was accomplished by leveling eroding fronts using bulldozers and subsequently seeding exposed soil with commercial grasses (*Poa* and *Festuca* sp.) and adding chemical fertilizers for a few years. Also, old trawler fish nets were placed on the eroding fronts to stabilize them. Eroding fronts were also leveled down simply by using a shovel, but this method is very labor intensive. Erosion of the remaining remnants containing the native ecosystems was halted by stabilizing eroding fronts. The total length of eroding fronts that were stabilized on Haukadalsheidi exceeds 50 km. Stabilizing and protecting vegetation remnants that contain the original heathland vegetation was a high priority, and it was accomplished by using the methods described above. The protection of fragmented vegetation patches is important in conserving native species diversity. Also, heathland fragments can act as nuclei for recolonization of native plants into barren areas. In fact, most heathland plant species (about 90%) can colonize barren lands without any soil amendments.

The fourth goal was to stabilize exposed soil lacking vegetation by harrowing hay stubbles (mulching) into the soil. This was followed by mechanized seeding of *L. arenarius*. Fifth, barren areas where little sand drift was taking place were typically revegetated by seeding a mixture of commercial grasses (*Poa, Festuca*, and *Descampsia* sp.) and spreading chemical fertilizers using low flying aircraft to stabilize the ground and initiate primary succession. Soil of eroded lands is in fact a gravel or glacier moraine. Such soil is characterized by low levels of available plant nutrients, especially nitrogen and phosphorus. Another approach in revegetation of barren gravel areas is to use an introduced species (*Lupinus nootkatensis*). The critically low nitrogen level of the soil of barren areas was an ecological threshold and can be restored by cost-effective means by the use of this lupine. Results of small-scale testing suggested that this lupine species survived, produced seed, and extended its cover rapidly under the harsh climatic conditions provided that appropriate rhizobium inoculum was provided at the time of seeding.

Lupine seed coated with rhizobium inoculum was sowed mechanically in the barren ground. Also, lupines were raised in nearby nurseries and transplanted strategically on barren land that was less accessible by machinery. During transplanting seedlings were spaced out at 5- to 10-m intervals. Such transplanting strategy allows more area of barren lands to be covered by lupines. Bird populations, especially golden plover and whimbrel, have increased their populations in lupine stands. Birds undoubtedly facilitate dispersal of native seed into the barren lands, especially crowberry (*Empetrum nigrum*) and blueberry (*Vaccinium uva-ursi*). Willow shrubs typically colonize lupine stands and are expected to outcompete and shade out the lupine in few decades. Succession has been demonstrated on sites at lower latitudes where the lupine is eventually replaced by native vegetation. Natural succession of plant communities in the arctic is very slow, and restoration efforts are greatly needed in order to facilitate this process.

The final goal was to improve the hydrology by elevating the groundwater level of the restoration area. Ecological engineering involved building earthen dams to regulate water levels of Lake Sandvatn. Fluctuating water levels of Lake Sandvatn was a notorious problem, and low water levels have caused massive sand encroachment onto vegetated lands, which in fact was probably one of the main factors behind the drastic soil erosion (**Figure 14.7**). This work was accomplished by using bulldozers and nearby material (gravel and stones). By building these dams the size of the lake

(continued)

Case Study (continued)

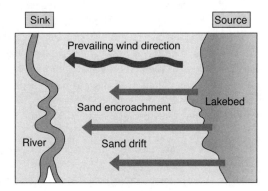

Figure 14.7 Schematic model showing the "erosion cell" of Haukadalsheidi. Drifting sand originates from the lakebed and is carried across the landscape by the prevailing wind.

was increased from about 250 hectares to 1,040 hectares. The source of drifting sand was subsequently inundated by water. The hydrology management of Lake Sandvatn has proved to be successful in curtailing drifting sand and fugitive dust and at the same time improved the groundwater level of the restoration area. The groundwater level was improved in nearby damaged wetlands, and this will undoubtedly facilitate their natural recovery. The largest restoration of such wetland is about 100 hectares in size, and it is now dominated by native cotton grasses (*Eriophorum* sp.).

Future restoration work will involve strategies in introducing birch into restoration sites. Native willows can act as native nurse plants and facilitate the establishment of birch trees. Native willows colonize restoration sites rapidly and form small shrubs. Enhancing establishment of birch on restoration sites could, therefore, involve strategic outplanting of container-based birch seedling on sheltered sites of willow shrubs. Ecosystem restoration should also involve restoration of key species at different trophic levels, such as the ptarmigan, snow owl, and the arctic fox.

The restoration methods mentioned above have been adjusted to the local conditions using adaptive management based on results from small-scale trials. These trials have, for instance, aimed at optimizing fertilizers use by evaluating different types of fertilizers and rate and timing of their use. Also, timing of seeding, sowing depth, and other factors influencing plant establishment have been similarly tested. In this work few keystone species have played a pivotal role; *L. arenarius* in stabilizing drifting sand, *Festuca rubra* and *Deschampsia beringensis* in stabilizing barren ground, and *L. nootkatensis* in enhancing the nitrogen budget of the soil. Adaptive management is important in a restoration that is in fact pioneering and where limited ecological information is available to guide appropriate restoration methods.

The cost of the initial restoration work including material for protective fences and timber sand-drift fences was provided by a governmental agency (The Sand Stabilizing Service). The Soil Conservation Service later oversaw this work with more emphasis on revegetation of barren lands by using aircraft in spreading seed of commercial grasses along with fertilizers. Maintaining the protective fences was an annual duty; due to high snow banks protective fences were mended each spring. Applying chemical fertilizers using ground-based machinery and extending restoration sites was also conducted annually. A fund for building the earthen dams at Lake Sandvatn was provided generously by one of the largest bank in Iceland. Also, funds for the aircraft seeding were provided by a special national campaign. The involvement of the local community was critical in the success of this project. Collectively, this restoration effort can be used as a model system for other restoration projects of severely degraded lands.

14.2 Case Study

Decision Making in Ecological Restoration
Stephen D. Murphy, Environment and Resource Studies,
University of Waterloo, Waterloo, ON, Canada

Introduction and Context to Decision Making in Ecological Restoration

As a transdisciplinary subject, ecological restoration presents challenges because many of us were trained in—and find it hard to escape from—disciplinary silos and experiences. I am trained and educated as a biologist with a reductionistic focus in coursework (a lot of biochemistry and physiology) and a bit more of a holistic framework in terms of my doctorate (community ecology with biochemical ecology as a main theme), but there was not much explicit grappling with why I chose to focus my energy on certain aspects or what the decision-making process should be once I stated my conclusions from my research. This is not a bitter complaint. In fact, I considered myself well served by my education and certainly had some exposure to the philosophy and decision-making process that influences and stems from scientific research. Nonetheless, it meant that I had to adapt further to influence decisions because as most of us learn, the science itself will not sway anyone without appropriate processes of consultation and explicit consideration of the context of the research and the implications to society beyond the science.

This can frustrate scientists sometimes because the whole scientific framework is designed to produce results that are as objective and, therefore, unbiased as is humanly possible; hence, there is an underlying assumption that the science should speak for itself and be translated instantaneously into policy and action. Of course, this does not happen, much to scientists' consternation. We are asked to explain ourselves and our rationale. Given that we are not often trained to be great communicators, this is a challenge. And there may be resentment that we even get asked to do this because it seems like our integrity and the integrity of science is questioned.

(continued)

Case Study (*continued*)

Sometimes it is, and that is actually part of science as a process, though that can be grossly abused—witness the cultural wars over scientific discussions of everything from evolutionary theory to stem cell research to aquaculture (fish farming in a loose sense) to mercury pollution to global climate change (mislabeled as "warming" because not all areas will warm). Scientists may become tired of obfuscation in the guise of sheen of honest skepticism. Scientists may fear they have to turn too much into advocates, though I argue that advocacy is not a failing as long as you are forthright about the amount and quality of scientific evidence in support of your scientific and policy decisions. There is indeed a danger of confusing "ought" with "is" even as a scientist, but the process of science should force us to confront that question and focus on what "is" happening rather than what "ought" to be. There is, however, room to ask what "might" happen (i.e., as the formal statement of hypotheses).

In some respects scientists really want to be decision makers by stealth. I do not mean this conspiratorially. I mean we want our scientific evidence to be accepted by society and decision makers. We forget that few really understand the scientific process of conditional hypotheses and conclusions that must be reassessed and refined as new evidence gathers, thereby eliminating most images of scientists as people who deliver irrefutable facts.

It disturbs people when they learn a scientific conclusion is falsified; over the last 50 years I contend that public confidence in science as a driver of society and its decisions has eroded. One part of this was overconfidence by scientists and a collective amnesia that science is not immutable and trying to defend the indefensible. Another part was the public conflating science and the unintended consequences of technological applications—concluding from visible examples of pollution like the Cuyahoga River catching fire or from smog-choked cities or from contaminated lands that science was out of control. Again, this is a conflation of science with technology and decision making, but there is the grain of truth that stems from the inability of science to help influence decisions that created these situations.

This conflation remains a problem but did, ironically, benefit the emerging field of ecological restoration because it demonstrated the need for restoring sites or using the approach to mitigate environmental problems. So, ecological restoration has been driven by a focus on the scientific process. As a scientist I appreciate the need to focus on the methods required to initiate, monitor, and complete a successful ecological restoration project. There are experimental design issues to consider, and using appropriate statistical methods is important to the scientific integrity of a project. There are hours spent considering the scale of focus (genetics, populations, ecosystem processes, biomes), of sampling in the field, or setting up the actual lab or field experiments. There is the exhaustion of just doing science well.

But all science is for naught if we fail to appreciate how social systems influence the application of science and how cultural demands—and cultural aspects of ecological restoration or the broader environmental management issues—intersect science. Indeed, Higgs, Light, Naveh, and many others have written extensively about this (see the Further Reading section).

Studying Decision Making in Ecological Restoration 1999–2006

Objective

Since the studies initiated by then-graduate student Soonya Quon (in 1999 and 2001), I have continued the constant evaluation of how restoration projects actually succeed or fail in terms of planning and decision making. Between January 1999 and February 2006 I used Quon's conceptual framework, written in collaboration with Larry Martin and myself, to interview participants in 46 additional ecological restoration projects in and around the Region of Waterloo.

Methods

Because of time constraints due to my work on the biophysical aspects of ecological restoration, my sample university is essentially bounded by the Counties of Wellington, Perth, Oxford, and Brant and the Regional Municipalities of Waterloo and Halton. The sampling followed that of Quon in that cases were chosen to represent

- The variety of land use categories that exist Waterloo Region
- The variety of terrestrial and aquatic rehabilitation activities found
- The variety of locations and jurisdictions (urban, rural, different municipalities)
- The variety of perspectives of different organizations or agencies involved in rehabilitation (e.g., municipal, ministry, conservation authority staff, community groups)
- The complexities or controversies arising from diverse political or community opinions

My sample universe is less comprehensive than that of Quon in that she was able to locate all 79 projects within the defined boundaries of the Region of Waterloo. In contrast, though I am aware of most projects via my role in the Society for Ecological Restoration Ontario and as an academic, I can say only that snowball sampling has left me with a count of 202 projects that can, in any sense, be considered ecological "restoration, rehabilitation, or reconciliation" in the four counties and two regions. I would not claim this counts all of them, although it is likely to be a representative sample because of my roles and contacts with most personnel involved in any form of ecological restoration. Hence, I can say that I've followed Quon's methods for about 20% of the 202 projects beyond those originally cited in that publication. Following this method exactly, I sought a factual description of the project and the practitioners' perceptions of each project by asking 61 structured questions on the following:

- Profile of the case (jurisdiction, size, type, and scope of rehabilitation activities, types of habitats being rehabilitated, date begun, and progress at time of interview)
- Conception and boundaries of the project (why the project began, was the project part of a larger strategy, was the project assigned a priority from a list of projects, were related studies performed)
- Project planning (relative formality, rigidity, and method of choosing objectives, existence of preplanning activities)
- Project organization (participants and their roles, leadership and its effectiveness, decision-making processes, outside consultation (especially with other projects)

(continued)

Case Study (*continued*)

- Community response (support, consultation method)
- Implementation (costs, funding, in-kind sources, permits or authorizations needed)
- Evaluation (method and frequency of monitoring [if any], did practitioner consider project to be successful and why)
- Practitioners' perceptions of opportunities and obstacles to ecological rehabilitation in the Waterloo Region
- Recommendations for fellow practitioners

In-person and telephone interviews with practitioners associated with the additional projects were conducted from 1999 to 2006. Lasting approximately 90 minutes each, interviews were tape recorded (when permitted by the practitioner) to verify written responses and clarify responses.

Pragmatism is universal, but idealism is not dead. Given that Quon found three near-universal criteria that were applied by practitioners, I examined the correspondence in interview answers versus these three criteria:

- Obtaining political/social support
- Promoting projects and changing attitudes about projects
- Securing sufficient and persistent funding to maintain a project's life

Every project of the new ones I examined followed the pragmatic principles. The responses in the intervening years, however, followed ideal criteria raised by Quon more closely. Following are results of how many projects met those ideal conditions for "good" ecological restoration projects—with a nod and deference to Eric Higgs because he has a much more comprehensive approach to this concept of "good ecological restoration" than I do here. Examining **Table 14.1**, the trends are likely related to a larger, broader sample than possible during Quon's Master's thesis project, increased sophistication of practitioners related in turn to increased accessibility to information and communication via Web sites. Indeed, all projects highlighted the fact that they used the Web to send and receive information and ideas. The trends are as follows:

- Contextual criteria are more closely considered, although too many projects still fail to do adequate pilot studies or consultation with the public.
- Ecological criteria are taken seriously, except for connection with landscapes.
- Related to the second point, few projects do consider the landscape around them, reexamine priorities in light of these or other unexpected challenges, or truly prioritize projects within a landscape.
- Adaptiveness is improving, but the lack of experimentation and, therefore, any quantitative analysis is distressing.
- Communication, despite my optimism about the Web and statements from practitioners, is still poor between projects; this reflects time constraints, competition and proprietary concerns of private consultants and landowners, and occasionally a worry about admitting failures—with a concomitant inability for others to learn lessons because the failures are hidden. In fact, some of these may be a classic file drawer problem (i.e., Type II statistical errors of seeing failure where success may eventually exist, but the lack of quantitative analysis moots this).

Table 14.1 Results from Interviews of 46 Projects Considered as Some Form of Ecological Restoration and a Comparison of Quon's Sample of 11 Projects (Interviews Completed 1997) with Murphy's Sample of 46 Different Projects (Interviews Performed 1999–2006)

Criteria	Quon in 1997	Murphy in 1999–2006
Contextual		
Conduct biophysical, socioeconomic, and attitudinal studies before planning	18	43
Clearly identify the origins of the problem leading to the need for rehabilitation	55	87
Include early public consultation, involving neighbors and stakeholders in decision making; establish constructive contact with skeptics and opponents	27	59
Obtain early institutional (political) staff support and cooperation for a project	64	89
Understand the policy context of the site (e.g., relevant land-use planning policy or environmental policy) to be aware of how ecological rehabilitation projects are officially and unofficially supported, protected, or managed	91	96
Ecologically appropriate		
Setting ecologically defensible bounds for the project	18	67
Using local genotypes of species native to the region	100	100
Establishing corridors to allow for immigration from nearby sources of genotypes	100	100
Using rehabilitation projects to physically or dynamically reconnect landscapes	27	24
Interconnected		
Work within large area objectives set for the desired ecological state of the site	27	26
When projects cannot include all degraded areas within the set boundaries, identify and address those external factors that can hinder project success	45	54
Prioritize projects according to consistent criteria, e.g., environmental importance of projects, feasibility of site rehabilitation, social value of sites	18	39
Use the many available sources of information to design goals and approaches and standardize reporting procedures to allow for comparisons to other projects	36	63
Adaptive		
Flexibility of plan, budget, time to adapt to surprises during the project	55	65
Monitoring plans designed, a priori, to establish baseline inventories and allow for time series comparisons of ongoing results; this helps acquire long-term funding and support	27	43

(continued)

Case Study (continued)

Table 14.1 (continued)

Criteria	Quon in 1997	Murphy in 1999–2006
Replicable experimentation to increase the reliability and validity of observations project objectives that are measurable within realistic time constraints	0	9
Communicative		
Incorporating appropriate sources and types of information for goals and project planning mechanisms to communicategoals, resolve conflicts, and administer projects	36	41
Effective leadership, preferably someone with relevant rehabilitation expertise and an ability to resolve conflicts and negotiate bureaucratic difficulties	91	96
Broad dissemination of project results via meetings and media	0	30

Projects were designed and implemented by a total of 151 practitioners associated with the 46 projects. Values are expressed as percentages (% of projects that exhibited the criterion), but note unequal sample sizes (11 vs. 46).

The lessons here can be stated briefly. There needs to be quality control on projects, and there is a role and a vital need for some prioritization and planning of projects. This does not mean wresting control and discouraging any enthusiastic local or nodal restoration. It means that volunteers and professional practitioners could be guided to prioritize, and it definitely means there needs to be cross-jurisdictional and large-scale planning of public projects that receive public funds (Census Metropolitan Area, regional or county, and provincial and federal in terms of Ontario's situation). But who will do this? The answer that emerges is that academic, government, and private practitioners are recognizing this need but rather than wait for political decisions are creating tools and forums where exactly these large-scale databases and planning may occur. In Ontario and across Canada I see a role for organizations like the Society for Ecological Restoration to have its boards and members (of local chapters and international headquarters) to be part of the leaders—but not to do this without true collaboration and transactive planning. If the biophysical data that drive or arise from ecological restoration are collated, shared and analyzed via statistics like multivariate approaches or meta-analysis, and the landscape implications and priorities are negotiated, then ecological restoration will fulfill its mandate of real repair of structure and processes. Again, this does not mean the trampling or death of local enthusiasts; small-scale, low-cost initiatives are welcome, and the practitioners can simply advise about the basics that avoid discouraging results. So too will a more comprehensive application of the principles from Quon—there will be an avoidance of frustration, disappointment, déjà vu, and a deflating "when we will we ever learn?"

Key Terms

Adaptive management   359
Aftercare   362
Assessment   362
Coordinated resource management and planning   355
Ecological management unit   355
Ecosystem regression   362
Model project   357
Monitoring system   353
Pilot study   358
Reference site   354
Trial-and-error   359

Key Questions

1. What is a reference ecosystem and how is it important for ecological restoration?
2. Describe the main features of coordinated resource management and planning.
3. What are model projects and how are they used in ecological restoration?
4. Describe the difference between the trial-and-error approach and adaptive management.
5. Define ecosystem regression.

Further Reading

1. Holling, C. S. (ed.). 1978. *Adaptive environmental assessment and management.* New York: John Wiley & Sons.
2. Kraft, M. E. 2004. *Environmental policy and politics.* White Plains, NY: Pearson Longman.
3. Sutherland, W. J. 1995. *Managing habitats for conservation.* Cambridge, UK: Cambridge University Press.

Glossary

Adaptive management Small-scale trials that act as a monitoring system that gives feedback information to allow beneficial modification for the restoration procedure.
Aftercare Active restoration efforts that are required to counteract unfavorable ecosystem regression.
Alternative stable states Existence of different but persistent ecosystem states.
Arbuscular mycorrhizal fungi Soil fungi that form mutualistic symbiosis with most plants and greatly enhance phosphorus uptake.
Area threshold Minimum area of a critical habitat where a species can survive and reproduce.
Assessment Evaluation of ecosystem functioning and integrity.

Beach nourishment Offshore dredging and pumping sand onto the shore.
Bioaccumulation Increase in toxic chemical concentration in different organisms through several trophic levels.
Bioaugmentation Inoculation of contaminated soil with naturally occurring or genetically engineered microorganisms.
Biodegradation Catalysis of an organic toxic chemical to another one that is usually less toxic.
Biodiversity Variety of organisms at genetic, species, and ecosystem levels.
Biological control Use of biological agents to control invasive species.
Biological legacy Species that survive disturbances and influence recovery processes.
Biomanipulation Strategic manipulation of fish populations to reduce the level of phytoplankton.
Bioremediation Use of microorganisms to metabolically convert contaminants to much less harmful substances.
Biotreatability Potential effectiveness of bioremediation in a given situation.

Captive breeding   Rearing and assisting breeding of animal species in special facilities.
Chelates   Complex of an organic ligand and a metal.
Clear-cutting   Wide-scale felling of trees.
Clonal offset   Propagation of the smallest viable rhizome fragment of dune grasses.
Connectivity   Movements of wildlife between patches within a matrix.
Connectivity thresholds   Artificial or natural barriers in the landscape that prevent movement of organisms.
Coordinated resource management and planning   Interdisciplinary team effort to solve problems that arise during the restoration process.
Corridor   Habitat-connecting patches that allow movement of wildlife across the landscape.

Decomposition   Gradual transformation of intact organic matter to amorphous matter mainly through microbial activities.
Deforestation   Conversion of forests to agricultural and other uses.
Desertification   Land degradation leading to desertified landscape.
Disturbance   Event that disrupts ecosystem structure and function.
Disturbance regime   Sum of all disturbances (natural and anthropogenic) affecting an ecosystem.
Dune reconstruction   Reshaping coastal sand to mimic natural fore or back dunes.

Ecological constraints   Conditions that maintain certain ecological states.
Ecological drift   Random fluctuations in species composition within ecosystems.
Ecological engineering   Use of engineering structures (e.g., drainage systems, levees, or dams) to enhance ecological functioning of an ecosystem.
Ecological filters   Environmental conditions that select species that can colonize a particular site.
Ecological resilience   Ability of an ecosystem to return to its original state after a disturbance.
Ecological resistance   Amount of disturbance needed to cause a regime shift in an ecosystem.
Ecoregion   Vast areas of different ecosystems but within a similar climate zone.
Ecosystem engineers   Species that alter ecosystem structure and function and maintain spatial heterogeneity in ecosystems.
Ecosystem regression   Gradual degeneration of a particular function of a restored ecosystem.
Ecosystem threshold   Ecosystem conditions where small disturbances lead to regime shift.
Edge effect   Open, exposed habitat that results from fragmentation.

Endemic Native species that are only found in one location.
Eradication Elimination of non-native species.
Eutrophication Excess amounts of macronutrients (N and P) entering an ecosystem.
Evapotranspiration Transfer of water from soil via roots through plant leaves into the air.

Fragmentation Change of intact habitat into smaller disconnected patches.
Functional group Species demonstrating similar physiological activities.

Hard release Process transferring animals from a source to a release site without any conditioning.
Hypoxia Loss of oxygen from water mainly due to decomposition of organic matter.

Invasive Non-native species that spreads rapidly in its new range where it tends to decimate native species and affect ecosystem function.
Invasiveness Capacity of non-native species to invade a new range.

Keystone species Species that is essential for the structure and function of the ecosystem.

Land degradation Drastic reduction in potential productivity of lands.

Matrix Dominating ecosystem of the landscape.
Metapopulation Network of distinctive populations that are connected by effective migration.
Mineralization Gradual transformation of organic matter to inorganic nutrients through microbial decomposition.
Minimum viable population Number of individuals that are needed for a continuous survival of a particular population.
Monitoring system Permanent plots within a restoration site that can indicate decline or progress following restoration efforts.
Monoculture Dominating stand of a single species.
Mulching Placing chopped straw onto the ground or incorporating it into the soil.

Native species Species that originate in a particular habitat.
Naturalized Non-native species that have adapted to local environmental conditions.
Nexus species Keystone species that determine the assembly trajectory but are not necessarily part of the final state.
Non-native species Species that have been introduced outside the range where they evolved.
Nurse plants Commercial plants used to establish a quick cover on the ground.

Patches   Discrete habitat within the landscape where local population survives.
Perturbation   Induced disturbances that shift an ecosystem to an alternative state.
Phytoplankton   Free-floating microscopic algae.
Phytoremediation   Use of plants to extract contaminants from soil.
Pilot project   Implementation of an entire restoration project on a small-scale.
Piscivorous   Fish-eating fish.
Planktivorous   Plant- or algae-eating fish.
Primary succession   Changes in plant communities with time on sites that do not have intact soil

Recalcitrant   Compounds resistant to biodegradation.
Reclamation   Assisted recovery of a degraded site so that it is habitable for wildlife.
Re-creation   Installing important physical or landscape features in the ecosystem.
Reference site   Intact site representing the predisturbance or "pristine" state of an ecosystem that is used as a reference in restoration projects.
Reforestation   Reestablishment of a forest that does not necessarily contain the same structure as the predisturbance forest.
Rehabilitation   Assisting ecosystem recovery to a functional state that is not necessarily the same as the predisturbance or the "pristine" state.
Reintroduction   Introduction of animals into a habitat where they once existed but became extirpated.
Restoration   Assisted recovery of a degraded ecosystem to its predisturbance or "pristine" state.
Revegetation   Establishment of vegetation (native or non-native) on a barren or degraded site.

Safe site   Microenvironment in the soil suitable for seed germination.
Sea level rise   Predicted rise in average global sea level due to global warming
Secondary succession   Changes in plant communities with time on sites that contains intact soil and even vegetation.
Seed bank   Dormant but viable seeds that are stored in the soil and germinate when suitable conditions prevail.
Seed rain   Primary dispersal of seed usually by prevailing wind.
Soft release   Process of transferring animals from a source to a release site with conditioning efforts.
Species richness   Number of species found within a particular area.
Stepping stones   Network of patches that facilitate movement of organisms across the landscape.
Succession trajectories   Various pathways of community changes on a particular site with time.

Translocation Introduction of animals into a habitat where they have not existed before.
Transplanting Outplanting container-based plants in a certain pattern on restoration sites.
Tropical rain forest Productive forests with high biodiversity that occur around the equator.

Watershed Drainage area including wetlands, rivers, and streams for a particular aquatic system.
Wetlands Terrestrial lands where soils are inundated with water over most of the year.

Xenobiotic Unnatural or synthetic compounds.

Zooplankton Microscopic aquatic invertebrates.

INDEX

Abandoned Mine Reclamation Fund, 364
abiotic factors
   ecosystem threshold, 113
   in resistance to invasion, 164
abiotic filters, 114
abruptness of disturbances, 37
Acadian forest, 284–288
acetylcholinesterase, 261–262
acid drainage, 239
acidification, 17, 325
acid rain, 3, 17, 270–271
   lakes and, 325
   legislation on, 363
action plans, 357–358
action thresholds, 361
active restoration, 163, 358
   of forests, 272–275
   of rivers, 330–332
   of wetlands, 320–321
adaptive management, 359–360
   in assembly, 113
   definition of, 352
adder (*Vipera berus*), 65–66
aerial photographs, 95–96, 360–361
Africa
   aquatic ecosystem degradation in, 319
   deforestation in, 4, 280
   desertification in, 183, 184, 366–367
   gold mine restoration in, 252–257
   invasive species in, 212
   land degradation in, 177
   population growth in, 2–3
   sand dunes in, 212
age structure analysis, 96
agriculture. *See also* grazing
   abandoned farmlands and, 89
   deforestation and, 266, 280–281
   desertification and, 181–185
   erosion from, 3
   eutrophication from, 323
   forests and, 267–270
   Great Lakes and, 326
   greenhouse gases from, 22, 23–24
   in land degradation, 177
   organic, 189
   pathogen resistance in, 13
   salinization and, 189
   shifting cultivation in, 280–281
   soil carbon sequestration and, 30–31
   soil conservation and, 185–189
   tragedy of the commons and, 10–11, 183, 267
   wetlands destruction and, 317, 321
   yields in, 30–31

air purging, 250
algal-biomass energy, 10
alligators (*Alligator mississippiensis*), 322
alternative stable states, 106–111
   regime shifts and, 107–110
   restoring, 110–111
Amazon rain forest, 4, 266, 280
   agriculture in, 267–268, 281
   destruction of, 281
   indigenous people in, 283
AMF. *See* arbuscular mycorrhizal fungi (AMF)
ammonification, 193
amphibians, 58
analysis
   age structure, 96
   gap, 134
   population viability, 61
   spatial, 357
analytical hierarchy process (AHP), 357
Animal and Plant Health Inspection Service, 365
ants, fire (*Solenopsis invicta*), 154, 160–161
aquaculture, 334–338
aquatic ecosystems, 314–351. *See also* lakes; rivers; wetlands
   case study on, 333–338
   degradation of, 316–319
   legislation on, 363–364
   restoration of, 319–333
Aral Sea, 318–319
arbuscular mycorrhizal fungi (AMF), 168, 193–195, 201–205
   in phytoremediation, 260
   in sand dunes, 209, 219, 224–225
   stockpiled soil and, 241–242
Area de Conservación Guanacaste (Costa Rica), 143–145
area threshold, 294
artificial insemination, 296
assembly, 101–130
   alternative stable states and, 106–111
   case studies on, 119–129

dune ecosystem, 220–221
   ecological filters and, 113–114
   ecosystem thresholds and, 113
   in forest restoration, 274
   regime shifts and, 107–110
   resilience and, 104–105
   rules of, 111–114
   unified neutral theory of biodiversity and biogeography and, 114–116
assessment, 362
auks, great (*Pinguinus impennis*), 60, 61–64
Australia, 134

bait stations, 160, 161
ballast water, 151, 157
barren gravel lands, 98–99
beach nourishment, 207, 214
bears, grizzly (*Ursus arctos horribilis*), 294, 304
beavers (*Castor canadensis*), 48
beech trees, American (*Fagus grandifolia*), 163
Belcher, Richard, 261–262
bensulide, hydroxylation of, 261–262
bicarbonate leaching, 28
Billington, Neil, 306–312
bioaccumulation, 244
bioassay, 243–244
biodegradation, 248
biodiesel, 10
biodiversity, 55–78
   agriculture and, 13, 281
   definition of, 56
   dense transplanting and, 275
   economic value of, 56
   ecosystem, 71–73
   extinction and, 60–64
   in forests, 275, 284–288
   genetic, 64–66
   geographical differences in, 57
   grassland, 51–53
   hybridization with invasive species and, 155–156

international conventions on, 367
metapopulations and, 291
microbial, 73, 197–200
phylogenetic, 119–129
population size and, 61
resilience and, 104–105
in resistance to invasion, 153–154
restoration of species, 66–71
role of natural disturbances in, 34
in sand dunes, 207
species diversity measures, 56
stability and, 72–73, 105
succession and, 83–84
threats to, 57–60
in tropical rain forests, 280
unified neutral theory of,
  114–116, 121
in wetlands, 315
biogeography, 101
equilibrium theory of, 102–104, 131
island, 131
unified neutral theory of biodiversity
  and, 114–116
biological control, 161–162
biological degradation, 25
biological legacies, 38–39, 271
biological monitoring, 243–244
biomanipulation, 325, 329
case study on, 338–347
Biophilia (Wilson), 281
bioremediation, 248–252
criteria for, 248
impeding factors in, 249
microbes in, 191, 197
of pesticides, 261–262
pros and cons of, 249
requirements for, 249–250
biostimulation, 250
biotic factors
ecosystem threshold, 113
in resistance to invasion, 153–154, 164
biotic filters, 114
bioventing, 251–252
birch trees (*Betula pubescens*), 268–270,
  274, 279–280

birds
California condor (*Gymnogyps
  californianus*), 298
critical habitats for, 292–293
eagles, bald (*Haliacetus
  leucocephalus*), 305
eagles, golden (*Aquila chrysaetes*), 305
European starling (*Sturnus vulgaris*),
  154
Florida scrub jay, 141–142
fragmentation and, 44
invasive species and, 156
ovenbird (*Seiurus aurocapillus*), 294
spotted owl (*Strix occidentalis*),
  89–90, 292
as succession indicators, 90
in U.K. beach restoration, 226–229
bison (*Bison bison*), 59–60, 298
Bob Marshall Wilderness (Idaho), 301
bogs, 315. See also aquatic ecosystems
bootstrapping hypothesis, 195
Bradshaw, Anthony D., 13–14, 239
Brazil, coastal salt march restoration in,
  333–338
breeding programs
captive, 294–299, 300
population size for, 294
Broads (U.K.), restoration of,
  328–330, 338–347
broom, French and Scotch (*Cytisus
  monspessulanus*) (*C. scoparius*),
  159–160, 220
Brown, Lester R., 18
brown snake (*Boiga irregularis*), 58
Buckely, Dan, 197–200
budgets, 356
buffer zones, 103–104
benefits of, 135–136
in forests, 272
in river restorations, 330–331
sand dunes as, 207
stepping stones in, 135–136
butterflies
Fender's blue (*Icaricia icarioides
  fenderi*), 136

Karner blue (*Lycaeides melissa samuelis*), 139–140

cabling, 93
camelthorn (*Alhagi pseudalhagi*), 159
Cape Wind Project (Massachusetts), 8
captive breeding, 294–299, 300
carbon dating, 96
carbon dioxide
  biodiversity and, 56
  in global warming, 4–6
  soil as source of, 23–24
  soil sequestration of, 21–31
  tropical rain forests and, 280
carbon sequestration, 17, 275–276
  bicarbonate leaching in, 28
  biodiversity and, 56
  case study on, 21–31
  chemical, 26
  geological, 26
  illuviation of clay-borne SOC in, 28
  oceanic, 26
  rates of soil, 29–30
  rates of soil carbon, 29–30
  in tropical rain forests, 280
carbon sinks, 182–183, 315
Carolina parakeet (*Conuropsis carolinensis*), 60
carrying capacity, 2, 294
case studies
  on carbon sequestration, 21–31
  on forest restoration, 284–288
  on invasive species, 154–174
  on kudzu, 165–169
  on mine restoration, 252–257
  on pale swallow-wort, 169–174
  on pesticide bioremediation, 261–262
  on phylogenetic structure, 119–129
  on phytoremediation of lead contamination, 257–260
  Pitcher's thistle restoration, 146–149
  on resilience and restoration, 116–119
  on sand dunes, 222–234
cation exchange capacity (CEC), 190
Cavender, Nicole, 119–129

Cavender-Bares, Jeannine, 119–129
Cedar Creek Natural History Area (Minnesota), 73
cedar trees (*Cedrus libani*), 277–279
Central and Southern Florida Project for Flood Control and Other Purposes, 322
Channel Islands (California), 304–305
chelate-assisted phytoremediation, 246–248, 258–260
chemical controls, 159–160, 161, 168
chemical pollution, 57, 58–59
chemical sequestration, 26
chemical soil stabilizers, 219
chestnut blight fungus (*Cryphonectria parasitica*), 156–157
chestnut trees, American (*Castanea dentata*), 156–157, 271, 272
China
  acid rain and, 17
  bamboo forest restoration in, 296
  carbon dioxide emissions in, 4
  desertification in, 183
  energy consumption in, 7
  soil erosion in, 178
  water supplies in, 2
chlorofluorocarbons (CFCs), 118
chlorosis, 237
cholinesterases, 261–262
Clean Air Act of 1963, 363
Clean Water Act of 1972, 363–364
clear-cutting, 44–45, 266–267
Clements, Frederic, 80
climate change, 4–6
  assembly and, 116
  desertification and, 181–185
  ecological restoration and, 17
  fires from, 43
  Gaia Model on, 11
  sand dunes and, 212–213
  sea level changes and, 212–213
  stability vs., 56
climax communities, 102, 112–113
climax vegetation, 80
clonal offsets, 216, 218–219

clover, purple prairie (*Dalea purpurea* Vent.), 74–77
"Coast 2050: Towards a Sustainable Coastal Louisiana," 364
coastal erosion, 213
colonization
 alternative stable states and, 106–111
 biological legacies in, 38–39
 equilibrium theory on, 102–104
 of forests, 271–272
 on islands, 87–88
 of keystone species, 92–93
 in landscape restoration, 143
 metapopulation and, 137, 148–149
 monitoring, 361
 in river restorations, 331
 of soil microorganisms, 193
Columbia River, 44
communication, 356
community ecology, 101, 102–103, 119
competitive exclusion principle, 120
Comprehensive Environmental Response, Compensation, and Liability Act of 1980, 364
Comprehensive Everglades Restoration Plan, 324
condor, California (*Gymnogyps californianus*), 298
conflicts of interest, 355
connectivity, 72
 corridors in, 134–135
 definition of, 132
 endangered animals and, 291, 293
 equlibrium theory on, 103–104
 landscape, 133–136
 matrix restoration and, 134–135
 Pitcher's thistle and, 148–149
 stepping stones in, 135–136
 thresholds of, 133, 141
Connell, Joseph, 80, 82
conservation genetics, 74–77
Conservation Reserve Program, 178–179, 188–189, 365
constraints. *See* ecological constraints
containment, 161, 197

context, 375–380
coordinated resource management and planning (CRMP), 355
coral reefs, 5, 323
core habitats, 72, 135–136
corridors, 72, 134–135, 272–273
Costa, César S. B., 333–338
Costa Rica, 143–145, 280
countouring, soil, 186, 188
cover crops, 216
Cowles, Henry, 80
coyotes, 304
critical habitats, 292–294, 297, 299
 Everglades as, 323–324
 reintroduction to, 300
CRMP. *See* coordinated resource management and planning (CRMP)
cropping methods, 181, 186
cutting, 93
cypress trees, bald (*Taxodium distichum*), 275–277

Damang Gold Mine restoration (Ghana), 252–257
dams, 44, 281, 319, 330
Dauphin Island (Alabama), 222–225
DDT (dichlorodiphenyltrichloroethane), 58–59, 305
debt swaps for nature, 283
decarboxylation, 193
decision making, 375–380
decomposers, 192
decomposition, 176–177, 192
deforestation, 265–266
 desertification and, 181
 Easter Islands, 3
 extinction and, 60
 nutrient cycling and, 44–45
 succession management and, 91–92
 of tropical rain forests, 4, 265, 266, 267–270, 280–282
denitrification, 193
dense transplanting, 275
desertification, 181–185
 causes of, 181, 182

definition of, 181
effects of on humans, 183–184
human actions in, 184–185
international agreements on, 366–367
regional effects of, 183
slowing, 185
Desertification Control Programme Activity Center, 184
deterministic model of succession, 101, 209
Diamond, Jared, 111
diazotrophic bacteria, 195
dinitrogen fixation, 195–196
direct seeding, 273–274
disease
  global warming and, 6
  invasive species and, 156–157
dispersal barriers, 110–111, 291
dispersal vectors, 243
disturbance regimes, 35, 293
disturbances, 34–39
  abrupt changes, 36
  cascading effects of, 34
  definition of, 34
  duration, abruptness, and return interval of, 37, 38
  fire, 39–43
  fragmentation and, 43–44
  frequency of, 36–37
  human-caused, 35
  hybridization and, 59
  magnitude of, 37
  mega-, 37–38
  nutrient cycling and, 44–47
  primary sites from, 83
  reducing or inducing, 91–92
  regime shifts and, 107–110
  resilience and, 104–105
  resistance after, 105
  restoration of natural, 39
  in sand dunes, 210–213
  scale of, 80
  severe, 38–39
  soil, effects of, 199–200
  stress, 35
  succession and, 79, 88–89, 91–92
DiTommaso, Antonio, 169–174
diversity. *See* biodiversity
dormancy, 69
Douglas, Marjory Stoneman, 321
dredging, 324, 329, 338
drift nets, 332–333
drivers, 48
droughts
  in Africa, 4
  desertification and, 181–185
  mycorrhizae and, 194
drought-tolerant plants, 95
ducks, Labrador (*Camptorhynchus labradorius*), 60
dune-building fences, 214–215
dunes. *See* sand dunes
Dust Bowl, 178, 183–184
Dyer, Andy, 51–53

eagles
  bald (*Haliacetus leucocephalus*), 305
  golden (*Aquila chrysaetes*), 305
early detection systems, 244, 361
Easter Islands, 3
ecological constraints, 105–106
ecological drift, 114, 115
ecological engineering, 16
ecological filters, 113–114, 121–123
ecological genetics, 74–77
ecological management units, 355
ecological resilience. *See* resilience
ecological resistance, 105, 164
ecological services. *See* ecosystem services
economics
  of biodiversity, 56
  of biological control, 161
  of invasive species, 151, 168
  of passive restoration, 145
  project funding and, 356
  of river restoration, 331–332
  of sand dunes, 207
  of soil erosion, 178–179

of soil nutrient loss, 190
of wetlands, 315
ecosystem engineers, 48–49
ecosystem regression, 361, 362
ecosystems. *See also specific ecosystems*
  degradation of, 3–12
  diversity of, 71–73
  fire-adapted, 40–42
  functioning of, 33–54
  future of, 11–12
  global warming and, 5–6
  management models for, 10–12
  matrix, 132
  negative feedback loops in, 4
  predisturbance state of, 14–15
  regenerative capacity of, 11
  stability of, 72–73
ecosystem services
  biodiversity and, 56
  definition of, 12
  value of, 12–13
  of wetlands, 315
ecosystem thresholds, 113
ecotourism, 279
ectomycorrhiza, 193
EDDS (ethylene-diamine-dissuccinate), 248
edge effects, 72
EDTA (ethylenediaminetetraacetic acid), 247–248, 258–260
education, on invasive species, 159–160
EGTA (ethyleneglycotetraacetic acid), 248
Egypt, desertification in, 182
elephants (*Elephantidae*), 49, 299
elk
  Cervus canadensis, 303–304
  Cervus elaphus, 357
Emergency Wetlands Resources Act of 1986, 364
emigration, 143
endangered species, 291–313
  captive breeding of, 294–299
  case study on, 306–312

critical habitat restoration and, 292–294
  hybridization and, 306–312
  metapopulation and, 137
  reintroduction of, 299–305
  succession and, 89–90
  translocation of, 299
Endangered Species Act, 40
endemic species, 154
energy consumption, 6–10
engineered systems, 13, 329
enrichment planting, 274
environmental impact assessment, 356
Environmental Protection Agency (EPA), 364
equilibrium theory of biogeography, 102–104, 131
eradication, 158–161
erosion, dune, 211, 213. *See also* soil erosion
An Essay on the Principle of Populations (Malthus), 2
ethanol, 10
Europe
  desertification in, 183
  wetlands degradation in, 316
eutrophication, 4
  in the Everglades, 323
  in lakes, 324–325, 328–330
  regime shift and, 108–110
  restoration from, 106
  succession management and, 94
  in wetlands, 317
evapotranspiration, 155
Everglades (Florida), 17, 118, 321–324
excavation, 248
expansion strategies, 272
extinction, 1, 60–64, 143. *See also* endangered species
  definition of, 60
  equilibrium theory on, 103–104
  fragmentation in, 43–44, 291
  genetic diversity and, 64
  habitat destruction in, 291

metapopulation and, 136, 137
rate of, 56, 60–61
threshold of, 133
vulnerability to, 61

facilitation model of succession, 80, 82, 209
facilitative effects, 278
false-color images, 221
famines, 184
Farm Bill of 1985, 365
Federal Noxious Weed Act of 1974, 365
Federal Noxious Weed List, 166
feedback
  in adaptive management, 360
  ecological constraints and, 106
  negative loops in, 4
  top-down, 33–34
fences, dune-building, 214–215
fens, 315
ferret, black-footed (*Musteles nigripes*), 296–298
fertilizer
  chemical, 219–220
  in mine restoration, 242, 243
  monitoring effects of, 95
  nitrogen restoration with, 191
  in primary succession, 84
  in revegetation, 45
  in sand dune restoration, 216, 219–220
  seed production and, 67
  in succession management, 93–94
50/500 rule, 295
filters, ecological, 113–114, 121–123
fire disturbances, 39–43
  biodiversity and, 57
  ecosystems adapted to, 40–42
  fire suppression and, 42–43
  grasslands and, 51–53
  invasive plants and, 155, 163–164
  prescribed burning and, 41–42, 92, 142
  regime shifts and, 107–108
  succession management and, 91–92
  suppression of, 163–164

fish
  biomanipulation and, 339–347
  eutrophication and, 318
  Great Lakes, 326–327
  hybridization with non-native, 306–312
  invasive species of, 58, 59
  keystone predators, 50
  minnow, sheepshead (*Cyprinodon variegatus*), 155–156
  Pecos pupfish (*Cyprinodon pecosensis*), 155–156
  piscivorous, 318
  planktivorous, 50, 318
  river fragmentation and, 44
  salmon, Chinook, 35
  salmon, North Atlantic (*Salmo salar*), 332–333
  sauger (*Sander canadensis*), 306–312
  sheepshead minnow (*Cyprinodon variegatus*), 155–156
  swordtails, 59
  trout, 331
  walleye (*Sander vitreus*), 306–312
fisheries, 11
fish ladders, 330
Five Star Restoration Program (EPA), 364
flagship species, 291
Florida
  dune restoration in, 207
  Everglades, 17, 118, 321–324
  invasive species in, 151, 154
  Kissimmee River restoration, 330, 347–351, 358
  longleaf pine in, 41–42
  phytoremediation of lead contamination in, 257–260
  regime shifts in, 107–108
  scrub jay, 141–142
  wetlands degradation in, 316
Florida Panther National Wildlife Refuge, 324
food security, 30–31, 183–184

Food Security Act of 1985, 186, 364
forests, 265–289
  on abandoned farmlands, 89
  canopy thickness and invasiveness, 164
  case studies on, 284–288
  clear-cutting and selective cutting, 44–45, 266–267
  core habitats in, 72
  deforestation and, 265–266
  degradation of, 266–271
  edge habitats in, 72
  fire suppression and, 42–43
  fragmentation of, 44, 287, 288
  human settlement and, 267
  hurricanes and, 36–37
  nutrient cycling and, 44–45
  pollution and, 270–271
  prescribed burning of, 41–42
  prevalence of, 265
  regeneration niches in, 275–280
  regeneration potential of, 285–288
  regime shifts in, 107–108
  restoration of, 271–280
  succession in, 81, 84–86, 89
  tropical rain, 265
forest zone of sand dunes, 207
fossil fuels, 6
  acid rain and, 17
  in global warming, 4
  greenhouse gases from, 21–22
  oil consumption, 7
founder effects, 74
founder populations, 297
fox, Channel Island (*Urocyoon littoralis*), 304–305
fragmentation, 43–44
  biodiversity and, 57–58
  connectivity and, 133–136
  definition of, 43, 57
  ecosystem diversity and, 71–72
  extinction due to, 43–44, 291
  in forests, 272–273
  of forests, 287, 288
  landscape, 132, 133–136
  tolerance to, 137
French Polynesia, 155
frogs, coqui (*Eleutherodactylus coqui*), 151
functional groups, 34, 48–49, 120
  drivers and passengers in, 48
  ecosystem engineers, 48–49
  in ecosystem functioning, 71
  phylogenetic diversity and, 123–124
funding, 356

Gaia Model, 11
gap analysis, 134
gene flow, 143
genetic diversity, 64–66. *See also* biodiversity
  captive breeding and, 295, 297
  invasive species and, 155–156
  in sand dunes, 216
  selective cutting and, 266–267
  tallgrass prairie case study on, 74–77
genotypes, 64–66. *See also* biodiversity
  metal-tolerant, 239, 240–241
  in restoration projects, 66–67
  seed storage and, 68
geographical information systems (GIS), 42, 134, 221, 294, 302, 357
geological sequestration, 26
geomorphology, 330–331
geothermal power, 10
germination, 68–69, 216
  in forests, 273–274
  improving, 217–218
  seed priming for, 69–70
Ghana, gold mine restoration in, 252–257
Glacier Bay (Alaska), 83
Glacier National Park, 301
Global Biodiversity Outlook, 367
global positioning systems (GPS), 141
global warming. *See* climate change
goals
  in restoration ecology, 14–15, 34

setting, 353–354
unclear/unrealistic, 356
Goat Island Bay Reserve (New Zealand), 11
Gondwana Link (Australia), 134
Gould, Stephen J., 120
Grand Coulee Dam, 44
The Grapes of Wrath (Steinbeck), 183–184
grasses
  African jaragua (*Hyparrhenia rufa*), 163–164
  cheatgrass (*Bromus tectorum*), 155
  commercial, 242
  loss of vigor in, 210
  lymegrass (*Leymus arenarius*), 216, 217
  marram (*Ammophila areanaria*), 212, 216
  metal-tolerant, 240–241
  for mine restoration, 240–241, 242
  in sand dunes, 212
  sea oats (*Uniola paniculata*), 212–213, 222–223
  tussock (*Agropyron cristatum*), 162
grasslands
  biodiversity and stability in, 73, 74–77
  dispersal barriers in, 111
  ecological constraints in, 105–106
  fire and, 39–40, 51–53
  invasive species in, 155
  niche preemption in, 162–163
  tallgrass prairie case study, 74–77
Grasslands National Park (Canada), 162
grazing, 1
  desertification and, 181, 182, 184–185
  dune disturbance through, 211
  eradicating invasive plants with, 159
  fire frequency and, 51–52
  forests and, 267
  kudzu control through, 169
  non-native animals and, 155
  restricting/preventing, 358
  succession and, 89, 90–92
  tropical rain forests and, 281

Great Lakes, 17
  invasive species, 160
  restoration of, 325–327
Great Lakes Fishery Commission, 327
Great Smoky Mountains National Parks, 162–163
greenhouse gases (GHGs), soil sequestration of, 21–31
Greipsson, Sigurdur, 97–99, 368–375
Gulf of Mexico, 222–225, 320
Gunderson, Lance, 116–119
Gustafson, Danny, 74–77

habitat fragmentation. *See* fragmentation
Habitat Suitability Model, 297–298
Haiti, deforestation of, 3
halophytes, 189
Hardin, Garrett, 10–11
hard release, 301
Hawaii
  coqui frogs in, 151
  extinction on, 60–61
  invasive species on, 154–155, 156
heather (*Calluna* sp.), 91–92, 98–99
  heathland restoration, 368–375
  regime shift and, 108
herbicides, 93, 160, 161
  bioremediation of, 261–262
historical contingency, 106
historical reference data, 354
Holling, C. S., 117
horizons, soil, 178
Horse Hollow Wind Energy Center (Texas), 8
horses, Prewalski (*Equus prewalski*), 298
Hovsepyan, Anna, 257–260
Hubbell, Stephen, 114
humans
  aquatic ecosystem degradation and, 319
  carrying capacity of, 2
  in deforestation, 265–266
  in desertification, 183–185
  disturbances caused by, 35
  effects of on ecology, 1–2

Everglades destruction by, 322–323
in land degradation, 177
population growth of, 2–3, 184–185, 265–266, 283, 366
sand dune disturbance by, 211
water supplies and, 2
humification, 27–28
hunting, 1, 57, 59–60
endangered animals and, 291
eradication of invasive species through, 160
in the Everglades, 322
extinction from, 57, 59–60, 61, 63
Hurricane Katarina (2005), 37
hurricanes, 36–37, 210, 223, 224
hybridization, 57, 59, 155–156, 331
case study on, 306–312
in restoration projects, 157
of salmon, 332
hydrogels, 240
hydrological cycles, 4, 47–48
biodiversity and, 56
deforestation and, 266
invasive species and, 155
reforestation and, 279
restoration of, 34
wetlands restoration and, 319–320
hydropower, 7, 10, 281, 319
hydroseeding, 70, 242
hyperaccumulators, 246
hypoxia, 318, 320
hysteresis, 108, 118–119

Ice Ages, 15
ice caps, melting of, 5
Iceland
birch woodlands in, 268–270, 274, 279–280
heathland restoration in, 368–375
immigration, 143
inbreeding, 295
inbreeding depression, 61, 74
reversing, 64–66
incineration, 248
income, 3

India
desertification in, 183
energy consumption in, 7
indicator species, 193, 361–362
indigenous people, 267, 283
individualistic model of succession, 101, 103
inhibition model of succession, 82
inoculation, 219
in landfarming, 250
Rhizobia, 196
insecticides, 161
insects
animals in controlling, 91
invasive, eradication of, 160–161
integrated pest management, 160
integrated restoration management, 162–163
intermediate zone of sand dunes, 207
international conventions, 366
International Union for Conservation of Nature and Natural Resources (IUCN), 291, 297, 366
invasiveness, 153
invasive species, 150–175
biodiversity and, 58
case study on, 165–169
chemical and biological control of, 161–162
containment of, 161
control methods for, 157–162
definition of, 150
effects of on ecosystems, 154–157
eradication of, 158–161
in the Great Lakes, 326–327
hybridization with, 306–312
invasion process of, 152–154
kudzu (*Pueraria montana*), 165–169
legislation on, 365
monitoring, 361
naturalization of, 153
pale swallow-wort (*Vincetoxicum rossicum*), 168–174
prevention measures against, 157–158
resilience and, 104–105

restoration to constrain, 162–164
in sand dunes, 211–212, 221
susceptibility to, 153–154
traits of, 153
in tropical rain forest restoration, 282
understory species, 39
ion enhancement, 216
irreversibility, threshold of, 124
irrigation, 94–95, 189
desertification and, 181
lake degradation and, 318–319
islands
equilibrium theory of biogeography, 102–104, 131
extinction rate on, 60–61
invasive species on, 154–155
primary succession on, 84–88
unified neutral theory of biodiversity and biogeography on, 114–115
isolation
Florida scrub jay and, 141–142
tolerance to, 137

Jamaica, bauxite mining in, 282
Janzen, Dan, 143–145
jay, Florida scrub (*Aphelocoma coerulescens*), 141–142

Kellogg Biological Station Long-Term Research site (Michigan), 200
keystone species, 34, 49–50
definition of, 49
in ecosystem functioning, 71
establishing and maintaining, 358
introducing or removing, 92–93
resilience and, 104–105
in restoration of species diversity, 66
in succession management, 92–93
transitional/nexus, 49
wolves, 303–304
Kinney, Patricia, 165–169
Kissimmee River restoration (Florida), 330, 347–351, 358
Krakatau, 84–86
kudzu (*Pueraria montana*), 165–169

Kuwait, crude oil lakes in, 251–252

lag growth, 152–153
Lake Chad, 319
Lake Michigan, 146–149
lakes
degradation of, 318–319
fragmentation in, 44
invasive species in, 151
regime shift in, 108–110
restoration of, 324–330, 338–347
roles of, 316
lampreys, sea (*Petromyzon marinus*), 326–327
land degradation, 177–178
landfarming, 250–252
land management, 10–12
carbon sequestration rates and, 29–30
degraded land purchases and, 18
erosion and, 181
forest degradation and, 267–270
in river restoration, 331
landscape conversion, 71
landscape ecology, 131–149
connectivity in, 133–136
definition of, 131
metapopulations in, 136–142
reference sites in, 132
restoration in, 143–145
spatial heterogeneity in, 131
landscapes
connectivity in, 132
definition of, 132
fragmentation of, 132
matrix in, 132
patches in, 132
land use practices, 188–189
leaching, salinization and, 189
lead (Pb) contamination, 257–260
legislation, 66, 353, 363–367
endangered animals and, 291–292
international conventions, 366
on mine reclamation, 239
legumes
non-native, 155

in nutrient cycling, 45
Rhizobia and, 195–196
liberal (passive) restoration. *See* passive restoration
line intercept method, 95
livestock
  in Africa, 3
  desertification and, 184–185
  overgrazing and, 3
  tropical rain forests and, 281
  wolves and, 302–303
living dead, 137
Long-Range Transboundary Air Pollution Agreement, 363
long-term management, 219–221
loss of vigor, 210
lupines
  Karner blue butterfly and, 139
  Kincaid's (*Lupinus sulphureus* sp. *kincaidii*), 136
  in restoration, 373
  in sand dunes, 220
lymegrass (*Leymus arenarius*), 216, 217

MacArthur, Robert, 102, 114
macronutrients, 177, 208–209, 250
Malthus, Thomas, 2
management. *See also* land management
  ecosystem, models of, 10–12
  long-term, 219–221
  of restoration projects, 352–381
  of sand dunes, 219–221
  of succession, 82, 91–95
management blocks, 142
management models, 10–12
maple trees, sugar (*Acer sacharum*), 163
marine reserves, 11
marram grass (*Ammophila areanaria*), 212, 216
marshes, 315, 333–338
matrix, 132, 133–134
McEachern, A. Kathryn, 146–149
mechanical controls, 168
mega-disturbances, 37–38
mercury pollution, 17

metal chelates, 246–248, 258–260
metals, heavy, 201–205, 238–239. *See also* mines
  hyperaccumulators of, 246
metal-tolerant genotypes, 239, 240–241
metapopulation, 132, 136–142
  case study on, 146–149
  dynamics of, 137
  endangered animals and, 291
  networks of, 133
  restoration projects, 138–142
miconia (*Miconia calvescens*), 155
micronutrients, 236–237
microorganisms, 178
  arbuscular mycorrhizal fungi, 193–195
  in bioremediation, 248–252
  case study on, 197–200
  diversity of, 73
  inoculation of, 196, 219
  invasive species and, 168
  in landfarming, 250
  mutualistic species and, 50
  in phosphorus cycling, 46
  in sand dunes, 209, 219
  in soil, 191–196
  soil disturbance and, 199–200
  stockpiled soil and, 241–242
migration
  from global warming, 5
  population growth and, 3
milestones, 354
mineralization, 46, 190, 193
mines, 236–264
  arbuscular mycorrhizal fungi and, 201–205
  bioremediation of, 248–252
  case studies on restoration of, 124–127, 201–205, 252–257
  grasses for, 242
  heavy metals from, 238
  landfarming, 250–252
  legislation on, 364
  monitoring restoration of, 243–244
  native plants for, 242–243
  phytoremediation of, 245–248

pollution from, 238–239
reclamation laws on, 239
stockpiled soil and, 241–242
tropical rain forest destruction
 by, 281–282
uranium, 6–7
waste restoration of, 239–244
wetland wastewater systems and, 16
minimum viable population, 61
minnow, sheepshead (*Cyprinodon variegatus*), 155–156
Mississippi River Delta, 320
Mississippi River watershed, 314–315
model projects, 357–358
monitoring, 360–362
 biological, 243–244
 of containment, 161
 of critical habitats, 293
 of endangered animals, 298, 300
 eradication efforts, 158
 in forest restoration, 271
 landfarming, 250
 in mine restoration, 243–244
 parameter selection for, 361
 proxy techniques in, 360–361
 of reintroduction, 300
 of sand dunes, 221, 226–229
 succession, 95–96
monitoring systems, 353
monocultures, 281
Montana, river restoration in, 331–332
Montreal Protocol, 118
moose (*Alces alces*), 6
Mount St. Helens (*Washington*), 88
mowing, 159
mulch
 erosion and, 181
 hydrological cycling and, 48
 in mine restoration, 240
 on sand dunes, 219
 in soil conservation, 186, 188
multispecies approach, 66
Murphy, Stephen D., 375–380

mussels, zebra (*Dreissena polymorpha*), 151, 327
mutualistic species, 34, 50
mycorrhiza, 178, 193–195
 arbuscular mycorrhizal fungi, 168, 193–195, 201–205, 224–225, 241–242
 in biodiversity and stability, 73
 invasive species and, 168
 in mine tailing restoration, 201–205
 as mutualistic species, 50
 phosphorus cycling and, 46
 in phytoremediation, 260
 stockpiled soil and, 241–242

National Environmental Policy Act of 1969, 363
National Invasive Species Council, 365
National Park Service, 301, 302–303, 305
National Priorities List, 364
native seeds, 67
native species, 150
 in mine restoration, 242–243
 for sand dunes, 216–219
 in tropical rain forest restoration, 282–283
natural disturbances, 34
 in biodiversity, 34
 definition of, 35
 reintroducing, 39
naturalized populations, 153
N cycle. *See* nitrogen (N) cycle
N-dimensional niches, 120
negative feedback loops, 4
Nepal, population in, 3
networks, metapopulation, 133
neutral theory, 121
New Zealand, aquatic ecosystems in, 11, 316
nexus species, 49, 93, 112–113, 355
niche preemption, 162–163
niches, 121
 assembly, 111
 N-dimensional, 120
 regeneration, 275–280
 succession and, 87–88

Nigeria, population growth in, 2–3
Nile perch (*Lates niloticus*), 58
nitrification, 193
nitrogen (N) cycle, 44–46, 83
  dinitrogen fixation and, 195–196
  in sand dunes, 208–209
  soil in, 178
  soil restoration and, 190–191
nitrogen-fixing species, 49
nitrogen oxides (NOx), 17, 363
nonequilibrium paradigm, 36
Non-Indigenous Aquatic Nuisance Prevention and Control Act of 1990, 365
non-native species
  after fires, 39
  edge effects and, 72
  eradication and control of, 358
  hybridization with, 57, 59
  introductions of, 151
  invasive, 150–175
  unified neutral theory of biodiversity and biogeography on, 115–116
  wetlands restoration and, 321
North American Waterfowl Management Plan of 1986, 364
North American Wetlands Conservation Act of 1989, 364
North Atlantic Salmon Fund (NASF), 332–333
Northwestern Hawaiian Islands Marine Natural Monument, 11
nuclear power, 6–7, 10
nucleated succession, 98
nurse plants, 93, 320
nursery-raised plants, 70–71
nutrient cycling, 33, 34, 44–47
  assessment of, 362
  biodiversity and, 56
  biostimulation and, 250
  deforestation and, 266
  metals in, 236–237
  microorganisms in, 191
  in mine restoration, 243
  mutualistic species in, 50

  soil in, 176–177
  soil restoration and, 190–191
  succession and, 83
  time scales in, 200
nutrient exhaustion, 94, 106, 321
nutrient management, 93–94

oak trees (*Quercus*), 107–108, 121–123, 141–142, 278
oats, sea (*Uniola paniculata*), 212–213, 222–223
oceanic sequestration, 26
organic farming, 189
organophosphates, 261–262
osmotic priming, 69–70
otters, sea (*Enhydra lutris*), 50
ovenbird (*Seiurus aurocapillus*), 294
overhunting, 57, 59–60, 61, 63
ozone layer
  depletion of, 3
  hole in, 20
  restoration of, 18, 19–20

panda, giant (*Ailuropaoda melanoleuca*), 295–296, 299
panthers, Florida, 322, 323–324
passengers, 48
passive restoration, 15
  of forests, 271–272
  landscape, 143–145
  of rivers, 330
  of wetlands, 320
patches, habitat, 132
  metapopulation and, 137, 138–139
  mycorrhizae and, 194–195
  spacing between, 138–139
Pavlovic, Noel B., 146–149
P cycle. *See* phosphorus (P) cycle
pedological pool, 23
permanent plots, 95–96, 360
Perrow, Martin, 225–234
perturbations, 117. *See also* disturbances
pesticides, bioremediation of, 261–262
pH, 70
  bioremediation and, 250

in lake restoration, 325
in succession management, 94
pharmaceutical industry, 56
phosphate stripping, 329
phosphorus (P) cycle, 45–47, 178
   regime shift and, 108–110
   in sand dunes, 208–209
   soil nutrient restoration and, 191
phosphorylation, 193
photovoltaic (PV) cells, 8, 9–10
phylogenetic scales, 198–199
phylogenetic structure, 119–129
phytodegradation, 245, 258
phytoextraction, 245, 258–259
phytoplankton, 317. See also eutrophication
phytoremediation, 189, 202–203
   case studies on, 257–260
   definition of, 245
   hyperaccumulators in, 246
   metal chelates in, 246–248, 258–260
   of mines, 245–248
phytostabilization, 245–246, 258
pilot studies, 358
pine trees
   Australian (Casuarina equisetifolia), 164
   longleaf, 40–42, 107–108
   ponderosa, 42–43
   regime shifts in, 107–108
   white (Pinus strobus), 271–272
pioneer zone, 207
piscicides, 331
planktivorous fish, 50, 318
planning, 355–357
plovers, ringed (Charadrius hiaticula), 226–229, 231–233
point intercept method, 95
polar bears (Ursus maritimus), 6
pollinators, 50, 91
pollution, 1. See also mines
   aquatic ecosystems and, 325
   arbuscular mycorrhizal fungi in removal of, 201–205
   biodiversity and, 57, 58–59

chemical, 57, 58–59
forests and, 270–271
heavy metal, 201–205
soil, sources of, 177
polymerase chain reaction–based methods, 96, 192
population age structure analysis, 96
population growth, 2–3, 184–185
   deforestation and, 265–266
   international agreements on, 366
   tropical rain forest restoration and, 283
population viability analysis, 61
possum, mountain pygmy (Burramys parvus), 134–135
predators, 34
   invasive species and, 153, 162
   as keystone species, 49–50
   reintroduction and, 305
   release from natural enemies hypothesis of, 153
   translocation and, 299
   wolves, 303–304
Predatory Animal and Rodent Control Service, 301
predisturbance state, 14–15
prescribed burning, 41–42, 92
   eradication of invasive plants with, 160
   management blocks in, 142
   of undesirable plants, 93
prevention measures, 157–158
primary productivity, 33, 34
primary sites, 83–88, 94
   case study on, 97–99
primary succession, 83–88
   mycorrhizae in, 195
   in sand dunes, 206
priming, seed, 69–70
productivity, ecosystem, 57
proxy techniques, 360–361
public relations, 361, 362
pupfish, Pecos (Cyprinodon pecosensis), 155–156

quarantines, 159–160

rainfall
　desertification and, 182–185
　sand dune disturbance and, 213
Rajapuna desert (India), 181
Rakata, 85–86
reclamation
　definition of, 15–16
　laws on, 239
recolonization, biological legacies in, 38–39. *See also* colonization
re-creation, 16
reference sites, 354
　in alternative stable states, 110–111
　landscape features and, 132
reforestation, 278–279
regeneration niches, 275–280
regime shifts, 36, 105, 107–110
　ecosystem thresholds and, 113
　resilience and, 117–118
rehabilitation, 16
reintroduction, 299–305
"release from natural enemies" hypothesis, 153
release strategies, 300–301
remote sensing technologies, 221
renewable energy, 7–10
reservoirs, 330
resilience
　assembly and, 104–105
　assessment of, 362
　case study on, 116–119
　definition of, 117
　succession and, 83–84
resistance, 105
　abiotic, 164
　biotic, 164
　increasing, 164
Resource Ratio Hypothesis, 83
restoration ecology
　approaches in, 15–16
　biological legacies in, 38–39
　cost of, 18

　definition of, 14
　endpoint selection in, 110
　future of, 18
　goals in, 14–15, 34, 120
　outlook for, 13–20
　predisturbance state in, 14–15
　resilience to, 105, 118–119
　scales of, 16–18
　succession in, 82–91
　value of, 12–13
restoration project management, 352–381
　action plans in, 357–358
　adaptive, 359–360
　aftercare in, 362
　assessment in, 362
　case study on, 368–380
　decision making in, 375–380
　goal setting in, 353–354
　legal framework for, 363–367
　monitoring in, 360–362
　planning in, 355–357
return intervals, 37
revegetation
　definition of, 15
　goals for, 354
　nutrient cycling and, 45–47
Rhizobium bacteria, 178, 191. *See also* symbiotic microorganisms
rill erosion, 180
rivers. *See also* aquatic ecosystems
　degradation of, 319
　fragmentation in, 44
　non-native species introduced to, 151
　regime shift in, 108–110
　restoration of, 330–333, 347–351
　roles of, 316
　salinization of, 189
　water quality in, 316
Roberts, Andrew, 222–225
Rocky Mountains, 17

safe sites, 91
salinization, 155, 181, 189
salmon

Chinook, 35
fragmentation and, 44
North Atlantic (*Salmo salar*), 332–333
salt marsh restoration, 333–338
sand dunes, 179–180, 185, 206–235
  beach nourishment and, 207, 214
  biodiversity in, 207
  case studies on, 222–234
  disturbances in, 210–213
  ecological processes in, 208–209
  fences for, 214–215
  formation of, 207–208
  human disturbances of, 211
  invasive species in, 211–212
  long-term management of, 219–221
  loss of vigor in, 210
  native plants for, 216–219
  reconstruction of, 213
  restoration strategies for, 213–219
  sea level changes and, 212–213
  succession in, 206
sand stabilization, 185
sauger (*Sander canadensis*), 306–312
scarification of seeds, 69
sea-cliff gulls, 87–88
sea level, 5, 212–213
seals
  common (*Phoca vitulina*), 229, 233–234
  grey (*Halichoerus grypus*), 229, 233–234
sea oats (*Uniola paniculata*), 212–213, 222–223
sea otters (*Enhydra lutris*), 50
secondary succession, 88–89, 271–272
sedimentation, 319, 330
seed banks, 88–89
seeds
  direct seeding, 273–274
  dispersal of, 66, 91, 273
  dormancy of, 69
  evaluating, 68
  germination of, 68–70, 216, 217–218, 273–274

  harvesting, 67, 216, 274
  hydroseeding, 70, 242
  native, 67, 216
  pretreatment of, 69, 217–218
  priming, 69–70
  processing and storage of, 67–68
  production of, 67
  safe sites of, 91
  in sand dune restoration, 216–218
  for sand dunes, 67, 216
  sowing, 70
  viability of, 68–69
  in wetlands restoration, 320
selective cutting, 266–267
self-regulating mechanisms, 11
sequence introduction, 111–113
sequestration. *See* carbon sequestration
Shannon-Wiener Index, 56
sheep, 369
sheep, bighorn (*Ovis canadensis*), 140–141
sheep, overgrazing, 155
sheet erosion, 180
shrimp farms, 334–338
SIC. *See* soil inorganic carbon (SIC)
siltation, 330
Simpson Index, 56
single-species approach, 66
sink populations, 133
site preparation, 213
Slatyer, Ralph, 80, 82
snails, 60
SOC. *See* soil organic carbon (SOC)
Sodbuster program, 365
soft release, 300–301, 302
soil, 176–205
  acidification of, 17
  aggregation of, 191
  amelioration of, 91
  amending, 70
  biodiversity and, 56
  carbon sequestration in, 17, 21–31
  compaction of, 267
  conservation of, 178, 185–189
  definition of, 176

degradation of, 24–25
erosivity of, 181
horizons, 178
humification of, 27–28
landfarming, 250–252
lead (Pb) contamination
 phytoremediation for, 257–260
microorganism diversity in, 73,
 191–196, 197–198, 197–200
mine waste restoration, 239–244
non-native species' effects on, 155
nutrient cycling in, 44–47
nutrient restoration in, 190–191
oxygen levels in, 250
pollution sources for, 177
salinization of, 155, 189
stockpiled, 241–242, 256
succession and, 83, 91
washing, 248
water impervious, 48
soil erosion, 3, 178–181
deforestation and, 267, 268, 270
desertification and, 181–185
factors affecting, 181
land degradation in, 177–178
legislation on, 365
nutrient cycling and, 44–47
reducing, 181
rill, 180
sedimentation and, 319, 330
sheet, 180
siltation and, 330
SOC depletion and, 24–25
soil conservation and, 185–189
water in, 180
wind in, 179–180
soil inorganic carbon (SIC), 23, 28
soil organic carbon (SOC), 23, 24–26, 28
soil organic matter (SOM)
erosivity and, 181
land degradation and, 177
nutrient restoration and, 190–191
in primary sites, 83
solar power, 8, 9–10

SOM. *See* soil organic matter (SOM)
source populations, 133, 138–139
spatial analysis, 357
spatial heterogeneity, 131
species
biological legacies, 38–39
compatibiity of, 113
connection and, 137
diversity of, 56
functional groups, 34, 48–49
indicator, 193
invasive, 39, 58
keystone, 34, 49–50
mutualistic, 34, 50
regime shifts and, 36
sequence of introduction, 111–113
in succession, 80, 82, 92–93
spurge, leafy (*Euphorbia esula*), 161–162
stability
alternative states of, 106–111
assembly and, 105
assembly rules and, 111, 112
phylogenetic diversity and, 124
stacking approach to succession,
 89, 282–283
starlings, European (*Sturnus vulgaris*),
 154
Steinbeck, John, 183–184
stepping stones, 72, 135–136
in community assembly, 103–104
in forests, 272–273
sterile insect technique, 160
St. Helena, 155
stochastic processes, 114
stockpiled soil, 241–242, 256
stratification of seeds, 69
stratospheric ozone layer. *See* ozone layer
stress, 35, 43
strict restoration, 14, 15
subarctic coastal ecosystems, 87–88
succession, 79–100
alternative stable states in, 106–111
animals in, 89–91
assembly models of, 102

assembly rules vs., 111–114
biological legacies in, 38–39
case study on, 97–99
in critical habitats, 293
cyclic, 91–92
definition of, 79
deterministic model of, 101, 209
facilitation model of, 80, 82, 209
factors affecting, 85
in forests, 81, 84–86, 89, 271–272, 286–287
individualistic model of, 101, 103
inferring, 96–97
inhibition model of, 82
on islands, 84–88
keystone species in, 92–93
management of, 82, 91–95, 354
mathematical models of, 96–97
microbial, 199–200
in mine restoration, 243
monitoring, 95–96
mycorrhizae in, 195
nucleated, 98
primary, 83–88
processes in, 82–91
in sand dune restoration, 220–221
in sand dunes, 206, 209
secondary, 88–89
stacking approach to, 89, 282–283
theories of, 79–82
tolerance model of, 82
trajectories of, 80, 106
in tropical rain forest restoration, 282–283
wave approach to, 89
sulfur oxides (SOx), 17, 363
Superfund, 364
Surface Mine Control and Reclamation Act of 1978, 364
Surface Mining Control & Reclamation Act of 1988, 242
surface stabilizers, 240
surrogate sites, 66–67
Surtsey, 86–88

swallow-wort, pale (*Vincetoxicum rossicum*), 169–174
swamps, 315
Sweden
  adders in, 65–66
  human population, 2
  mercury pollution in, 17
swordtails, 59
symbiotic microorganisms, 73, 178, 191, 193–195. *See also* mycorrhiza *and* rhizobium
  invasive species and, 152–153, 167–168
  in sand dunes, 209

Tahiti, 155
taxonomical surveys, 95
Tay, Charlotte, 252–257
teprachronology, 96
  terms, little (*Sterna albifrons*), 226–231
terrestrial carbon sequestration, 26–27
thermal desorption, 248
thistles
  Pitcher's (*Cirsium pitchen*), 146–149
  star (*Centaurea solstitialis*), 151, 155
thresholds
  action, 361
  area, 294
  connectivity, 133, 141
  ecosystem, 113, 124
  extinction, 133
  of irreversibility, 124
tillage, 181, 186–188
Tilman, David, 73, 83, 105
timber management strategies, 285–288
tolerance model of succession, 82
top-down control mechanisms, 49–50
topsoiling, 202
Toth, Louise A., 347–351
tourism, 279, 331–332
toxicity, from metals, 237–238
toxicity tests, 244
trade, international, 151
tragedy of the commons, 10–11, 183, 267
translocation, 299
transplanting, 211, 218–219, 273

dense, 275
in tropical rain forests, 282–283
travel, international, 151
trees. *See also* forests
American beech (*Fagus grandifolia*), 163
American chestnut (*Castanea dentata*), 156–157, 271, 272
Australian pine (*Casuarina equisetifolia*), 164
bald cypress (*Taxodium distichum*), 275–277
birch (*Betula pubescens*), 268–270, 274, 279–280
black willow (*Salix nigra*), 277
cedar (*Cedrus libani*), 277–279
longleaf pine, 40–42, 107–108
oaks (*Quercus*), 107–108, 121–123, 141–142, 278
pioneer, 282–283
ponderosa pine (*Pinus palustris*), 42–43
slash pine, (*Pinus elliottii*), 40
sugar maple (*Acer sacharum*), 163
transplanting, 273–274
white pine (*Pinus strobus*), 271–272
willow (*Salix*), 277, 279–280
trial-and-error, 359
Tropical Forest Conservation Act of 1998, 283
tropical rain forests (TRFs), 280–283
destruction of, 4, 265, 266, 267–270
restoration of, 282–283
threats to, 280–282
turkeys, wild (*Meleagris gallopavo*), 135
Turnau, Katarzyna, 201–205
tussock grass (*Agropyron cristatum*), 162

ultraviolet radiation, 3
U.N. Conference on Desertification, 184, 366–367
U.N. Convention on Biological Diversity, 291–292, 367
U.N. Earth Summit (1992), 184, 366, 367
unified neutral theory of biodiversity and biogeography (UNTB), 114–116

U.N. International Conference on Population and Development, 366
United Kingdom
Broads restoration in, 327–330, 338–347
sandy beach restoration in, 225–234, 327
United States
energy consumption in, 6
extinction rate in, 60–61
hurricanes in, 37
longleaf pine savanna in, 40–42
soil erosion in, 178–179
wetlands destruction in, 316–317
wind energy in, 8
U.N. Millennium Goals, 30
UNTB. *See* unified neutral theory of biodiversity and biogeography (UNTB)
uranium mining, 6–7
urbanization, 317
U.S. Fish and Wildlife Service, 301, 365
U.S. Forest Service, 42
U.S. Global Assessment of Soil Degradation, 177
U.S. Public Health Service, 365
U.S. Soil Conservation Service, 178–179, 365

vapor stripping, 248
vegetation
acidification of, 17
climax, 80
eradication of invasive, 158–160
metal-tolerant, 239, 240–241
metal toxicity in, 237–238
in mine waste restoration, 240–241, 242–248
monitoring, 95–96
nursery-raised, 70–71
phylogenetic structure of, 119–129
restoration of species diversity, 66–67
revegetation, 15, 45–47, 357
salt-tolerant, 189
sand dune, 207–208

seed harvesting and, 67
succession of, 79–89, 93, 96–97
vehicle emissions, 17
volunteers, 356

walleye (*Sander vitreus*), 306–312
Warming, Eugene, 80
waste remediation, 56
water. *See also* aquatic ecosystems; hydrological cycles
ballast, non-native species in, 151, 157
erosion by, 180
human usage of, 2
imbibition of for germination, 69
New York City, 13
pollution of from mines, 239
in succession management, 94–95
wetlands and quality of, 315
watersheds, 314–315, 325–326
wave approach to succession, 89
weather
disturbances caused by, 35, 36–37
global warming and, 4
weed management, 93, 362

wetlands. *See also* aquatic ecosystems
definition of, 315
degradation of, 316–318
legislation on, 363–364
regime shift in, 108
restoration of, 319–324
wastewater systems in, 16
The Wilds (Ohio), 124–127
willow trees (*Salix*), 277, 279–280
black (*Salix nigra*), 277
Wilson, Edward O., 12, 102, 114, 281
wind energy, 7–8
wind erosion, 179–180
desertification and, 181
preventing, 186, 188
wolf, gray (*Canis lupus*), 17, 301–304

Yellowstone Park
fires in, 42
gray wolf reintroduction to, 17, 301–304

zooplankton, 318, 324–325, 329
zoos, 295, 297, 298–299